An Introduction to the Physical Basis of

SOIL WATER PHENOMENA

An Introduction to the Physical Basis of
# SOIL WATER PHENOMENA

E. C. CHILDS

*Reader in Soil Physics, Cambridge University,*
*Hon. Director, Agricultural Research Council*
*Unit of Soil Physics*

A WILEY—INTERSCIENCE PUBLICATION
John Wiley & Sons Ltd
London  :  New York  :  Sydney  :  Toronto

Library of Congress catalog card No. 69-17251

SBN 471 15581 0

Printed in Great Britain by
C. TINLING AND CO. LTD.,
LIVERPOOL, LONDON AND PRESCOT

# Preface

Certain aspects of soil are discussed in University courses in more than one Faculty. Its systematic study is a feature of Departments of Agriculture, but selected topics are of interest also to engineers, geographers, botanists and microbiologists among others. Among the most important of soil phenomena are those of a physical nature, which are in the main associated with physical properties which are dependent upon the interactions between the solid particles and the liquid in the pore space. Thus a part of the study of soils lies in the field of physics, and of quite advanced and difficult physics at that.

A prime difficulty in the presentation of a discussion of soil water phenomena will now be apparent. The interested parties cover a very wide spread of backgrounds, from physicists and engineers who will be entirely familiar with the tools of quantitative logic and will require to go deep, to biologists who from natural inclination and undergraduate training will be accustomed to quite other approaches. With the latter in particular the topic has gained a reputation of being difficult and obscure.

The object of this book is ambitious. It is to provide for both categories of reader without boring the one or stranding the other. The exposition starts from the most elementary concepts, demanding no previous familiarity, and proceeds in logical sequence to cover the essence of the present state of knowledge. It is in fact developed from a course of lectures delivered over some years in the University of Cambridge School of Agriculture to post-graduate candidates for the Diplomas in Agriculture and Agricultural Science. These students were drawn primarily from successful candidates for the Natural Sciences Tripos and commonly had a biological bias.

While mathematical treatments are recognized to be necessary to the attainment of the objects of the book, it is also recognized that they must form a serious break of continuity of exposition for a reader who does not follow them easily. Hence where possible and desirable, particularly in the earlier part of the book, the detailed mathematical treatments have been included in separate numbered Notes, only the essential results of which are quoted in the continuous text. Thus at a first reading these Notes may be entirely ignored if desired. On the other hand a reader who is already familiar with the field in a general way might well find himself giving the Notes more attention than the body of the text.

Finally it must be said that in the opinion of the author a course of lectures has only one advantage over a book, and that is that a lecturer may be interrupted when his standards of clarity fall below what is acceptable. The author remembers an occasion when he was having an 'off day', and after ten minutes had to confess to his class that what he had said did not convince even himself, and begged leave to start again. This remark was well received, and the author is emboldened now to express the hope that no reader will fail to draw his attention to short-comings of this kind.

No book of this nature is completed without great assistance in many ways from colleagues over a long period. This one is no exception. A good deal of the subject matter itself has been developed by collaborators in one capacity or another over the years, whether as research students, staff colleagues or distinguished visitors. Our discussions have not always been harmonious but invariably fruitful. It will be no difficult task to recognize these collaborators in the list of more recent references, but it is neverthe-less fitting to acknowledge their contributions here.

Dr. E. C. Youngs and Professor N. Collis-George occupy special places, since they made notable contributions at critical times.

Other colleagues at various times include Dr. G. D. Towner, Dr. A. Poulovassilis, Dr. D. H. Edwards, Mr. T. O'Donnell, Dr. A. J. Peck, Dr. D. E. Smiles and Dr. C. N. Evans. Discussion with distinguished visitors has been particularly rewarding. Among these must be especially mentioned Dr. J. R. Philip, Professor G. B. Bodman, Dr. Wilford Gardner, Professor Paul Day, Dr. P. J. Bruijn and the late Professor W. R. van Wijk.

Lastly, no words of mine can adequately express my debt to my wife, who prepared the draft for the press. But for her this book would most likely not have been completed.

E. C. CHILDS.

# Contents

A*

## 12   The movement of water in the soil profile

## 13   Surface infiltration

## 14   The flow of groundwater

# CHAPTER 1

# The structure of materials

## 1.1 The packing of particles

To introduce a discussion of the soil and its moisture by remarks upon the ubiquity of crystalline forms in natural materials may appear to be to take a detached and remote point of view. However, that approach to any subject is not necessarily the most perspicuous and economical of words, which is the most uncompromisingly direct. The subject of this book is the neighbourly relations between the solid and liquid parts of the soil, and some initial discussion of the constitution of these parts separately can hardly fail to be illuminating. Nor will it occasion surprise nowadays when it is seen that this discussion turns upon the orderliness of the arrangement of atoms or ions which constitutes crystal structure, since advances in the technique of X-ray diffraction and in the interpretation of diffraction patterns, which comprise the chief means of elucidating crystal structures, are steadily reducing the ranks of those materials which fail to present evidence of some degree of such structure.

The manner in which particles pack together may be illustrated by reference to some visible types of structure of common experience. If some coarse sand is poured into a graduated measuring vessel, it will be seen to occupy a certain apparent volume, say $V_1$. When the vessel is tapped on the bench top, the sand will be seen to settle and to occupy a smaller volume. Continued tapping will ultimately reduce this apparent volume to a minimum, say $V_2$. The reason for this may seem to be self evident, but it is worth-while putting it into precise physical terms as a prelude to discussing the reasons, not quite so self evident, of the types of packing which are more commonly met in real materials. The forces acting on each particle of sand are gravity, acting downward; the repulsion of neighbouring particles which prevents any closer proximity than "touching"; and the imposed shaking force. There is no necessity here to discuss at length the mutual repulsion. Let it suffice to say that it is our common experience of hard bodies that the repulsion between them is negligible until they *do* touch, and then suddenly increases so that very large forces are required to squeeze them appreciably closer together. Each sand particle has a tendency to

1

descend to the bottom of the containing vessel; gravitational force or weight is an expression and a measure of this tendency, which is an example of a general rule that the potential energy of a system tends to become as small as possible. In this case potential energy is gravitational and is measured by the height of the body. Not every particle can in fact reach the bottom of the vessel, since for a particular grain, gravity is balanced by the repulsion of the grains underneath which got there first. However, the potential energy of the mass of grains is at a minimum when the centre of gravity of the mass is as low as possible. Since the walls of the vessel prevent the spread of the sand which would bring the centre of gravity right down, and since the centre of gravity of a cylindrical mass is half way up the axis, the lowering of the centre of gravity is accompanied by an even greater lowering of the "ceiling" of the mass of sand grains, i.e. by a reduction of the volume to the minimum possible. This cannot be done by the sand grains deforming each other, but only by a reduction of the space between grains, and the mass is said to be closely packed.

If one repeats the performance, but this time substitutes for the sand a large number of spherical particles of uniform size, such as graded lead shot, one finds that the state of closest packing is associated with a particular regular pattern of shot. This is illustrated by the perspective drawing in Figure (1.1). The shot are arranged in layers, and in each layer

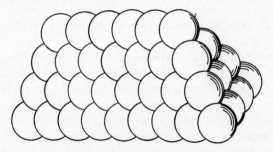

Figure 1.1 Perspective view of uniform spheres in cubic
close packing.

the centres form a mesh of equilateral triangles; each sphere is surrounded by six others, each of which it touches. The pattern of a single layer is shown in Figure (1.2), drawn from a photograph, and this also shows that the pattern is not perfectly continuous except in blocks of limited size. Any three spheres whose centres are at the apexes of one of the equilateral triangles enclose a nest in which a sphere of the second layer may rest, this layer having approximately the same pattern as the first. Since, upon inspection of Figure (1.2), it will be seen that there are twice as many nests as spheres, only one-half of the nests of the lower layer will accommodate a sphere of the upper layer. The resulting pattern is shown in Figure (1.3), where the occupied nests are marked *o*. It is clear that a sphere occupying a site *o* prevents another occupying the nearest neighbouring nests, and these unoccupied sites are marked *e*. It will be seen that the two different sorts of site have quite different geometry. At *o* a space between three spheres of the base layer is blocked up by a sphere of the upper layer, whilst at *e* such a space between three lower spheres lies under a similar space between three spheres of the upper layer, so that a straight rod could be poked through the pair of layers without meeting an obstruction. The geometry of these different sites will be treated in detail in

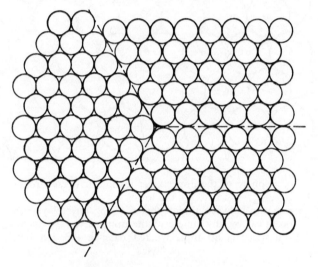

Figure 1.2 Single layer of uniform spheres showing mosaic of close packed regions.

Section 1.2, and its significance will become evident in Section 2.2. Finally, there are sites marked *o'*, at each of which a sphere of the lower layer fits into a nest between three spheres of the upper layer. These sites are similar to the *o* sites, but are inverted. Examination of Figure (1.3) shows that there are two sites of type *o* or *o'* to each site of type *e*.

One may now build on a third layer, to find that there are two alternative arrangements of the spheres in this layer. They may be fitted into nests above the *e* sites, or above the *o'* sites, but not both. In the former pattern the clear passage through the lower pair of layers is blocked up by a sphere of the third layer, while in the latter arrangement the through-way persists. There is clearly no difference between the closeness of packing in the two patterns, for both are closest packing. For reasons which we need not discuss here, but which may be found expounded in standard works on crystal structure, the former arrangement is called cubic close packing and the latter hexagonal close packing. In Section 2.5 this difference will be referred to briefly in passing.

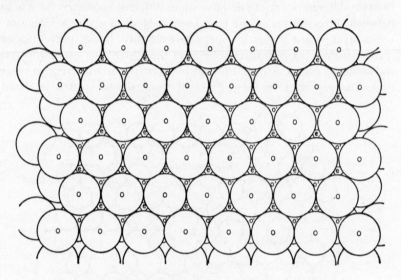

*o*, a sphere of the upper layer resting between three of the lower

*e*, a site between three spheres of the lower layer unoccupied by any sphere of the upper layer. At such a site there is an unobstructed passage through both layers

Figure 1.3 One layer of close packed uniform spheres superimposed on another.

## 1.2 The geometry of close packed spheres

It will be seen in due course that prime importance is attached to the shapes of the cavities enclosed by spheres at the sites *e* and *o* of Figure (1.3). Each of the latter is formed by a tetrahedron of spheres, three of those of the base layer supporting one of the neighbouring layer; whilst each *e* site is characterized by three spheres of the base layer lying under three differently oriented spheres of the upper layer. A single group of each type is shown both in plan and in perspective in Figures (1.4) and (1.5). In Figure (1.4) it is seen that the centres of the four defining spheres form the apexes of a regular tetrahedron each side of which has length 2*R*, where *R* is the radius of each sphere. The essential features of the geometry of this tetrahedron are listed below the diagram, but we may draw special attention to the fact that the centre of the tetrahedron lies at a distance of $1{\cdot}225R$ from each of the apexes. Since the surface of each sphere only extends to a distance $R$ from the appropriate apex, it follows that a small sphere of radius $0{\cdot}225R$ could be fitted into the $o$ cavity, with its centre coinciding with the centre of the tetrahedron. Such a small sphere would be imprisoned

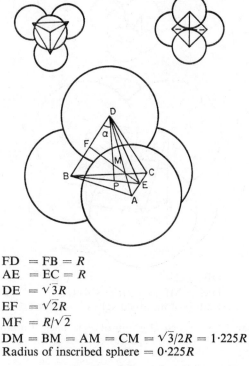

FD $= $ FB $= R$
AE $= $ EC $= R$
DE $= \sqrt{3}R$
EF $= \sqrt{2}R$
MF $= R/\sqrt{2}$
DM $= $ BM $= $ AM $= $ CM $= \sqrt{3}/2R = 1{\cdot}225R$
Radius of inscribed sphere $= 0{\cdot}225R$

Figure 1.4 Tetrahedral group of close packed uniform spheres.

within the cavity, since the largest sphere which can pass through the orifice between the three spheres forming any one face of the tetrahedron has a radius of only $0{\cdot}157R$.

A similar examination of Figure (1.5) shows that the centres of the six spheres grouped at an $e$ site lie at the apexes of a regular octahedron, each edge of which has length $2R$. The centre of this octahedron lies at a distance of $1{\cdot}414R$ from each apex, so that there is room for a sphere of radius

0·414$R$ in each cavity at a site $e$. Any one face of the octahedron is again formed by three mutually neighbouring spheres enclosing an orifice passable only by a sphere of radius smaller than 0·157$R$, so that the enclosed sphere of radius 0·414$R$ is imprisoned.

Before leaving this topic, one may note in Figure (1.3) that between two layers of close packed spheres there are just as many tetrahedral (*o* or *o'*) cavities as there are spheres, but only half as many octrahedral (*e*) cavities.

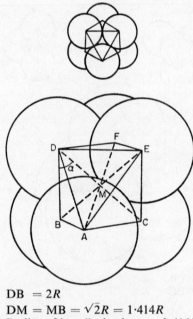

DB $= 2R$

DM $=$ MB $= \sqrt{2}R = 1\cdot414R$

Radius of inscribed sphere $= 0\cdot414R$

Figure 1.5 Octrahedral group of close packed uniform spheres.

### 1.3 The nature of the particles constituting matter

The particles which, when packed together, form a recognizable material are either atoms or ions (or, in some cases, molecules) and these are not simple "hard" spheres of the kind exemplified by the lead shot discussed in Section 1.1. It is now universally accepted that the atom is a more or less well-defined space containing a distribution of minute charged elementary particles. Those contributing the greater part of the mass are concentrated in a positively charged nucleus at the centre; whilst the negative charge is contributed by the electrons distributed through the remaining space. The space is otherwise empty. In neutral atoms the total negative charge just balances the nuclear positive charge. In cations the electrons are deficient in number so that the net charge is positive, while in anions the electrons are excessive in number. Such a concept seems at first sight to be quite incompatible with a discussion of matter which requires atoms to be thought of as hard spheres, yet the compatibility is demonstrable.

Suppose one considers the forces between, for example, a cation and anion in mutual proximity. Provided the separating distance is not too small, each ion behaves as a body with a net charge, and, the charges being of opposite sign, the force between them is one of attraction. Since the bulk of each ion consists of empty space, there would appear to be no reason why the approach of one to the other should not continue until the two were intermingled; but in such a case they can no longer be regarded as separate entities with net charge. In particular, if the approach were to continue until the nuclei were very close together, with but little electronic charge between to screen one from the other, these nuclei, of large and like charge separated by small distance, would repel each other with great force. There is thus some distance of separation at which the attractive force gives place to repulsion. If they are constrained to approach closer, repulsion tends to restore the separation, while if separated more, the tendency is to restore proximity. This is the same as the behaviour of a pair of elastic spheres pressed into contact. The ions may thus be regarded as a pair of spheres, the distance between them being the sum of their radii. In particular, if two ions of the same species are neighbours in a close packed cluster around another ion of opposite sign of charge, the distance between their centres is equal to double the radius of either. Distances between ions in crystals, as revealed by X-ray diffraction (see Section 1.5), provide very important evidence for assigning sizes to atoms.

Ions of one sign of charge will cluster more closely round another of opposite charge if the attractive force is greater. Thus a small cation of high charge will attract a cluster of anions more closely than will a large cation of low charge. This will be reflected in the X-ray diffraction pattern, the analysis of which will reveal a smaller distance between anion centres and therefore a smaller anion radius in the former case than in the latter. Hence the size of an ion is not invariable, but will depend upon the structure in which that ion finds a place. This needs to be borne in mind when reference is made to a list of atomic radii, such as that presented in Section 2.1.

This discussion of the meaning of particle radius has been introduced by reference to ions because the electric force of attraction holding ions in contact is readily understood. Such ions take part in the building of the structures of crystals of simple inorganic salts, exemplified by rocksalt (NaCl) and sylvite (KCl), where characteristic anions and cations are present, and also in quite complicated minerals such as the alumino-silicates. The force of attraction between neighbours is then simply the electrostatic attraction between charges of unlike sign, and is known as the electrovalent bond.

If, however, the material is an element (for example, diamond), the particles clearly cannot be ions, since each would be an ion of the same sign of charge contributing to a total charge born by the total mass. Such materials are commonly formed instead by the binding together of the constituent uncharged atoms by a mechanism known as the covalent bond; each of the neighbouring atoms contributes electrons to a shell common to both. In the special case of metals these shared electrons are regarded as being common to the whole bulk rather than to immediate neighbours only, and are mobile. The covalent bond is typical of organic compounds.

Sometimes structures are built of molecules which, themselves owing existence to covalent or electrovalent bonds between their constituent atoms or ions, are not bound into the structure by such bonds. In some such molecules the "centre of gravity" of the positive charges does not coincide with that of the negative charges, and for the purpose of calculating the force between neighbours each molecule may be regarded as an electric dipole, that is to say, as a body with two equal and opposite charges separated by a fixed distance. Such a molecule is called a polar molecule, and the product of the magnitude of each charge and the separating distance is called the dipole moment. The most important molecule of this kind, from our point of view, is the water molecule, the structure of which, together with the structure of water in bulk, will be discussed more fully in Chapter 3. A polar molecule tends to present a positive end to the negative end of a neighbour, and the resultant force between the neighbours is one of attraction.

Figure (1.6) is presented in order to compare the forces between ions with those between dipoles. Diagram (a) shows two ions of charge $+e$ and $-e$ respectively, each of radius $R$, with centres separated by distance $D$; diagram (b) shows two dipoles similarly of radius $R$ and separation $D$, the dipole moment being $ed$ corresponding to charges $+e$ and $-e$ at distance $d$ in each dipole. Under each diagram is worked out the relationship between the forces between particles and the relevant properties of the particles. Taking a particular case as an example (i.e. the case $d = R$), Figure (1.7) is presented to show the forces between the ions (curve a) at various degrees of separation; likewise for the dipoles (curve b). For this purpose the unit of distance has been taken as the length $R$ and that of force as the quantity $e^2/R^2$. It will be seen that although the force between dipoles may be great when the separating distance is small, it decreases much more quickly as the separation increases than does the force between ions. This is shown also by curve (c) which plots the ratio of the forces between ions and dipoles at different distances.

Yet another type of bond is that between uncharged molecules which are

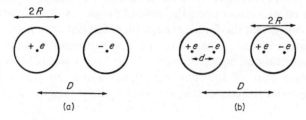

(a) force of attraction $= e^2/D^2$

(b) force of attraction $= \dfrac{2M^2}{D^2}\dfrac{3D^2-d^2}{(D^2-d^2)^2}$, where $M = ed$

Figure 1.6 Interaction between (a) ions and (b) dipoles.

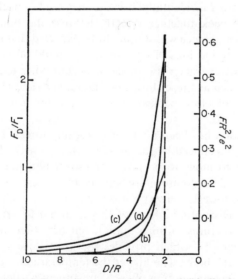

Figure 1.7 Force between (a) ions and (b) dipoles as a function of separation. (c) Ratio of force between dipoles to force between ions.

not even permanent dipoles. This is known as the Van der Waals force. Examination of Figure (1.6b) shows that a dipole exerts forces on the individual charges of another dipole which tend to separate those charges still further and thus to increase the dipole moment. In a similar way the charge distribution of a non-polar molecule may influence, and be influenced by, another non-polar molecule so that each becomes a temporarily induced dipole during the period of proximity. The result is a force of attraction between the molecules. Since the magnitude of the dipole moment is itself dependent upon the mutual proximity of the partners, it will be self evident that the force between them falls off even more rapidly as the separation increases than is the case with permanent dipoles. The temporary induction of dipole moments is no longer regarded as contributing a major part of the Van der Waals force, but the quantum mechanical treatment necessary to expound modern views is beyond the scope of this book.

### 1.4 The packing of atoms, ions and molecules; the solid, liquid and gaseous states

Owing to their light weight in comparison with bond strengths, atoms, ions and molecules do not pack together under gravity, as does lead shot. The latter pack together incidentally, because they tend to pack downward. Particles of matter pack together directly because the bond forces act directly between them. Hence whilst it might be expected that the clustering of ions of one sign round ions of opposite sign should often result in the formation of groups characteristic of close packing, i.e. tetrahedral and octahedral groups such as those shown in Figures (1.4) and (1.5), extensive continuous masses of close packed structure are not necessarily to be expected.

So far little mention has been made of disrupting forces. In the case of lead shot these are introduced by the shaking which is necessary to permit the individual shot to move to positions of lowest potential energy. The packing state achieved is a compromise between the closest possible packing with all particles in lowest energy positions and loose packing due to the necessary shaking. Disrupting effects of this kind are present also in the atomic and ionic structures forming matter, but here they are due to the movement of the particles which provides the kinetic energy constituting the heat content of the body. The higher the temperature of the body, the more violent is the thermal motion of each constituent particle. The nature of the structure achieved must therefore depend upon the violence of these thermal motions in comparison with the strength of the bonds between the particles. If the force between particles remains strong even when the separating distance is large (i.e. if the bonds are electrovalent or covalent), the thermal motions must be very violent before one particle is removed from the range of attraction of its neighbours. If, however, the attractive force dies away quickly as the separation increases (as in the case of polar, or even more markedly Van der Waals bonds), the particles may be shaken apart by correspondingly less violent thermal motions, that is to say, at lower temperatures.

Now consider a hypothetical structure of close packed particles, in which an external force is applied in an endeavour to slide one layer over the next below. Such a force is recognized to be a shear force. The act of sliding requires the spheres of the upper layer to rise out of the nests of the lower layer, to climb over the divides between nests, and to fall again into the nests next encountered; that is to say, the upper layer must become more separated from the lower, against the force of attraction, before resuming the state of closest packing. It is readily seen that an appreciable force must be applied in the direction of sliding in order to effect the shear failure; it is rather like raising a body up an inclined plane by applying horizontal force. The material thus possesses shear strength which is a characteristic property of solids. If, however, the temperature is raised sufficiently, the spheres of one layer may be bounced out of the nests of the neighbouring layer for a sufficiently great proportion of the time, so that the externally applied sliding force is not called upon for this purpose, and the shear failure will set in with but a negligibly small applied force. Most structures are mosaics of small crystals, so that the picture of the transmission of a shear failure through the mass is more complicated than the mere sliding of one continuous layer of the structure over another. Nevertheless it remains true that a structure which cannot be sheared without a disturbance which extends the bond lengths appreciably will not shear with less than a certain threshold stress; and that a sufficient increase of temperature will so extend the mean bond lengths as to change the structure sufficiently to permit shear without further extension caused by the shear stress itself. Materials which have no shear strength are characteristically liquids or gases, and thus it is that a sufficient rise of temperature will transform a solid into a liquid. The stronger the bonds between particles, the higher is the temperature at which the transition from the solid to the liquid state takes place. Materials of great bond strength are therefore hard solids at ordinary temperatures, and have high melting points. Examples of these are minerals with electrovalent or, sometimes, covalent bonds. Where the bond strength is weaker, the material is characteristically a soft solid of low melting point. Ice is an example of such a material; the water molecules are held together by polar bonds.

The very concept of structure implies that the constituent atoms have assigned sites, and such a state is the less self evident the more violent is the thermal motion. Some semblance of structure is demonstrable in liquids, as will be seen in Chapter 3 when water is discussed. At sufficiently high temperatures, however, each atom tends to behave as an independent individual, buffetting and being buffetted by its agitated neighbours. It has little inclination to cling to partners of chance encounters and tends to move freely about such space as is available to it. The material will be a random distribution of particles with random motions, obeying those laws which follow as a consequence; these laws are the gas laws and the material will be a gas. If the bonds between the constituent atoms are of negligible strength, this state can persist at extremely low temperatures, as in the case of the inert gases of which helium and argon are examples.

### 1.5 The evidence for the structure of materials

Nothing has yet been said about the experiments which demonstrate the orderly packing of atoms and ions which constitutes structure, and it is not the purpose here to go into detail on this subject. However, something must be said if the state of knowledge is to be critically appreciated. It will be sufficient to discuss the evidence which is provided by the analysis of X-ray diffraction, which is perhaps the most powerful method of elucidating structure.

If a beam of X-rays of suitable wavelength (say between 1 and 2 angstrom units, this unit being $10^{-8}$ cm) from an X-ray tube fitted with a pin-hole collimator system impinges upon a small fragment of solid material, each of the constituent atoms is disturbed by the radiation, and the result is a number of scattered radiations, one from each disturbed particle. If the material has a lattice structure, i.e. an orderly repetitive pattern of particles of the kind described in the preceding sections, there may be directions in which the weak waves from each of a particular series of regularly spaced particles reinforce each other to produce a relatively powerful scattered beam. One must say "may be" because an analysis of the situation shows that such reinforcement is not achieved unless the piece of material is presented to the incident beam at a particular angle. Hence, if one places the piece of material at the centre of a cylindrical strip of photograph film, fires the X-rays at it along one of the radii through a gap in the film, and then turns the material about in all possible directions, each time the angle of incidence suits a particular set of atoms there will be a more or less powerful scattered beam emerging in a direction which amounts to reflection at planes in which those atoms are found to lie. The end result is a pattern of lines on the developed film, the positions of which indicate the angles of reflection of the X-rays. From an analysis of the diffraction pattern recorded on the film, and taking into account the intensities of the scattered beams and the wavelength of the X-rays, exponents of the art can specify the structures of quite complicated crystals, including most of the known minerals.

81°    279°

Figure 1.8 X-ray powder photograph of kaolinite (from a photograph by courtesy of Dr. R. M. S. Perrin).

Regularity of structure was early held to account for the occurrence of certain types of solid materials, such as crystals of specific shapes, and X-ray analysis naturally began with obviously crystalline materials. Nowadays lattice structures are commonly referred to as crystal structures, and materials exhibiting such structures are referred to as crystalline, whether or not they occur as visibly recognizable crystals. Suppose that one were to take an undoubted crystal and grind it to powder so fine that, whilst each particle might be a crystal capable of reflecting an X-ray beam if correctly presented to it, it would be impossible to perform the necessary manipulations to secure this correct presentation. One could still secure the X-ray evidence, since a quantity of the powder has so many microscopic crystals with random orientations that proportions of the crystals will lie in all the directions necessary to produce all the reflected beams at once. These beams will constitute hollow cones of X-rays intersecting the cylindrical photographic film to give blackened curved lines. The result is known as a powder photograph, and an example is shown in Figure (1.8).

If the powder consists of very fine particles indeed, such as those constituting clays, the lines of the powder photograph become ill-defined. The reason for this is simple. Since the scattering of X-rays in well-defined directions amounts to reflection in planes of atoms, the structure must present well-defined populated planes if there is to be a well-defined reflected beam. Two or three atoms do not constitute a well-defined populated plane; hence if grinding is carried to the point where no single fragment is large enough to contain planes well populated with atoms, no well-defined coherent reflection is possible. It may be said that the grinding has been carried to the point at which structure has become well-nigh obliterated, and however powerful a tool X-ray diffraction may be, it cannot demonstrate the presence of a structure which does not exist. If only a powder sample is available, there being no other form of the material in nature, the X-ray evidence must be imperfect and the structure inevitably more or less uncertain; but such evidence is still generally accepted to be among the best available. We shall meet this situation when discussing the structures of the clay minerals, and it is a useful corrective to have at the outset an appreciation of the strength or weakness of the evidence upon which discussion will turn.

CHAPTER 2

# The structures of some soil minerals

## 2.1 The ions forming common minerals

SINCE soil is a product of the weathering of rock minerals, it may be expected to contain a variety of minerals characterizing the parent rock, as well as some others which are formed in the weathering process and have no existence in the unweathered rock. It is not proposed to enter into a systematic discussion of the composition of the mineral part of the soil such as would cater for the needs of the mineralogist. For this one may turn to specialized monographs. The discussion will rather be confined to a few of the more significant and more abundant minerals as illustrations of the way in which structure influences significant properties and the roles played by minerals in soils.

Soil minerals fall for the most part into chemically differentiated groups, such as oxides, hydroxides, and the particularly well-populated group known as the alumino-silicates. The ultimate particles of which these are built are ions, the binding being mainly of the electrovalent type. The species of ions occurring are relatively few, and one need hardly consider more than those listed in Table 1 (Evans, 1964), which also indicates in each case the commonly accepted radii. The valency of each ion is shown by the superscripts in the column of symbols, in the form of the number and sign of the electric charges. For example, silicon occurs as a quadrivalent cation; it takes its place in a structure as a particle with four positive units of charge, the unit being the magnitude of the charge on an electron. The hydroxyl ion is a monovalent anion; and so on. The cations are listed in descending order of abundance (i.e. silicon is the commonest ion, aluminium less so, and so on), and similarly for the anions.

Inspection of Table 1 reveals some striking features. In the first place, the anions are of more uniform size, irrespective of species, than are the cations. With the exception of the chlorine ion, which is not very abundant, the anions may be regarded without serious error as of quite uniform size. As regards the oxygen and hydroxyl ions this is, of course, no accident, since the latter is but an oxygen ion with a hydrogen nucleus (i.e. a proton) embedded in it. We shall see later that the water molecule is also of the

14

same size, being an oxygen ion with two protons embedded in it. The fact that fluorine has also the same size is, however, fortuitous. By contrast, the cations vary widely in size according to species. Secondly, the anions are, as a group, considerably larger than the cations, of which only the potassium ion approaches the anions in size. The potassium ion is itself remarkable in that its size is almost exactly the same as that of the oxygen, hydroxyl and fluorine anions. Thirdly, the anions are of low valency whilst the cations include polyvalent species. Silicon, for example, is quadrivalent, having a charge equal in magnitude to four electronic units.

*Table 1*
*Radii of ions abundant in common minerals*

| Ion species | Symbol | Radius (in angstroms) |
|---|---|---|
| Silicon | $Si^{4+}$ | 0·41 |
| Aluminium | $Al^{3+}$ | 0·50 |
| Ferric iron | $Fe^{3+}$ | 0·64 |
| Magnesium | $Mg^{2+}$ | 0·65 |
| Calcium | $Ca^{2+}$ | 0·99 |
| Potassium | $K^{+}$ | 1·33 |
| Sodium | $Na^{+}$ | 0·95 |
| Hydroxyl | $OH^{-}$ | 1·40 |
| Oxygen | $O^{2-}$ | 1·40 |
| Chlorine | $Cl^{-}$ | 1·81 |
| Fluorine | $F^{-}$ | 1·36 |

## 2.2 Pauling's rules of structure

Consider a tetrahedral group characteristic of close packed structure, in which the four spheres are similar ions, say hydroxyl ions. Referring to Section 1.2 one recalls that a small sphere of radius $0·225R$ can be accommodated in the cavity between the four spheres, where $R$ is the radius of each of these spheres. Substituting in this expression the value of the radius of the hydroxyl ion, given in Table 1, one finds that the radius of the sphere within a tetrahedron of hydroxyl ions is 0·3 angstrom. Similarly the sphere which can fit inside the cavity between the six hydroxyl ions of an octahedral group has a radius of 0·55 angstrom (i.e. $0·414R$ as given in Section 1.2). These radii are of the order of magnitude of the radii of the commonly occurring cations, silicon and aluminium respectively. One is therefore led to conclude that the chief geometrical requirement of a structure is that it should be able to accommodate the large anions, for the cations are able to fit into the cavities between the anions. Moreover, since the anions are all of the same size, and such uniformity of size favours the formation of close packed structure, the frequent occurrence of the tetrahedral and octahedral elements of this structure is to be expected. Examination of the many structures now elucidated shows that the expectation is indeed fulfilled, to a degree which permits of the enunciation of a general rule of structure, constituting one of a number of such rules associated with the name of Linus Pauling (1929). This particular rule, known as the co-ordination rule, is that the large anions cluster round the small cations to form groups of regular geometrical pattern, and that the number so clustering about a single cation is determined by the relative magnitudes of the radii of the respective ion species. Thus the relative lengths of the silicon and oxygen radii ordain that four of the latter form a regular tetrahedron round one of the former, and the coordination number is said to be four. Similarly, six hydroxyl ions tend to form a regular octahedron round an aluminium ion, and the coordination number in this case is said to be six.

This rule is not one of the more rigid laws of chemistry. It may often be broken, but not too flagrantly and not in too wholesale a manner. Thus it will often be found that aluminium occurs in a structure in fourfold co-ordination, i.e. it has elbowed its way into a tetrahedral group of oxygen ions. The opposite kind of replacement, where a silicon ion rattles about in an octrahedral group too big for it seems to be not at all common. However, the important fact is that such misplaced ions will not occur generally, but in only a small fraction of the tetrahedral groups present, perhaps to the extent of one in four; and further only those cations which are but a little oversize are to be expected to occur with the wrong coordination number. Proposed structures which flout the rules too blatantly are regarded with suspicion. To summarize this rule briefly so far as it concerns the ions of importance in soil minerals, it is to be expected that silicon will occur characteristically with a coordination number of four (i.e. it will lie at the centre of a tetrahedral group of close packed anions), and aluminium, ferric iron and magnesium are to be expected at the centres of octahedral groups, with a coordination number of six. Larger cations than these require larger cavities than are to be found between close packed oxygen or hydroxyl ions.

There are several other rules of structure, but only one more, of first importance, will be discussed. This is Pauling's valency rule. It may be illustrated by reference to the structure shown in Figure (1.3). Suppose that the two sheets of close packed spheres are in fact sheets of hydroxyl ions, each ion having one electronic unit of negative charge. At the end of Section 1.2 attention was drawn to the feature that there are twice as many spheres in this structure as there are octahedral cells. Therefore, if each octahedral cell contains a divalent cation, i.e. an ion with two electronic units of positive charge, the structure will be electrically neutral as a whole. A suitable cation, with the required valency and coordination number (six), would be the magnesium ion. It will now be shown that the structure is electrically neutral in quite small detail.

If one examines any one of the cations in the octahedral cells, it will be seen that it is equally distant from each of the six anions which surround it. One may regard it as being divided into six parts, each one-sixth part of its charge being devoted to binding it to each of the different members of the surrounding group of six anions. Similarly, any one of the anions is equally distant from three surrounding cations, and one may therefore regard it as being divided into three parts, each of which provides the charge binding the anion to a different one of the three cations. Hence any one bond between an anion and a cation may be regarded as a link between a positive and a negative charge, the positive charge being one-sixth of the charge on a cation (i.e. one-sixth of two positive units) and the negative charge being one-third of the single negative charge on a hydroxyl ion. The net charge of each mutually attracting pair, the charge associated with the bond, is zero, and the structure is electrically neutral, bond by bond. The cation with two positive units of charge lies between six anions, each of which devotes only one-third of a unit charge to that particular octahedral group, i.e. the two positive units are just neutralized by the two negative units contributed to the group by the anions. This detailed neutrality constitutes the valency rule. Like the coordination rule, it is not obeyed rigidly. For example, in a coordinated group which requires a quadrivalent silicon ion for its electrical neutrality, one may find instead a trivalent aluminium ion, the consequence being a deficiency of one positive unit of charge. Again such contraventions of the rule may be neither flagrant nor wholesale, and the resulting charge deficiency must in some way be made good by the accommodation in the structure, at not too great a distance, of an additional ion contributing the charge necessary to restore electrical neutrality.

It will not have escaped attention that the hypothetical structure which has been used here to illustrate the valency rule obeys also the coordination rule. The magnesium ion is of the size to occur with coordination number six in combination with hydroxyl ions, and its charge is just neutralized by its share of the surrounding anion charge. It is therefore a structure which might be expected to exist, and one may reasonably ask, "Is it found in nature?" The answer is that it is found as the mineral brucite, and also as a component of certain clay minerals which will be found to occupy a great deal of attention in this book.

## 2.3 The structure of some common minerals: quartz

The way has now been prepared for the systematic examination and discussion of a few dominant soil minerals, beginning with some forms of silica, $SiO_2$. The most widely known form is, perhaps, quartz. Originating as crystals of various sizes in "acidic" igneous rocks such as granite, it is highly resistant to chemical change and forms a large proportion of those soil particles large enough to be seen with the unaided eye or with the help of a hand lens or low magnifying power. The structure is based on the tetrahedral group of four oxygen ions round a silicon ion, and the arrangement of these tetrahedra in the structure is shown in plan view in Figure (2.1), in which the tetrahedra present that aspect shown in the upper right-hand

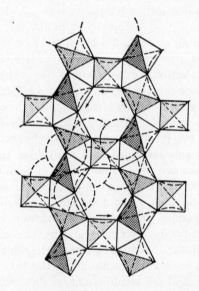

Figure 2.1 Plan view of silica tetrahedra spirals in quartz structure. Each apex of a tetrahedron is the centre of an oxygen ion.

detail of Figure (1.4). Figure (2.2) is an attempt to give a perspective view of the basic elements of the structural pattern. This latter diagram shows how the oxygen ions form interlocking spirals. The cylinders depicted are imaginary surfaces on which lie the centres of the oxygen ions, these centres lying on curves which have the form of a double threaded screw. Six ions in succession comprise a complete revolution of the spiral, and such a complete spiral is indicated by the ions lettered *A* to *G*. Any one cylinder is surrounded by six others in a hexagonal pattern, and at those places where a spiral of ions round one cylinder crosses a similar spiral on a neighbouring cylinder, a tetrahedral group of four oxygen ions is formed round a silicon ion. Such a group is shown at *ABHJ*, each spiral contributing two ions to the tetrahedron. This group is common to two cylinders and provides the link binding these two together; similar linking groups bind any one cylinder to each of its six surrounding neighbours. Such other groups are shown at *DEKL* and *FGMN*. Whilst the centres of all the oxygen ions lying on both spirals of each of five cylinders are shown, the actual ions are shown only for one complete spiral and those illustrating linkages to other spirals. Any attempts to show more would result in pictorial chaos.

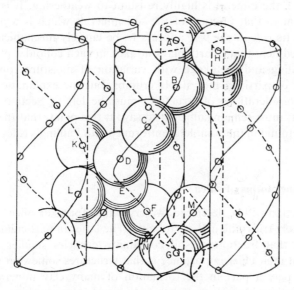

Figure 2.2 Perspective view of spiralling oxygen ions in quartz structure. Neighbouring spirals cross and touch to form a tetrahedral group.

Figure (2.1) is provided to show more completely the linkages between two complete double spirals and between these and all the neighbouring spirals, the tetrahedra shown being those with apexes at the centres of the ions of the groups shown in Figure (1.4). It is to be noted that every apex of each tetrahedron is joined to an apex of another tetrahedron, i.e. every oxygen ion is common to two different tetrahedral groups. Thus half of the charge on the anion may be regarded as devoted to one group and the other half to the other. In any one tetrahedral group there are thus eight negative units of electric charge, since oxygen is divalent; half of this is to be regarded

B

as devoted to the group, the other half to other groups. The resulting four units of negative charge therefore just neutralise the four positive units of the silicon cation at the centre of the tetrahedron, and both of Pauling's rules are satisfied.

Every bond in the structure is of great strength, since it is an electrovalent bond between a small ion (silicon) of high charge and a negative neighbour. The smallness of the ion implies a short distance of separation between the mutually attracting charges and consequently great strength of the attractive force. The bonds continue the structure in all directions equally, and the end result is a mineral of great hardness, without cleavage planes, which is highly resistant to the action of disruptive forces, whether physical or chemical; i.e. the mineral is highly resistant to weathering. It is therefore not abundant in that fraction of the soil minerals which is a product of weathering. Its role in the soil appears to be entirely mechanical. It may be mentioned here that quartz plays no part in such colloidal phenomena as soil swelling and shrinkage with variation of moisture content, and aggregation of particles into stable crumbs, but the explanation of this, based on considerations of structure, must be left to Section 2.10 and Chapter 4. It must remain sufficient to say, at this point, that the explanation is to be found in the complete conformity with Pauling's rules displayed by quartz.

### 2.4 Some other forms of silica

In this section will be described two crystalline forms of silica other than quartz, namely tridymite and cristobalite. The latter in particular has been reported in that fraction of soil containing particles smaller than two microns, but their chief importance is that structures somewhat similar to cristobalite play a part in the formation of many clay minerals. These structures approach much more nearly to simple close packing of spheres than does quartz, in which only the elementary tetrahedral group can be traced. If one looks again at Figure (1.3) and recalls that there are as many tetrahedral cells as there are spheres in the two sheets, it is at once obvious that no structure can exist in which the spheres are oxygen ions and in which each tetrahedral group contains a silicon ion. This is because there would be as many silicon ions, each with four positive units of charge, as there would be oxygen ions with only two negative units each. Pauling's valency rule would be flagrantly flouted and the structure would be highly charged electrically. The number of tetrahedral cells needs to be reduced drastically without an equally severe reduction of the number of anions; and this may be done as follows.

In a single sheet of close packed spheres let alternate spheres in alternate rows be removed, so that the gaps in any one gappy row lie between those in the incomplete rows on either side. The resulting pattern is shown in Figure (2.3). The removal of these spheres reduces the number of nests,

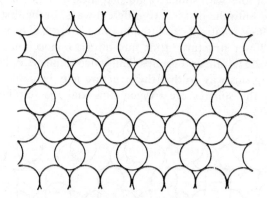

Figure 2.3 Gappy close packed oxygen ions forming one layer of a cristobalite structure.

Figure 2.4 Honeycomb arrangement of oxygen ions in second layer of a cristobalite structure. This layer is not close packed.

but every one of those remaining can accommodate a sphere in a second sheet, each such sphere completing a group as shown in Figure (2.4). It will be noticed that these spheres of the second sheet are *not* close packed; the distance between centres of neighbours is $4R/(3)^{\frac{1}{2}}$, i.e. $2·31R$, instead of $2R$ as in close packing, $R$ being the radius of a sphere. This diagram will be referred to in Sections 2.7 to 2.10 during a discussion of the micas and clay minerals. For the present, consider the structure resulting from filling alternate nests only, as in Figure (2.5), where $a$, $b$, $c$ and $d$ are examples of occupied nests and $e$, $f$, $g$ and $h$, are examples of unoccupied nests. A second pair of sheets of ions with similarly labelled sites, $a'$, $b'$, $c'$, $d'$, $e'$, $f'$, $g'$ and $h'$, can be fitted onto the first in two different ways. In the first, by a translation only without rotation, $e'$ is fitted over $a$, $f'$ over $b$, $g'$ over $c$ and so on, spheres of the lower structure fitting into hitherto unoccupied nests of the upper to complete tetrahedral groups. Unlimited numbers of layers may be superimposed in this way so that the structure may be continued in three dimensions. If the spheres are oxygen ions and the tetrahedral groups

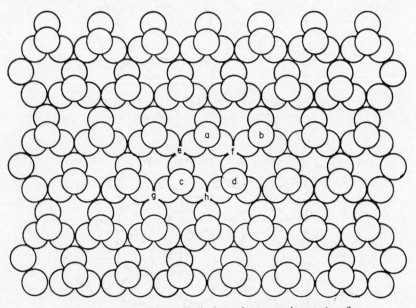

Figure 2.5 As Figure (2.4), but alternate sites only of
second layer occupied, as in a true cristobalite structure.

contain each a silicon cation, we have the mineral cristobalite. In the second type of superimposition the upper pair of layers is turned clockwise through $60°$, so that $h'$ fits over $c$, $e'$ over $a$, $f'$ over $d$, and so on. The reader may confirm for himself the difference between the two structures by making two tracings of Figure (2.5) and placing one over the other in the two possible ways described. This second structure, with oxygen and silicon ions as before, is the mineral tridymite. As in the case of quartz, each oxygen ion is shared between two tetrahedral groups and each tetrahedral group includes a silicon ion, so that Pauling's rules are complied with perfectly.

The minerals described in this section are of more interest by reason of their associations with other structures than in their own right. For example, the occurrence of the pattern of Figure (2.4) will be described in Sections 2.7 to 2.10, and this pattern is often referred to in that connection, for brevity and not entirely correctly, as the cristobalite layer. Again, in Chapter 3 comparisons will be drawn between the structures of water and quartz, and between ice and tridymite, and for that purpose it may be recorded that the specific gravities of quartz, tridymite and cristobalite are 2·66, 2·30 and 2·27 respectively.

## 2.5 The felspars

In the felspars the tetrahedral group of four oxygen ions clustered round a silicon ion is again the basic building unit, and in the illustrations to this section the group is shown as a tetrahedron derived by joining the centres

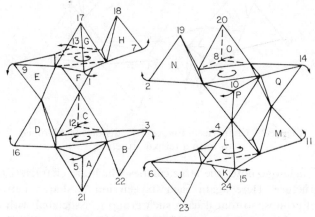

Figure 2.6 Chain arrangement of silica tetrahedra as in a felspar structure.

of the oxygen ions by straight lines; for clarity not a single ion is shown as such. In Figure (2.6) are shown four such tetrahedra, *A*, *B*, *C* and *D*, with their bases in the same plane and joined by their corners so that they lie round a square. The apexes of neighbours *A* and *B* point downwards, those of *D* and *C* upwards. A second group of four tetrahedra is similar (shown at *E*, *F*, *G* and *H*), but the upward and downward pointing roles are reversed, so that the two sets of tetrahedra join together via the apexes of *C*, *D*, *E* and *F*. This process continued indefinitely both upwards and downwards results in a chain. A similar array of tetrahedra, *J*, *K*, *L*, *M*, *N*, *O*, *P* and *Q*, constitutes a mirror image of the first chain.

These chains have now to be distorted by twisting in alternate senses as indicated by the arrows, so that apexes *1* and *2* approach each other, as do also *3* and *4*, to provide linkages between the two chains. The same distortion turns apexes *13*, *14*, *15* and *16* to join the pair of chains to similar pairs on each side, whilst *5*, *6*, *9*, *10*, *7*, *8*, *11* and *12* provide links with chains to the rear and in front. The chains of Figure (2.6) when joined together in the manner described present the pattern shown in Figure (2.7), the structure being continued in three dimensions by the conjunction of other similar units via the apexes *c,c'*, *d,d'*, and *b,b'*.

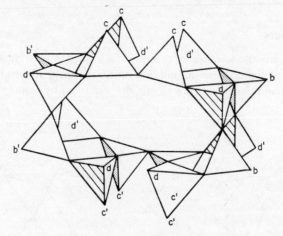

Figure 2.7 Chains as in Figure (2.6), distorted and joined
to form a felspar structure.

At this stage one notes the large cavities which are characteristic of the felspar structure. Here again each oxygen ion is shared between two tetrahedral groups, so that if each such group is associated with a silicon ion, one would have complete conformity with Pauling's rules. One would then have to enquire into the stability of a large cavity of the kind shown. In fact, however, one tetrahedral group in four contains not a silicon ion but a trivalent aluminium ion. Here then is an example of an exception to Pauling's rules, for both are flouted. Aluminium is typically associated with octahedral anion groups, and has only three units of positive charge whereas four units, as in silicon, are required for the electric neutrality of the structure. The occurrences of ions of one kind where the purity of the structure requires others are called isomorphous replacements. Felspar provides an example of the phenomenon. The contravention of the valency rule demands that an adventitious ion shall be introduced somewhere into the structure to make good the deficiency of positive charge of the aluminium ion. This is effected in orthoclase felspar by the siting of mono-valent potassium ions in the cavities of the structure, and in the plagioclase felspars by various proportions of sodium and calcium cations. These adventitious cations serve both to maintain the balance of electric charge and to keep the cavities open. The evidence of the latter is that the cavities are rather more distorted when they contain small ions than when the large potassium cation is enclosed.

Perhaps the most noteworthy feature of the felspar structure is the comparative weakness of the binding of the potassium ion. Firstly, it is a large ion, and secondly it has only one unit of charge: low charge and large separation from anions result in a low bonding force. Lastly the ion has no fewer than ten neighbouring oxygen anions sharing the attraction, six at a distance of 2·85 angstroms and four at 3·1 angstroms. When a crystal of felspar is placed in water, the potassium ions in surface cavities may receive accessions of energy by collision with water molecules or ions in solution sufficient to separate them from the parent body, when the intervention of water dipoles may so reduce the restoring force that the potassium ion becomes one of the free ions in solution. This process is known in elementary chemistry as dissociation. Further, it seems plausible that in such an event, there being no ion to keep the cavity open, there should ensue a collapse of the structure at the surface of the crystal, fresh surfaces being thus exposed for further chemical reaction. In the present state of knowledge this proposed mechanism can only be regarded as a plausible speculation about the nature of the weathering of the felspars, but it is an experimentally established fact that felspars *do* weather rapidly in warm moist climates, and that soils resulting from such rapid weathering *are* rich in potassium ions which are available to vegetation. Furthermore, ground felspar suspended for long periods in water *does* break down into sludge with the release of potassium ions. The sludge, on analysis, is found to contain a proportion of clay minerals whose structure has no resemblance whatever to that of the original felspar, and must therefore be the result of the complete collapse of the felspar followed by recrystallization into quite new forms.

## 2.6 Layer lattices: brucite and gibbsite

Brucite has already been described at the end of Section 2.2 as exemplifying Pauling's rules. It consists of two sheets of close packed hydroxyl ions in which each octahedral cell contains a magnesium ion, as shown in Figure (2.8). The bonds between magnesium and hydroxyl ions tend to extend the structure as a layer which is electrically neutral, so that there is no strong electrical attraction between one such layer and another. A structure of this kind is known as a layer lattice. By restricting the use of the word "layer" to mean a structure which extends indefinitely in only two dimensions, one distinguishes the "layer" from the "sheets" of ions of which it is composed. A crystal of brucite consists of a pile of these electrically neutral layers loosely bound together. The binding force is not quite zero, since only a uniform sheet of negative charge could entirely screen the sheet of magnesium cations of one layer from the nearest sheet of hydroxyl anions of the neighbouring layer. The negative charges being in fact located on individual ions, there is a weak stray electric field binding the several layer lattices together. Hydroxyl ions of the lower sheet of an upper layer fit into the nests between the upper hydroxyl ions of a lower layer to continue the close packed structure. The weakness of the binding between successive compound layers in the pile accounts for easy cleavage along these planes.

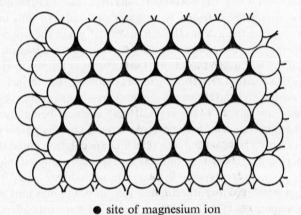

● site of magnesium ion

Figure 2.8 Two layers of close packed hydroxyl ions
with each octahedral unit containing a magnesium ion.
The brucite structure.

Gibbsite, which is also known as hydrargillite, is a similar structure in which the cations in the octahedral cells are aluminium instead of magnesium. Since each aluminium ion has a charge of three positive units as compared with the two units of charge on a magnesium ion, only two-thirds of the octahedral cells can be occupied, otherwise the layer lattice would accumulate high net positive charge. The occupied cells form a honeycomb pattern as shown in Figure (2.9), and from that diagram it can be seen that each anion is shared between only two octahedral groups, instead of between three as in brucite. Thus one-half of the total of six units of negative charge on the six hydroxyl ions of an octahedral group is associated with that group, i.e. the group contributes three negative charges which just neutralize the three positive charges on the enclosed aluminium ion; there is total conformity with Pauling's rules. The layer lattice is electrically neutral, and the gibbsite crystal consists of a loosely bound pile of such layers. The way in which one layer of the pile rests on that underneath it is somewhat different from that of the brucite structure, and when discussing the difference, we must introduce a rather special case of a dipole bond, which we shall need to refer to again when describing the clay minerals.

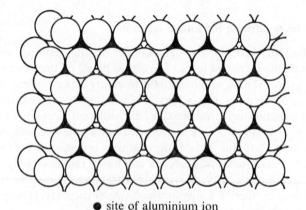

● site of aluminium ion

Figure 2.9 As Figure (2.8), but two out of three octahedral units are occupied by aluminium ions, the remaining units being empty. The gibbsite structure.

It was mentioned in Section 2.1 that a hydroxyl ion is but an oxygen ion with a proton embedded in it. The proton does not, of course, lie at the centre of the ion because the oxygen nucleus already occupies that place. The positions of such embedded protons will be described in further detail in Chapter 3, but it is enough here to note that the hydroxyl ion must be a dipole, the net negative charge being located at a different point from that occupied by the positive charge. Hence there is a tendency for the cation at the centre of a coordinated group of hydroxyl ions to turn the axis of the dipole into a preferred direction, or to polarize it. The greater the electric force exerted by the cation, the greater will be this tendency. Aluminium exerts a considerably greater force than does magnesium, since reference to Table 1 shows that it has both the larger charge and the smaller size. This is reflected in the fact that the anions cluster much more closely in gibbsite than in brucite, the radius of the hydroxyl ion in the latter being 1·61 angstroms and in the former only 1·40 angstroms. This discrepancy, and that between either value and the radius listed in Table 1, has been commented on already in Section 1.3. The consequence of this greater electric force appears to be that aluminium polarizes its surrounding hydroxyl ions more than does magnesium, with the consequence that the lower hydroxyl

B*

ions of one compound layer of gibbsite exert specific forces on the upper hydroxyl ions of the layer below. The ions therefore rest one upon the other, instead of in nests between the hydroxyl ions of the lower layer as in brucite. Gibbsite is not close packed except as regards the individual layer lattice. This specific binding force due to polarizing the proton embedded in a hydroxyl ion is called a hydrogen bond.

### 2.7 Mixed layer lattices: the kaolin minerals

It may be remarked that the structures of brucite, gibbsite, and the so-called cristobalite layer of Figure (2.4) are all based on the close packing of oxygen or hydroxyl ions, and one might therefore expect to find minerals in which layers of the different kinds are stacked together. To expect this is to forget that one and the same ion can have different sizes depending upon the company it keeps. It has been recorded in the previous section that the hydroxyl ion has a radius of 1·61 angstroms in brucite and 1·40 angstroms in gibbsite; in cristobalite the oxygen ion has a radius of only 1·26 angstroms, since silicon is a very strongly attracting cation. By the merest accident, however, there is a way in which two of these layers may condense to form a double layer lattice.

Reference to Figure (2.4) and to Section 2.4 shows that the upper sheet of oxygen ions forms a honeycomb structure in which the separation of neighbouring ions is $2·31R$, where $R$ is the oxygen radius in this structure, i.e. 1·26 angstroms. Thus the distance between centres of neighbouring ions of this sheet is 2·91 angstroms, and moreover the honeycomb pattern is derivable from a single layer of close packed type by the mere removal of one sphere in three. Compared with this, gibbsite is based on close packing with a distance of 2·8 angstroms between centres of neighbouring ions, a value which differs by only 4% from the spacing of the "cristobalite" honeycomb sheet. Now, if each of the ions in the latter sheet is an oxygen ion with two charge units, only one of which is required to contribute to neutralizing the charge of the silicon ion at the centre of the tetrahedral group, there will be a single unit charge available to contribute to the gibbsite layer. Hence it seems possible that the gibbsite and "cristobalite" layers could be welded together by these oxygen ions being common to both layers, substituting for the corresponding hydroxyl ions of the gibbsite layer. Such a condensation of one "cristobalite" layer with one gibbsite layer is in fact known to exist, for it provides the minerals of the kaolin group of clays.

There are various members of the kaolin group, such as kaolinite itself, halloysite, metahalloysite, dickite and nacrite. The members of the group differ from each other in the way in which successive layers (double layers as they are called from the mode of formation) are stacked in the crystal. Since a "cristobalite" layer is applied to one face only of the gibbsite layer, one face of the kaolin double layer consists of close packed hydroxyl ions appropriate to gibbsite and the other consists of oxygen ions proper to cristobalite. In the stacked pile of double layers constituting the kaolin crystal, each layer presents its hydroxyl surface to the oxygen surface of one neighbour and its oxygen surface to the hydroxyl surface of the other neighbour, as shown in Figure (2.10). The aluminium ions on one side of the

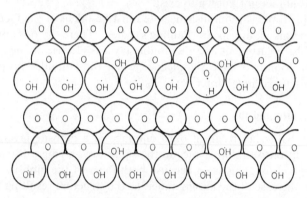

Figure 2.10 One layer each of gibbsite and "cristobalite" joined in the kaolinite structure. Two sheets of kaolinite double layer are shown, and one hydrogen bond between the two sheets is indicated.

hydroxyl sheet and the oxygen ions of the neighbouring double layer between them polarize the hydroxyl ions in the manner discussed in Section 2.6 with reference to gibbsite. The embedded hydrogen nuclei tend to take up positions between the oxygen and hydroxyl sheets presented to each other by neighbouring layers, as illustrated in Figure (2.10). They then provide $O^{2-}$—$H^+$—$O^{2-}$ links, or hydrogen bonds, between the layers. The spacing between layers and the relative positions of the layers with respect to each other, as revealed by X-ray analysis, are consistent with hydroxyl ions of the surface of one layer resting *on* oxygen ions of the neighbouring layer, as in gibbsite, instead of resting in nests between ions, as in brucite. This is, as has been already mentioned, to be expected when the confronting surfaces have attraction for each other in specific directions due to the polarization. Electron photomicrographs show, by the great opacity of kaolinite crystals to electron beams, that the crystals are commonly rather thick, the relatively strong binding between consecutive double layers holding many such layers together.

The gibbsite and cristobalite layers fit together sufficiently well to permit the existence of the mixed double layer, but not so well as to allow large crystals to grow. The slight cramping of the cristobalite layer and the stretching of the gibbsite layer, required to perfect the fit, result in internal bending moments which tend to produce curved crystals. These internal stresses place a limit on crystal size, which accounts for the fact that kaolin is indeed a clay. The fortuitous agreement in spacing does not extend to the cristobalite-brucite partnership. The internal stresses involved in a hypothetical double layer of this constitution would be very great, and such a mineral is in fact not known to exist.

Before this topic is relinquished, note must be taken of the fact that the kaolin minerals conform to Pauling's rules. There is no appreciable isomorphous replacement and the double layer is electrically neutral. This fact will be found to have significance when the colloidal properties of clays are discussed in Chapter 4.

## 2.8 Pyrophyllite and talc

If a cristobalite layer can fit onto one face of a gibbsite layer, another can clearly fit onto the other face in the same way, giving a three-layer or sandwich structure. This is so because both faces of the gibbsite layer are alike. As in the case of the kaolin minerals, the structure resulting from this condensation conforms to Pauling's rules in detail and is electrically neutral. It occurs in nature as the mineral pyrophyllite. The three-layer structure has oxygen ions on both faces, since the gibbsite layer, which normally has the hydroxyl ions, is the middle layer and the outer layers are the oxygen-containing cristobalite lattices. There is thus no question of polarization of surface ions, since there are no protons embedded in these ions to provide hydrogen bonds between successive triple layers. The stacked layers can therefore be held together only by stray forces.

Since the fitting of similar structures on opposite faces of the gibbsite layer removes the bending moments and the tendency to curvature of the crystals, there is *prima facie* a greater likelihood of brucite being able to take part in condensation with two cristobalite layers, in spite of poor conformity, than with one. In fact, such a three-layer lattice occurs in nature as the mineral talc. Neither pyrophyllite nor talc are common in soils, and have no intrinsic interest here, but their structures are basic to certain soil minerals of great importance.

## 2.9 The micas

Isomorphous replacements, which do not appear to occur to an appreciable degree in the kaolin minerals, are found in abundance in the three-layer lattices, and give rise to different classes of minerals. The structures of the micas are basically similar to those of pyrophyllite and talc, but with isomorphous replacements which consist, in most micas, of the replacement of silicon by aluminium to the extent of one silicon in four. Muscovite ("ruby" mica), for example, has the pyrophyllite structure with the above isomorphous replacements. There are no cavities in the layer lattice for the accommodation of adventitious cations necessary to redress the balance of charge which is upset by the ion replacement. These cations therefore occupy sites between the successive layers. In muscovite, and indeed in most micas, the cations are monovalent potassium ions which are located in the nests formed by the hexagonal rings of oxygen ions (the lower layer of the cristobalite structure shown in Figure 2.4) which constitute the faces of the mica layers. Since the cations are positive ions lying between negatively charged layers and having an attraction for both neighbours, they provide electrovalent bonds holding the pile of layers together to form a thick crystal. Nevertheless, the binding due to the potassium ions is not *very* strong, for reasons which were discussed in connection with the felspars. If the necessary force is applied, the micas cleave characteristically along planes between the layers, which hold together very much more strongly than they hold to each other.

Since the potassium ions occur between the layers, which are themselves very stable structures, their removal at exposed edges during weathering does not lead to a general collapse of the structure as it does in the case of the felspars. Whether for this reason, or because structures of mica type may be crystallized during weathering from the products of the breakdown of other minerals, mica structures are found even among the smallest particles which have been most exposed to weathering. A hydrated form of mica, known as illite, is one of the typical clay minerals.

## 2.10 The montmorillonite minerals

This group of minerals, among the most important in soils in that it is exceptionally colloidal in its behaviour, is, like the micas, based upon the pyrophyllite and talc structures. The isomorphous replacements are very varied from mineral to mineral, both in kind and in degree. There is no direct evidence on this point, and one has to proceed by trying to reconcile chemical analyses, which are very diverse, with the structure as revealed by X-ray analysis which, so far as it goes, is consistent with the three-layer lattice described in Section 2.8. The various ions found in the chemical analysis are allotted to their characteristic groups, so far as is possible, deficiencies being made good by the most plausible isomorphous replacements. Thus silicon ions are allotted to the tetrahedral groups, and any such groups left over are supposed to be filled by aluminium, with consequent negative net charge. When all tetrahedral groups are thus satisfied, the remainder of the aluminium is allotted to the octahedral groups in which it characteristically occurs. Deficiencies of aluminium are made good from among other ions found in the analysis and known to occur characteristically with a coordination number of six. Thus magnesium and iron are called upon to provide isomorphous replacements, and even calcium has been supposed by some to perform this role.

The balance of cations found in the analysis is assigned to the space between layers, thus neutralizing the negative charges due to the isomorphous replacements. Such neutrality of the final structure is not, of course, a proof of the correctness of the assumptions as to replacements, but only a check on the accuracy of the chemical analysis. The amount of inter-lattice ions thus derived should, however, agree approximately with the amount of the measured dissociable ions, of which more anon. When the assessment of isomorphous replacements is carried out for the various members of the montmorillonite group of minerals, it appears that montmorillonite itself has the pyrophyllite structure with magnesium replacing aluminium in a proportion of the octahedral groups of the gibbsite layer, whilst beidellite has the pyrophyllite structure with aluminium substituting for some of the silicon in the cristobalite layer. Nontronite is similar to beidellite, excepting that the "gibbsite" layer is a mixed aluminium-ferric hydroxide layer, and hectorite and saponite have the talc structure with aluminium invading tetrahedral groups in the cristobalite layers. Whatever the constitution of particular minerals, they all behave similarly in their physical properties and are known as a group as the montmorillonite minerals, montmorillonoid, or smectite. Indeed, there is no well-defined boundary between particular members of the group, all gradations between type members being known.

The montmorillonite minerals differ from the micas in that the adventitious ions between the charged triple layers are not strongly bound and do not serve to link layer to layer in a thick crystal. One can only speculate as to the reason for this. Possibly the sites of isomorphous replacements are so randomly distributed that the charge-balancing cations cannot occupy well-defined lattice points as they are known to do in the micas. Be that as it may, the fact remains that the inter-layer ions are readily dissociated from the minerals in aqueous suspension, with the proviso that ions of a few species are anomalous in this respect. The montmorillonite can then enter into ion exchange reactions, in the same way as any other dissociated salt. For example, when the dissociated ions are those of, say, potassium, and there is additionally ammonium chloride in solution, one can have the following reaction:

$$K \text{ montmorillonite} + NH_4Cl \rightarrow NH_4 \text{ montmorillonite} + KCl.$$

A reaction of this kind is called base exchange. An apparently thick crystal ("thick" in this context means of the order of 50 angstroms, since all the minerals of this class are clay minerals of very small particle size) is no more than a pile of unbound layers, the separation between layers depending upon the humidity of the environment. In very dry atmospheres the layers are close-piled, with a distance between corresponding sheets of neighbouring layers (the so-called basal spacing) of 9 angstroms. With increasing relative humidity, water is sucked between the layers, which separate to give a basal spacing of 12 angstroms; at ordinary room humidities of the order of 75%, the basal spacing increases to 15 angstroms with further uptake of inter-layer water. This feature, the so-called expanding lattice, is characteristic of the montmorillonite minerals.

When a montmorillonite crystal in aqueous suspension dissociates the inter-layer ions, it becomes a kind of ion itself. It is a fragment whose mass is great compared with that of ions of the elements, and is not even specific to the mineral species, this mass being more or less a matter of chance. An ion of this kind is called a micelle. The reason why the kaolin minerals do not dissociate ions to form micelles to the same extent is simply that the dissociable ions are not present in quantity, since the isomorphous replacements which bring them into being do not exist. The reason why the micas do not form micelles, in spite of the appreciable extent of isomorphous replacement, is that the adventitious charge-balancing ions are tightly bound and do not dissociate. This is a statement which will need some modification when we discuss other mechanisms of dissociation and charge development in Chapter 4.

# CHAPTER 3

# The structure of water

### 3.1 The nature of the water molecule

IT has been briefly mentioned in Sections 1.3 and 1.4 that the water molecule is an electric dipole, and that ice, for example, is a structure of neutral molecules bound together by dipole attraction. The matter must now be taken a little further.

From evidence provided by the absorption band spectrum of water vapour, Mecke and Baumann (1932) concluded that each of the two hydrogen nuclei in the molecule lies at a distance of 0·96 angstrom from the oxygen nucleus (i.e. the protons are inside the oxygen atom), whilst the H—O—H angle is 105°, which is not very different from the angle of 109° subtended at the centre of a tetrahedron by any two apexes. Bernal

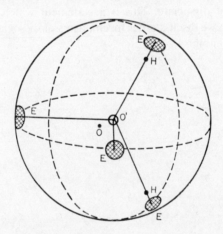

O . . . oxygen ion nucleus
E . . . tetrahedral distribution of electronic charge
H . . . protons

Figure 3.1 Structure of the water molecule after Bernal
and Fowler.

and Fowler (1933) continued the discussion by proposing a distribution for the negative electronic charge, basing their arguments on Mulliken's concept of molecular orbitals analogous to the electronic orbits in atoms. Their conclusions are illustrated by Figure (3.1). The molecule is represented by the sphere of radius 1·38 angstrom with centre at O', this being the "centre of gravity" of the negative charge distribution. The hydrogen nuclei (protons) are at the sites, H; the oxygen nucleus is at O, on that side of O' which is remote from the hydrogen nuclei, and on the bisector of the angle H—O—H. This siting of nuclei is in accordance with Mecke's conclusions. The outer electrons of the molecule provide a charge distribution, the chief feature of which is the localized concentrations at the regions marked E, which lie at the apexes of a regular tetrahedron. Of these four concentrations, two lie on the radii through O'H and reduce the effective positive charge of each hydrogen nucleus by about 50%; the remaining two concentrations provide regions of negative charge. To summarize, therefore, the water molecule may be regarded as a sphere upon the surface of which electric charge is distributed in such a way as to provide concentrations about four points which lie at the apexes of a regular tetrahedron, of which two apexes are charged positively and two negatively.

## 3.2 The packing of water molecules

Because of the charge distribution on the water molecule, one such molecule attracts another in a specific direction, the two tending to make

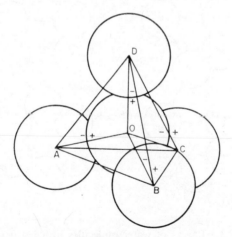

Figure 3.2 Tetrahedral grouping of four water molecules
round a fifth.

contact with the positive apex of one adjacent to the negative apex of the other. For this reason the molecules do not form close packed structure with typical tetrahedral and octahedral groups, but they *do* form tetrahedral groups of another kind due to the fact that specific directions of attraction of a molecule lie in the lines joining the centre to the points where charge is concentrated. The result is that a water molecule tends to be surrounded by a tetrahedral group of four other water molecules as shown in Figure (3.2). This is not the tetrahedral group of four close packed spheres depicted in Figure (1.4), but is a very open group of five spheres, one of which is at the centre of the tetrahedron formed by the other four. Nevertheless, because of the similarity between the groups of Figure (1.4) and Figure (3.1), the structures of water have points of similarity with those of silica minerals which are formed from groups such as that of Figure (1.4).

## 3.3 The structure of ice

If a layer of water molecules is arranged as shown in Figure (3.3a), and another as in Figure (3.3b) which is a reversal of the former in every respect, then the two can be superimposed with the molecule *D* falling on *O* of the under layer. The result is an array of tetrahedral groups of the kind shown

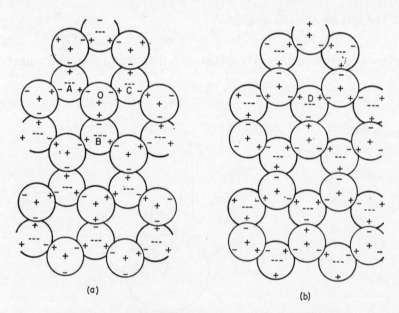

(a)                                              (b)

Figure 3.3 Layers of water molecules in ice structure. Layer (b) is superimposed on layer (a), and further layers continue the structure.

in Figure (3.2), an example being that formed by the molecules *DOABC*, lettered to conform with Figure (3.2). Other layers can be added to continue the structure indefinitely. This is the structure of ice as revealed by X-ray analysis, which does not, of course, give any evidence as to which apexes of the molecules have positive and which have negative charge. The charge pattern shown is hypothetical, and would require ice in the mass to have a large electric moment. It can be supposed that in nature this is eliminated by the different orientations of the elements of the mosaic of crystals which form the bulk.

A comparison between Figure (3.3a) and Figure (2.5) reveals a striking similarity, and the way in which two silica layers shown in Figure (2.5) are superimposed in the mineral tridymite is similar to the way in which two layers of water molecules are superimposed in ice. For this reason ice is sometimes said to have the tridymite structure.

### 3.4 The structure of liquid water

In Section 1.4 mention was made of the nature of the structure of matter in the liquid state, although from the very mobility of a liquid the lattice structure must be somewhat evanescent. The X-ray diffraction pattern of

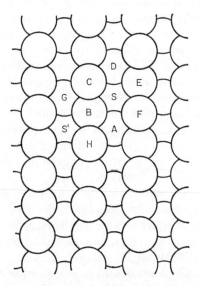

Figure 3.4 Quartzlike structure of water molecules in liquid water. Molecules *GBACH* form a distorted tetrahedral group.

liquid water is consistent with a mixture of random and lattice structures; the lattice structure is that shown in Figure (3.4). This has a striking similarity with the structure of quartz. Just as quartz, illustrated in Figure (2.1), presents spirals of oxygen ions, so does liquid water exhibit spirals of water molecules. The sequence of molecules *ABC* in Figure (3.4) is such a spiral, *DEF* being the second "thread" of the double threaded "screw". The imaginary cylinder on which these spirals lie is bound to neighbouring similar cylinders, as in quartz, by tetrahedral groups common to both cylinders formed by the crossing of two spirals of molecules. Thus the spiral *ABC* of cylinder *S* in Figure (3.4) crosses the spiral *GBH* of cylinder *S'* to form the group *GBACH* which is a somewhat distorted form of the basic water group of Figure (3.2).

### 3.5 Some properties of water

Much light is thrown on the properties of water by this elucidation of its structure. The openness of packing, whether of liquid water or ice, accounts for the relatively low density. Were the structure close packed, the density would be about doubled. Then again, the lower density of ice as compared with liquid water is consistent with the latter having quartzlike structure and the former having tridymitelike structure. The densities of quartz and tridymite were recorded in Section 2.4 just for the purpose of this comparison. The ratio of tridymite density to quartz density is 0·86, whilst the comparable ratio of ice to water is 0·91. Bearing in mind that water is an admixture of quartzlike and random structures, the agreement between these figures is significant. The maximum density of water is well known to occur at $4°c$, and this is explicable if one assumes that at that temperature water retains its quartzlike structure, but at lower temperatures there is some tendency to transpose to the tridymite ice structure of lower density. At higher temperatures than $4°c$ the density decreases for the normal reason, namely the expansion of the water mass due to thermal molecular motion expanding the lattice. Thus with a change of temperature in either direction from $4°c$ density decreases.

One further phenomenon involving the density of water is of importance. An ion or micelle in solution provides a centre of strong attraction which can modify the structure of water locally. The very open quartzlike structure gives way to close packed clusters round the ions. Since close packed water would have a density in the neighbourhood of 2 $g/cm^3$, the average density of water in an aqueous solution of ions or a suspension of micelles is greater than the normal value of unity. This fact invalidates certain methods for determining the density of micelle-forming materials, such as clay minerals, by displacement of water which is assumed to have the normal density. The attraction of ions and micelles for water is witnessed in other ways. In moving through the attractive field to take up clustered positions, the water molecules lose potential energy. The total energy in the system being conserved, there is a consequential increase of kinetic energy of the molecules, which is experienced as an increase of heat content and of temperature. This is known as the heat of solution, or, in the case of suspensions such as clay particles, the heat of wetting. Such heat of wetting is sometimes used as a method of qualitative determination of the area exposed by the particles for attraction of water molecules, i.e. of the fineness of division of the material. Conversely, if a crystal contains much water of crystallization which is bound with greater energy than is associated with the same molecules in quartzlike liquid water structure, those molecules in excess of the number clustering round ions will, in solution, gain potential energy by taking part in the open water structure, thereby causing a cooling effect. This may be demonstrated by using highly hydrated salts, such as photographers' hypo (sodium thiosulphate).

Whilst there are other properties which become explicable with the acceptance of a persistent structure for liquid water (a notable one is the anomalously high mobility of the hydrogen ion in aqueous solution, and the normal mobility of the ion of the heavy isotope of hydrogen), only one more need be mentioned. Examination of Figure (3.3a) shows a honeycomb network of hexagons similar to, and not very different in scale from, that of the "cristobalite" layer which lies at each surface of the layer lattices of the montmorillonite minerals. This has led Hendricks, Nelson and Alexander (1940) to put forward the speculation that the water between the layers of montmorillonite has the tridymitelike structure, matching the montmorillonite structure itself and serving to continue the building of the mineral in three dimensions. This might be held to account for the imbibition of inter-layer water causing an increase of spacing between the layers in well-defined steps of 3 angstroms at a time, as described in Section 2.10, instead of continuously.

## 3.6 Structures of other soil components

It would be very satisfactory if one could conclude this account of the atomic structures of soil constituents with a discussion of that colloidal organic fraction known commonly as humus. Unfortunately this must remain a task for the future. Such chemical evidence as there is goes to show that humus has some of the attributes of both lignin and protein materials. The proteins in particular are very difficult subjects to study which are only now beginning to yield to the methods of X-ray analysis.

The present situation in respect of humus is that there are two opinions as to whether this material gives a characteristic X-ray diffraction at all.

It does appear, however, that the groups, COOH, which occur in cellulose, lignin and protein, together with the groups, $NH_2$, which are characteristic of protein, persist in humus.

CHAPTER 4

# The equilibrium of particles
# suspended in water

### 4.1 The kinds of particles in suspension

THE word "suspension" is used in this chapter to mean a volume of water in which are distributed numbers of particles so small that they do not sink to the bottom of the containing vessel rapidly. This general definition obviously includes solutions of salts, in which the solute never settles out on the bottom. Hence "suspension" will be held to include "solution", but not vice versa.

The kinds of particles which need to be discussed include, naturally, water molecules themselves and the ions to which the dissociation of water molecules gives rise. These are commonly referred to as the hydrogen ion and the hydroxyl ion. The size of the latter has been referred to in Section 2.1, and from Section 3.1 it follows that the hydroxyl ion is an oxygen ion in which one hydrogen nucleus or proton is embedded. The so-called hydrogen ion, however, needs a little further discussion. A hydrogen atom which has lost its only electron in the process of becoming an ion is just a stripped proton of negligible size, even on the atomic scale. The properties of such an ion in solution, notably the mobility in an electric field, would be very different from those actually observed. It is now generally accepted that the proton is absorbed by a neutral water molecule, so that the hydrogen ion in solution is actually the ion $OH_3^+$, to which the name "oxonium ion" has been assigned. Nevertheless, for the sake of conformity it is desirable to adopt the established practice of using the name "hydrogen ion", reserving "proton" for the true ion or stripped hydrogen nucleus.

Further, suspensions will almost invariably contain neutral molecules of salts in solution, together with the ion species to which their dissociation gives rise. Indeed, very elaborate measures have to be taken to rid water of all dissolved salts. Other material in suspension may include small particles of insoluble matter, and these may be either electrically neutral or they may have electrical charges as a result of having dissociated loosely bound ions, as in the case of the montmorillonite minerals. Yet other

41

particles, the proteins, are able to absorb additional protons from hydrogen ions in solution in suitable circumstances (which will be explained in Section 5), thus becoming positively charged micelles.

From all this it emerges that the particles in a suspension may be numerous and of many different kinds. Owing to thermal motions, the particles will be colliding with each other, thereby giving or taking energy. Some neutral particles will be dissociating into component ions, and some ions will be recombining to form neutral particles. The measured concentrations of the various particles are states of equilibrium between the dissociations and recombinations. There is at the same time a condition of equilibrium of a different kind between the concentration of particles of a given species at a given point and the concentration of particles of that same kind at other points. Thus, for example, heavy particles tend to sink, under the force of gravity, to the bottom of the containing vessel, whilst the more or less violent thermal motions of these same particles tend to disperse them throughout the volume of fluid available to them. These two kinds of equilibrium are the subject of this chapter.

## 4.2 The equilibrium between neutral particles and their constituent ions

The bonds holding anions and cations together in neutral molecules will not be broken spontaneously. To separate the ions, force must be applied and work must be done, and the only source of such work available in the suspension is the energy of other particles, imparted during collisions. Since in dilute suspension by far the greater part of the total bulk is water, it is long odds on the partner in collision being a water molecule; the probability of any one neutral particle receiving a dissociating collision is therefore not dependent upon the numbers of other particles of the same kind present. It is the same for all particles of the same kind, and depends only upon the general violence of thermal motions (i.e. upon the prevailing temperature) and upon the strength of the bonds holding the constituent ions of that kind of particles together. The number of such particles per unit volume of suspension which dissociate per second will therefore simply be proportional to the number of neutral particles of the specified kind, per unit volume, available for dissociation. If this concentration be $n$, the relationship may be written:

$$\text{Rate of dissociation} = An \qquad (4.1)$$

where $A$ is a constant characteristic of the kinds of particles and of the temperature.

Now in the recombination of ions to form a neutral molecule of the species discussed in the preceding paragraph, it is self evident that only two kinds of particles of all those present can be concerned, namely the constituent ions of the specified neutral particles. One ion of each of these species must come together in a collision and energy must be given up by the pair to some other particle (most likely a water molecule), since otherwise the two ions would simply bounce apart again in an elastic collision. Since the recombination is due to a collision between an ion of one species with an ion of another, the probability of a particular individual ion of the first-named kind effecting a recombination will clearly be proportional to the likelihood of an appropriate collision, i.e. to the concentration of the second kind of ions. Let this concentration be $p$. Since each of the ions of this kind has the same probability of recombinations, the number of recombinations per unit volume per second will be proportional to the concentration, say $q$, of the first-named kind of ions. Hence the number of recombinations per unit volume per unit time is proportional both to $p$ and to $q$, and therefore to the product $pq$. Hence:

$$\text{Rate of recombination} = Bpq \qquad (4.2)$$

where $B$ is a constant characteristic of the kinds of particles and of the temperature.

The observed concentrations of both neutral particles and constituent ions remain constant at all times, so that the rate of removal of the former by dissociation is just equal to the rate of production by recombination of the latter, i.e. from Equation (4.1) and Equation (4.2),

$$An = Bpq$$

Since $A$ and $B$ are constant characteristics of the particles and of the prevailing temperature, the ratio $A/B$, which one may write as $K$, is another such constant, and the relationship between $n$, $p$, and $q$ becomes

$$pq = Kn \qquad (4.3)$$

which is the relationship sought. For a given amount of the specified material in suspension, $n$ can only increase at the expense of $p$ and $q$. In a highly ionized material $n$ is small and $p$ and $q$ are large, so that the constant $K$ must be large to satisfy Equation (4.3). Conversely, in a slightly ionized material with large $n$ and small $p$ and $q$, $K$ must turn out to be small. The constant $K$ thus reflects the readiness of the particles of the material to dissociate, and is called the dissociation constant. It is listed in standard chemical tables for a large number of chemical compounds.

It will readily be seen that the state of equilibrium between particles and constituent ions is stable. If disturbed, it will automatically tend to be restored. For if the number of neutral particles is, for some reason, momentarily increased and the concentration of ions decreased, it is seen from

Equations (4.1) and (4.2) that the rate of dissociation is also increased and the rate of recombination decreased. This condition persists for as long as the equilibrium state is thus disturbed. These changes of rate of dissociation and recombination are in the right direction for restoring the *status quo ante*.

Two features of Equation (4.3) and its derivation are noteworthy. No step in the argument requires reference to any particles in suspension other than the specified kind of neutral particles and the constituent ions, even though particles of many other kinds may be present. Hence the dissociation constant $K$ which emerges from the argument is in no sense dependent upon the presence or absence of particles other than those to which $K$ refers. The presence of such other particles does, however, account for the possibility that the constituent ions may be present in unequal concentrations, i.e. for the inequality of $p$ and $q$. Suppose, for example, that the salt to be considered is sodium chloride. It is clear that if no other chlorides or sodium salts are present, the only source of the constituent ion species, $Na^+$ and $Cl^-$, is the sodium chloride, and since each dissociation which produces one sodium cation also produces one chlorine anion, the concentrations, $p$ and $q$, of the constituent ions must be equal. If, however, among the particles present are some molecules of, say, potassium chloride, dissociation of this salt contributes to the concentration, $q$, of chlorine ions without contributing to the concentration, $p$, of sodium cations, so that $p$ and $q$ in the equation of equilibrium of sodium chloride molecules and ions are unequal. In fact, the increase of $q$, in consequence of the balance expressed by Equation (4.3), decreases the concentration $p$.

## 4.3 The dissociation of water molecules; the pH scale

The equilibrium of water molecules and the constituent hydrogen and hydroxyl ions is a very special case of the equilibrium discussed in the preceding section, since in aqueous solutions water provides the bulk of the molecules present. Furthermore, the dissociation constant of water is very low, so that the bulk of water is preponderantly in the form of neutral molecules. These two facts, taken together, lead to the conclusion that the number, $n$, of neutral molecules of water per unit volume remains sensibly constant in spite of quite wide variations of concentration of suspended particles in general and of hydrogen ions in particular. For example, in a solution of deci-normal hydrochloric acid there is about $0 \cdot 1$ g of hydrogen ions per litre (1000 g) of solution and but a negligible amount of hydroxyl ions. Allowing for the $3 \cdot 55$ g of chlorine ions, the solution consists of more than $99 \cdot 5\%$ by weight of undissociated water molecules. Soil solutions which are the subject of this book are almost invariably even more dilute than this, so that the content of neutral water molecules will range between

99·5% and sensibly 100%, a range so limited that $n$ may be regarded as constant. Hence the equation of equilibrium between water molecules and their constituent ions may be written, from Equation (4.3),

$$C_H C_{OH} = \text{constant} \qquad (4.4)$$

where $C_H$ is the concentration of hydrogen ions (the $p$ of Equation 4.3) and $C_{OH}$ is the concentration of hydroxyl ions (the $q$ of Equation 4.3).

The constant of Equation (4.4) may be determined, since it is possible to carry out experiments to measure it in a particular case. In the purest water obtainable (so-called "conductivity water") practically the only particles present other than neutral water molecules are the hydrogen and hydroxyl ions, and these will be present in equal concentrations. When an electric potential difference is applied to a pair of electrodes in this water, the movement of the ions gives rise to an electric current which may be measured. The speed or mobility of the ions is also measurable, and consequently the numbers of particles carrying the charges may be deduced. The total number being apportioned equally between hydrogen and hydroxyl ions, the concentrations of each are known. Such experiments give the result:

$$C_H = C_{OH} = 10^{-7} \text{ g equivalents per litre}$$

whence

$$C_H C_{OH} = 10^{-14} \qquad (4.5)$$

This product, being invarient, is now known for all aqueous dilute solutions.

The degree to which solutions exhibit acid properties is determined by the concentration of hydrogen ions, whilst alkaline properties are associated with a preponderance of hydroxyl ions. Pure water, which is neutral, has equal concentrations of each kind of ion, the concentration of each being $10^{-7}$ g equivalents per litre. Acid solutions have greater concentrations of hydrogen ion and, therefore, from Equation (4.5), smaller concentrations of hydroxyl ion, whilst in alkaline solutions the reverse is the case. It is not necessary to specify both $C_H$ and $C_{OH}$ when stating the degree of acidity or alkalinity, since from Equation (4.5) the one is related to the other. In practice, it is usual to state the hydrogen ion concentration, whether the solution is acid or alkaline. In acid solutions $C_H$ is greater than $10^{-7}$ and in alkaline solutions it is less. The range of hydrogen ion concentrations encountered is so great that it is a generally adopted convention to use the logarithm of $C_H$, with changed sign, as a measure of acidity. The sign is changed because otherwise, owing to the low concentrations generally encountered, the logarithm would almost always be negative. The resulting number designating the degree of acidity is called the pH of the solution, and the scale of acidity is called the pH scale. Thus the pH at neutrality is

7 ($C_H$ is $10^{-7}$, log $C_H$ is $-7$, pH is 7). Acid solutions have $C_H$ greater than $10^{-7}$, i.e. $C_H$ is $10^{-x}$ where $x$ is some number less than 7, so that pH for acid solutions is less than 7. By the same token the pH of alkaline solutions exceeds 7.

## 4.4 The influence of the pH of a solution on the electric charge developed by certain suspended micelles

Soil materials contain certain particles, the structures of which present hydroxyl ions at the surfaces. Some such ions, as in the clay minerals, are linked to silicon ions deeper in the crystal, whilst others, as in the organic soil material, are linked to carbon atoms in the structure. Such surface groups may be expressed by the symbols $\geqslant$Si—OH and $\geqslant$C—OH respectively; the bonds left unspecified indicate the structural linkage between the silicon or carbon atoms and the rest of the particle. Such $\geqslant$Si—OH groups can occur only at broken edges of the clay minerals, since at the surfaces of the layer lattices the silicon is associated with oxygen anions, not hydroxyl anions. If a layer lattice is broken, one could have a linkage $\geqslant$Si—O—Si$\leqslant$ in the complete layer becoming $\geqslant$Si—O$^-$ on one side of the fracture and $^+$Si$\leqslant$ on the other. In aqueous suspension a recombination of a proton from a hydrogen ion with the $\geqslant$Si—O$^-$ and of a hydroxyl ion with the $^+$Si$\leqslant$ would restore the similarity of the fracture surfaces, which would now present $\geqslant$Si—OH and OH—Si$\leqslant$ alike. On all exposed edges there is evidence that in suitable circumstances these surface hydroxyl ions can dissociate the proton to form a hydrogen ion in solution, leaving the surface with a negatively charged site. Such dissociation takes place, for example, in silicic acid and in organic acids. There are yet other materials, the proteins, the structures of which present surface amino groups, which one may symbolize in the form —NH$_2$. These groups can, in certain circumstances, admit an additional proton from hydrogen ions in solution to become —NH$_3^+$ groups, which thereby endow the structure with sites of positive charge. Just as negatively charged micelles associate with various cations in the solution, giving rise to base exchange phenomena, so do positively charged micelles associate with anions in solution, giving rise to anion exchange reactions. The circumstances in which these phenomena are to be observed have now to be examined.

In acid solutions the concentration of hydrogen ions is enhanced and that of hydroxyl ions is therefore depressed. Particles in suspension are therefore subject to an enhanced bombardment from the hydrogen ions, and consequently the rate of recombination between such ions and any negatively charged $\geqslant$Si—O$^-$ or $\geqslant$C—O$^-$ sites on the particles is enhanced. This leads, from the arguments of Section 4.2, to a depression of the degree of dissociation; i.e. the tendency will be for particles to have neutral $\geqslant$Si—OH or $\geqslant$C—OH sites rather than dissociated sites. But in those same conditions the enhanced bombardment encourages the addition of protons to amino groups to form charged —NH$^+$ sites. Conversely, in alkaline solutions the hydroxyl ion content is enhanced and the hydrogen ion content is depressed. The feeble bombardment of the particle surfaces by hydrogen ions depresses the rate of recombinations between $\geqslant$Si—O$^-$ or $\geqslant$C—O$^-$ groups and protons from the said hydrogen ions; hence the degree of dissociation is enhanced, and charged sites prevail at the expense of uncharged. At the same time the depression of recombination between neutral —NH$_2$ groups and protons to form positively charged —NH$_3{}^+$ groups results in a prevalence of the neutral groups. To summarize, the structure of the suspended matter determines whether the surfaces can, in any circumstances, have negative or positive charges, whilst the degree of acidity of the solution determines whether, given that possibility, the charges actually develop. Positive charges are encouraged in acid and discouraged in alkaline solution, whilst for negative charges the reverse is the case.

The discussion can be pursued quantitatively in the following way. Consider a particular group, say $\geqslant$Si—OH, the equilibrium, $\geqslant$Si—OH$\rightleftharpoons\geqslant$Si—O$^-$ + H$^+$, being governed by the dissociation constant $K_1$. Let the total concentration of groups, whether charged or neutral, be $C_t$, made up of $C_n$, the concentration of neutral groups, and $C_a$, the concentration of dissociated groups. Further, let the prevailing concentration of hydrogen ions in solution be $C_H$. Then the equation representing the equilibrium, i.e. Equation (4.3) as applied to the present case, is

$$C_H C_a = K_1 C_n \qquad (4.6)$$

Also

$$C_t = C_n + C_a$$

which, by substitution of the value of $C_n$ from Equation (4.6), becomes

$$C_t = C_a(1 + C_H/K_1)$$

or

$$C_a/C_t = 1/(1 + C_H/K_1) \qquad (4.7)$$

This gives the dependence of the ratio of dissociated to total groups upon the hydrogen ion concentration and dissociation constant. The same expression also holds for the dissociation of $\geqslant$C—OH groups, with the substitution of the appropriate dissociation constant, $K_2$, for this group.

The equilibrium, —NH$_2$ + H$^+$ $\rightleftharpoons$ —NH$_3{}^+$, is represented by the equation,

$$C_H C_N = K_3 C_c \qquad (4.8)$$

which is the form taken by Equation (4.3) in this case. Here $C_c$ is the concentration of positively charged $—NH_3^+$ groups, $C_N$ the concentration of neutral $—NH_2$ groups, and $C_H$ has its previous meaning. The dissociation constant for this particular equilibrium is $K_3$. The total concentration $C_T$ of groups of this kind is the sum of positive and neutral groups, i.e.

$$C_T = C_N + C_c$$

and by substituting for $C_N$ from Equation (4.8) as before, one arrives at the equation,

$$C_c/C_T = 1/(1 + K_3/C_H) \tag{4.9}$$

The total concentrations of groups of the two kinds considered, $C_t$ and $C_T$, are of course dependent only upon the structure of the suspended particles and the amount in suspension. For a given kind of particles in a given concentration in the suspension, $C_t$ and $C_T$ are constant. Equations (4.7) and (4.9) express the concentrations of charged groups as fractions of these constant totals. One can see that when $C_H$ is very small (alkaline solutions), the ratio $C_a/C_t$ approaches unity and $C_c/C_T$ approaches zero. This means that in these circumstances the whole of the structural hydroxyl groups are dissociated and negatively charged, whilst none of the amino groups are charged. In acid solutions $C_H$ becomes large, and Equations (4.7) and (4.9) show that then $C_a/C_t$ approaches zero and $C_c/C_T$ approaches unity. None of the hydroxyl groups are dissociated but the whole of the amino groups absorb protons to become charged. This is the qualitative result already mentioned.

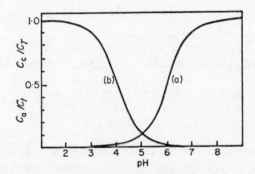

(a) development of negatively charged surface
(b) development of positively charged surface

Figure 4.1 Development of dissociation of surface active
material in suspension with varying acidity.

Curves may be drawn to show the course of development of charge with changing acidity, using Equations (4.7) or (4.9) with a selected hypothetical dissociation constant. Figure (4.1a) shows such a curve representing Equation (4.7) with a dissociation constant assumed to have the value $10^{-6}$. The acidity has been plotted on the pH scale. Similarly the curve of Figure (4.1b) represents Equation (4.9) with a dissociation constant having the value $10^{-4}$. Whilst these are assumed values of dissociation constants for the sake of illustration, it is reasonable to take $K_1$ less than $K_3$ since it might seem *prima facie* that the electrically neutral groups, —OH and —NH$_2$, are the normal stable forms. The former is undissociated and the latter may be regarded as the dissociated form of the —NH$_3$$^+$ group. That is as much as to say that the latter tends to dissociate more readily than does the former. However that may be, this is certainly observed to be the order of the dissociation constants of the two kinds of groups in that typical protein, gelatine.

One more type of dissociation equilibrium will be treated, namely that of materials such as ferric and aluminium hydroxides. These dissociate the hydroxyl ion complete, being bases usually, thereby acquiring positive charge. The positive micelles may then react with other anions in solution to take part in anion exchange. One may express the dissociation in the form, $> \text{Fe—OH} \rightleftharpoons \text{Fe}^+ + \text{OH}^-$, and the equilibrium may be expressed in the usual way by the equation,

$$C_\alpha C_{OH} = K_4 C_\nu \tag{4.10}$$

where $C_\alpha$ is the concentration of dissociated negatively charged groups, $C_{OH}$ is the concentration of hydroxyl ions in solution, $C_\nu$ is the concentration of undissociated groups, $C_\tau$ is the sum of groups dissociated or not, and $K_4$ is the appropriate dissociation constant. Also, from the equilibrium of dissociation of water (Equation 4.4),

$$C_H C_{OH} = \gamma$$

where $\gamma$ is a constant, usually taken to be $10^{-14}$. Equation (4.10) may, with the aid of this last equation, be written

$$C_\alpha \gamma / C_H = K_4 C_\nu \tag{4.11}$$

Also

$$C_\tau = C_\alpha + C_\nu$$

whence, substituting for $C_\nu$ from Equation (4.11), one arrives at

$$C_\alpha / C_\tau = 1/(1 + \gamma/K_4 C_H) \tag{4.12}$$

This is formally similar to Equation (4.9). Dissociation is encouraged in acid solutions and discouraged in alkaline solutions.

**4.5 Amphoteric properties; the isoelectric point**

The proteins have structural groups of both $\geqslant$C—OH and —NH$_2$ type and consequently both positive and negative charges are possible. As the acidity changes, both the trends of charge development shown in Figure (4.1) are to be expected. Whether the ranges of pH over which the charges develop overlap or not is determined by the two dissociation constants. If these are not very dissimilar, one has the events shown in Figure (4.2a), in which the assumption has been made that the concentrations of the two kinds of groups are equal. Here the negative charge begins to develop before the positive charge disappears, and one has a particular pH at which the two charges are equal in magnitude and the net charge is therefore zero. In Figure (4.2b) is shown the course of charge development when the dissociation constants are widely different. Here there is but negligible overlap and the individual charges, as well as the net charge, are effectively zero at a particular pH. Materials which behave in this way, exchanging bases in alkaline solutions and anions in acid solutions, are said to be amphoteric. In the state of zero net charge they are said to be isoelectric, and the pH at this state is the isoelectric point. It may be said at once that gelatine, which has been mentioned as illustrating the matter, behaves in the manner of Figure (4.2b). It is obvious that the properties of the material will differ markedly according to whether the behaviour is that of Figure (4.2a) or (4.2b). In the latter case there will be either base exchange or anion exchange, or neither, depending on the prevailing pH, whilst in the former case the isoelectric state will be a condition in which both base and anion exchange take place simultaneously. Other important differences will be discussed in Sections 4.9 to 4.11.

Since it is generally conceded that protein constituents enter into the structure of humus, amphoteric properties may be expected of humus. However, the matter has not so far been very thoroughly investigated, as a consequence, no doubt, of great technical difficulty both in isolating homogeneous humus constituents and in performing physico-chemical experiments on such material. Those clay minerals, however, which have appreciable charges, owe them in the main to isomorphous replacements in the structure which are in no way affected by the degree of acidity of the solution bathing them. Furthermore these charges are almost invariably negative, that is, of the same sign as those further charges developed in alkaline solution by the dissociation of protons from these structures. The effect of degree of acidity is, therefore, merely to increase or decrease somewhat the total negative charge, and no amphoteric behaviour is to be expected. In fact, amphoteric properties are the rare exception. Some tropical clays of low base exchange capacity have been reported as exhibiting the phenomenon of an isoelectric point, and the explanation has been proffered that the isomorphous replacements have been such as to result in a positive lattice charge. The development of negative charge in alkaline solution could then reduce the net charge, through zero, to negative values.

It has been suggested in recent years that kaolinite can develop positive

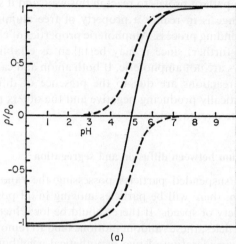

(a)

Figure 4.2 (a) Development of charge on surface active material with varying acidity. The dissociation constants of positive and negative sites are approximately equal. $\rho$ is the charge density expressed as a fraction of the maximum possible density $\rho_0$.

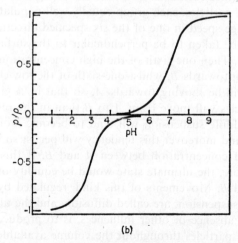

(b)

Figure 4.2 (b) As (a), but dissociation constants of positive and negative sites are well separated.

charge in acid solution by dissociation of the surface hydroxyl ions complete, as does ferric hydroxide, and certainly the phenomenon of anion exchange has been demonstrated in the finely ground material. Unfortunately only mineral ground so fine as to extinguish the characteristic X-ray diffraction pattern proves to react in this way, and it seems likely that the anion exchange is in reality a property of free aluminium hydroside released in the grinding process. Amphoteric properties of clay minerals will be discussed no further, since it may be taken as established that, as a general rule, clays are not amphoteric. If both anion and cation exchange is observed, these reactions are due to the presence of different minerals, some characteristically producing negative and the others positive micelles in the appropriate circumstances.

### 4.6 The equilibrium between diffusion and segregation

In a mass of suspended particles possessing the random motions of thermal agitation, there will be particles moving in all possible directions with a great variety of speeds. If there should be local inequalities of concentration of particles, these random motions tend to remove them, as may be seen from a consideration of two hypothetical neighbouring regions of different concentration. Let us label the regions $A$ and $B$. The velocity of a given particle may be resolved into components in the directions of the axes of rectangular Cartesian coordinates, and the particles into groups, each of which is regarded as containing particles moving in one of these directions only. Since each axis has positive and negative directions, there are six groups of particles, each group equally well populated and moving with equal average speed in one of the six specified directions. Let one of these directions be taken to be perpendicular to the surface between the regions $A$ and $B$. Then one-sixth of the high concentration of particles in $A$ will be moving towards $B$, whilst one-sixth of the low concentration of particles in $B$ will be moving towards $A$, so that in a given interval of time more particles will move from $A$ to $B$ than in the reverse direction. Thus the more dilute suspension will tend to become more concentrated and vice versa, and moreover this tendency will persist so long as there is any difference of concentration between $A$ and $B$. If this were the only tendency, therefore, the ultimate state would be equality of concentration in regions $A$ and $B$. Movements of this kind, regulated by concentration gradients in the suspension, are called diffusion, and the ultimate effect of diffusion, in the absence of other influences, is to produce uniformity of concentration of particles throughout the volume available to them.

Other influences are, however, coexistent with diffusion. All particles have some mass and therefore weight, and tend to sink to the bottom of the containing vessel. This is a segregating influence, the ultimate state due to this alone being a sediment sharply separated from a supernatant fluid which is free of particles. Such segregating forces may also be of an electrical origin. For example, an electrically charged clay particle will move but slowly, owing to its size, in playing its part in the general thermal agitation. In relation to the more mobile ions of atomic size in solution the clay particle will be a stationary centre of attraction or repulsion. The ultimate state, if this were the only influence, would be a concentration of ions at the clay surface sharply separated from ion-free solution, or, in the case of repelled ions, a similarly concentrated layer of ions removed as far from the clay surface as possible. When both diffusion and segregating forces operate, one has to enquire into the nature of the net result. Does one force overcome the other completely, or is there some form of distribution of particle concentration intermediate between complete uniformity and complete segregation? One has some guidance from the known distribution of gas molecules in a vertical column, the equilibrium of which is of the kind under discussion.

### 4.7 The equilibrium of particles subject to gravity and diffusion; osmotic pressure

As examples of gravitational segregation, one may consider the distribution of gas density or of the concentration of suspended particles in liquids in a vertical column. Both gas molecules and particles small enough to exhibit Brownian movement have thermal agitations opposing the gravitational segregation. The difference is that the gas molecules exist in space and experience only collisions with each other, while the particles are outnumbered by molecules of the suspending liquid and receive most of their thermal buffets in collision with those molecules. The random motions of the gas molecules, which are responsible for diffusion, are also responsible for the gross effect recognized as gas pressure. Other things being equal, the measured gas pressure is proportional to the concentration of molecules whose collisions with containing surfaces result in the phenomenon of pressure, so that the unobservable distribution of particle concentration may be taken as measurable in terms of the observable pressure distribution.

It is shown in Note 1, and is confirmed by observation, that the balance of diffusion and segregation in the case of a column of gas results in a pressure distribution of the form, numbered in the note as Equation (N1.6),

$$\ln(P_0/P) = Mgz/RT \qquad (4.13)$$

where $z$ is the vertical height of a point, at which the gas pressure is $P$, relative to a datum level at which the pressure is $P_0$; $\ln(P_0/P)$ is the natural logarithm of the ratio of these pressures; $M$ is the molecular weight of the gas concerned; $g$ is the acceleration of free fall under gravity; $T$ is the prevailing temperature on the absolute scale (i.e. degrees centigrade plus $273°$); and $R$ is the universal gas constant defined by the gas law:

$$PV = RT$$

In this expression $P$, $R$, and $T$ have their previous significance and $V$ is the volume occupied by one gram molecule of the gas in the given circumstances. Since a gram molecule of any gas comprises the same number, $N$, of molecules, irrespective of the chemical constitution of the gas, $V$ may equally well be regarded as the volume available to $N$ molecules. The constant $N$ is known as Avogadro's number and is given in tables of physical constants.

The interesting feature of Equation (4.13) is that the state of equilibrium is a pressure or concentration distribution, intermediate between uniformity throughout and complete segregation, as shown in Figure (4.3). This shows three different curves drawn with three different values of $M$, i.e. corresponding to particles of three different masses. The heavier the particle,

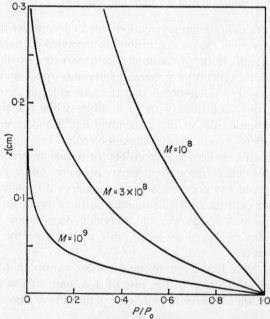

Figure 4.3 The vertical distribution of particles in equilibrium under gravitation and thermal diffusion. $M$ indicates the molecular weight.

the nearer does the equilibrium approach complete segregation, as might be expected from experience of suspended particles. It is next of interest to enquire how far suspended particles have concentration distribution of this kind. Perrin (1908) solved this experimental problem with his classical experiments with gamboge particles. The difficulty is, of course, that very large particles segregate completely, and since this is also in accordance with Equation (4.13), such a test is hardly crucial. On the other hand, particles of molecular dimensions are invisible and cannot be counted. It so happens that there is an intermediate range of particle sizes which, whilst large enough to be seen and counted in the field of view of a suitable microscope, are small enough to diffuse effectively and so to set up a concentration distribution intermediate between uniformity and segregation. The gamboge particles used by Perrin satisfied these requirements. The striking result was that when allowance was made for the buoyancy of the particles in the suspending liquid, the vertical distribution of concentration of particles was found to agree with the modified form of Equation (4.13), even to the extent of requiring the same value of gas constant $R$. The modified equation, derived also as Equation (N1.12) in Note 1, is,

$$\ln(n_0/n) = (Mgz/RT)(1-1/\rho) \qquad (4.14)$$

where $n_0$ and $n$ are particle concentrations at heights differing by $z$, $\rho$ is the specific gravity of the suspended particles, and the other symbols have their previous meanings. The "molecular weight", $M$, of the particles is in this case the mass of a particle expressed in atomic units, i.e. with the mass of the hydrogen atom as the mass unit. It is assumed that the suspending fluid is water with unit specific gravity.

The conclusion is that the distribution of suspended particles can be correctly computed by precisely the same process as for gas molecules; the particles in fact behave in this respect just as they would behave were they gas molecules occupying the same volume of space as is available to them in the suspension. They obey the gas law with the same value of the gas constant, $R$, as applies to true gases. They may be regarded as exerting, in suspension, a partial pressure which regulates their diffusion, and which may be calculated by assuming them to be molecules of gas. It is this partial pressure which is known as the osmotic pressure of the suspension due to the presence of these particular particles.

## 4.8 The hydrostatic pressure differences which accompany differences of osmotic pressure

If the concentrations of particles in a suspension can be maintained at different values at different parts of the suspension, in a state of true equilibrium (i.e. if the differences are not merely temporary and due to be removed in the normal course of diffusion), differences of hydrostatic pressure at the different parts arise in consequence. For example, if a solution is divided into two regions by a membrane which is permeable to the solvent but not to the solute, and the solution on one side of this membrane is more concentrated than that on the other, water diffuses preferentially from the dilute region, through the membrane, to the more concentrated region. Membranes of this kind are described as semi-permeable, and the whole apparatus is called an osmometer. The preferential diffusion of the water is called osmosis. The osmosis results in a rise of level of the solution on the more concentrated side, and this introduces an increase of hydrostatic pressure which tends to oppose the osmosis, so that eventually a state of equilibrium supervenes in which hydrostatic pressure and osmotic pressure are greater on one side of the membrane than on the other. It may be shown by thermodynamical reasoning that the difference of hydrostatic pressure is equal to the difference of osmotic pressure. In fact, semi-permeable membranes provide the basis of classical experiments designed to measure the osmotic pressures of solutions by direct measurement of the accompanying differences of hydrostatic pressure.

In the case considered in Section 4.7, the differences of concentration are maintained as a result of equilibrium between segregating and diffusive effects, and again it can be shown that at each point in the column of suspension there is a component of hydrostatic pressure due to the concentration of particles, and that this pressure is equal to the osmotic pressure. The derivation of this result is carried through in Note 2.

### 4.9 The equilibrium of particles diffusing in an electric field; the Gouy layer

Ions of atomic mass in solution are, of course, acted upon by gravity and have appropriate vertical distributions of concentration, but very tall columns of solution are required to demonstrate any sensible departure from uniformity of concentration, as a very rough calculation based on Equation (4.14) will show. Electrical segregating forces of the kind described in Section 4.6 do, however, result in marked variations of ion concentration within very small distances. A charged particle of a size which is very large compared with that of the ion acts as a relatively stationary centre which attracts ions of opposite sign of charge and repels ions of like sign. The ultimate equilibrium distribution of ions is a consequence of the balance of effects of just the same kind as were discussed in reference to a column of gravitating particles, but the calculations required to express this distribution in a form analogous to Equation (4.13) are much more complicated.

The difficulty arises from the fact that the force exerted on the ion by the electric charges present is not constant, irrespective of where the ion is located; the farther away from the centre of attraction or repulsion, the smaller is the electric force, and moreover other ions located between the one considered and the micelle exert their own repelling or attracting effect. The analysis is given in Note 3, following Verwey and Overbeek (1948). The results are illustrated by Figure (4.4), which presents the ion distributions for some selected cases.

Two distributions are involved, one for ions of sign unlike that of the charged surface, and another for ions of like sign. The former will depict a concentration of ions increasing with proximity to the attracting surface, the latter decreasing due to repulsion. The two distributions are not independent, since a cation and an anion are the constituents of a neutral

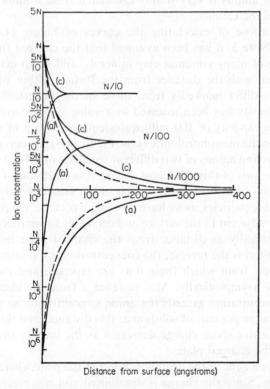

Distance from surface (angstroms)

Figure 4.4 The distribution of ions in the Gouy layer on a plane micelle. Full lines show monovalent ions, broken lines show divalent ions. Curves (c) indicate cations, curves (a) anions. The concentration of remote solution is indicated on the curves.

uncharged molecule, which will also have its separate existence and concentration distribution. The distribution of uncharged molecules is a simple one. Since there is neither repulsion nor attraction of an uncharged molecule, the charged surface exerts no segregating force whatever. Diffusion enjoys unhindered play and the neutral molecules become uniformly dispersed throughout the space available to them. Let this uniform concentration be $n_m$; let the concentration of constituent cations at a given point be $n_c$; and the concentration of constituent anions at that same point be $n_a$. Then by Equation (4.3),

$$n_a n_c = K n_m \qquad\qquad (4.15)$$

where $K$ is the dissociation constant of the molecule considered. The right-hand side of Equation (4.15) being constant for the given suspension, the distribution of anions is very simply calculated from a knowledge of the distribution of the cations.

For the purpose of calculating the curves of Figure (4.4) from the equations of Note 3, it has been assumed that the charged surface is flat, as in the case of many common clay minerals, and of an extent which is large compared with the distance from the surface within which the ion concentrations differ markedly from those of the unperturbed solution. The charge density has been assessed at a value which corresponds to a base exchange capacity of 100 milliequivalents per 100 g of dry mineral, which puts it in the montmorillonite class. Figure (4.4) presents distributions of both cations and anions of two different valencies in N/1000 solution and of monovalent ions of three different distant concentrations. One sees that the distribution of the attracted cations is of the same general form as that of the gravitating particles, as we have cause to expect; i.e. the concentration is a maximum adjacent to the surface and decreases to approach a constant value asymptotically as distance from the charged plate increases. The anion distribution is the reverse; the concentration is a minimum near the charged surface, from which these ions are repelled, and increases to a constant value asymptotically. At a distance $x$ from the charged surface the cation concentration exceeds the anion concentration so that there is net positive charge per cm$^3$ of solution at this distance, and this concentration of net positive space charge decreases as the locality becomes more remote from the charged plate.

There is thus a zone in the neighbourhood of the plate where there is net positive charge. Since the charge is distributed and not precisely located, the zone in which it resides is called the diffuse layer, or quite commonly the Gouy layer, after the first computer of this ion distribution (Gouy, 1910). The total charge in the system must clearly be zero, since the ions and micelle arise from the dissociation of initially neutral particles, and as much negative as positive charge must arise from such dissociation. The total positive space charge in the Gouy layer must therefore equal the negative charge on the surface of the micelle. At a sufficient distance from that surface the electric force on an ion must be zero, since the negative surface charge is completely screened by the intervening equal positive space charge. There is therefore, far from the surface, no segregating force even on the ions, which are therefore diffused uniformly at concentrations equal to those in solution undisturbed by charged stationary surfaces. The cation and anion concentrations therefore approach asymptotically the same value at points remote from the micelle, namely the value obtaining in undisturbed solution; this is shown in Figure (4.4). Lastly, it is seen that the thickness of the Gouy layer is less in solutions of divalent ions than in those of monovalent ions, for the same concentrations at remote distances, whilst for ions of the same valency, the Gouy layer thickness is less in concentrated than in dilute solutions.

The ion distribution found by these calculations is an impossible one at points very close to the charged surface. The calculation proceeds on the assumption that an ion is a point charge, and that therefore there is no upper limit to the possible charge concentration. In fact ions have radii of the order of $10^{-8}$ cm. They can approach the surface no nearer than this, and there is an obvious limit to their concentration. The simple Gouy theory can demand impossible concentrations at impossibly close distances from the surface. The finite size of an ion, rather than diffusion, opposes the segregating attraction of the cations and limits the ion concentration in the immediate neighbourhood of the charged surface; the closely adsorbed layer of ions so limited is known as the Stern layer. This layer will not be considered further, but it must be remembered in what follows that any deductions made on the basis of Gouy's analysis is subject to modification in the immediate vicinity of the micelle surface.

At each point in the Gouy layer at which the concentration of cations exceeds that of the anions, there is an osmotic pressure in excess of that in remote undisturbed solution. This may be shown by a development of Equation (4.15), from which

$$n_c n_a = n_0^2 = K n_m$$

This follows since both $n_c$ and $n_a$ take the value $n_0$ in the remote solution which is undisturbed by the micelle. Hence if $n_c$ exceeds $n_0$ by an amount $\Delta$, then $n_a$ must be less than $n_0$ by some other amount, say $\delta$, such that

$$(n_0 + \Delta)(n_0 - \delta) = n_0^2$$

or

$$n_0(\Delta - \delta) = \Delta \delta \qquad (4.16)$$

c*

The osmotic pressure in the solution is the sum of the partial osmotic pressures due to the contributory molecules and ions, and each partial pressure is determined only by the concentration of the relevant particles and not by their species. Thus the total osmotic pressure is proportional to the total concentration of particles. At the point in the Gouy layer the osmotic pressure $P_i$ is given by

$$\begin{aligned} P_i &= A(n_m + n_c + n_a) \\ &= A\{n_m + (n_0 + \Delta) + (n_0 - \delta)\} \\ &= A(n_m + 2n_0 + \Delta - \delta) \end{aligned}$$

where $A$ is a constant. Similarly the osmotic pressure $P_0$ in the remote solution is given by

$$P_0 = A(n_m + 2n_0)$$

Hence the pressure in the Gouy layer at the specified point exceeds that in remote solution by the amount,

$$\Delta P = P_i - P_0 = A(\Delta - \delta)$$

From Equation (4.16) the quantity $(\Delta - \delta)$ is essentially positive, since it equals the inherently positive ratio $\Delta\delta/n_0$. Hence the osmotic pressure at points within the Gouy layer exceeds that at points outside the influence of the micelle. Further, $\Delta$ may take very large values as the surface of the micelle is approached, but $\delta$ may not exceed $n_0$ since the ion concentration cannot be negative. Hence $\Delta - \delta$ and therefore $\Delta P$ increases as the micelle surface is approached. The difference may be calculated from Note 3, Equations (N3.6) and (N3.7), together with Equation (N3.2).

Just as in other cases of difference of osmotic pressures between regions in equilibrium, including the gravitational equilibrium described in Section 4.8 and Note 2, this difference of osmotic pressure is accompanied by a difference of hydrostatic pressure of equal magnitude. In this case, the hydrostatic pressure at the distance $x$ from the clay surface is due to the electrostatic attraction toward the surface which is exerted on the space charge in the Gouy layer beyond the distance $x$. This pressure varies with distance, and is a mutual interaction between the charged surface and the Gouy layer. At the charged surface itself the pressure is exerted on the micelle surface, but this is not a pressure which tends to move the micelle. It is analogous to the pressure which a person would experience were he to hug a sponge rubber cushion to his chest. Such a pressure would not tend to project him across a room. The micelle and its accompanying Gouy layer move together in equilibrium, and the pressure which tends to move this inseparable system is the zero pressure at the Gouy layer boundary remote from the surface of the micelle.

## 4.10 The interaction between two inter-penetrating Gouy layers

A single micelle does not constitute a mass of material with observable physical properties. Such physical properties are traceable to the mutual interactions between neighbours in such a mass. The interactions between two such neighbours in idealized circumstances will be discussed in this section.

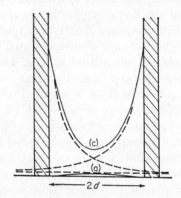

Figure 4.5 The distribution of ions in the Gouy layer between two parallel plane micelles. Broken lines indicate unimpeded Gouy layers.

Figure (4.5) presents a picture of two micelles with plane surfaces facing and parallel to each other. Each micelle is of the type considered in the preceding section, and each will tend to develop its own Gouy layer ion distribution as indicated by the broken lines. As is seen, the separation, $2d$, between the surfaces is taken to be insufficient to provide space for the free development of each layer, and the ion distributions must be of the kind shown, with the cation concentration nowhere falling to the value $n_0$ appropriate to remote undisturbed solution and the anion concentration nowhere rising to this value. The distributions must be symmetrical about the medial plane distant $d$ from each surface, as shown.

The mutual interactions between the micelles arise from the persistent difference between the cation and anion concentrations at the medial plane. The osmotic pressure and the accompanying hydrostatic pressure differences between a region in the Gouy layer and the region outside the influence of the micelles occur wherever there is an inequality between these concentrations, and this remains true at the medial plane where the cation concentration $_dn_c$ still exceeds the anion concentration $_dn_a$. At this plane, however, the hydrostatic pressure cannot be merely a mutual effect between the micelle and its accompanying Gouy layer, because from the condition of symmetry the gradients of the ion concentrations and of the electric potential must be zero. That is to say, there can be no electric force one way or the other at the medial plane. That half of the Gouy layer which is on one side of the plane just balances electrically the surface charge of the micelle on that side, and similarly for the other side. The medial plane is completely screened electrically from each micelle surface. On each side of it the space charge builds up increasing hydrostatic pressure by interaction with the corresponding micelle as that surface is approached, but at the medial plane itself, where the space charge urge toward either surface is zero as it passes through its change of direction, the hydrostatic pressure due merely to self-contained interaction is zero. Hence the hydrostatic pressure which is undoubtedly experienced because of the inequality of the cation and anion concentrations must be imposed from outside the space charge between the plates, and is consequently effective in tending to separate the micelles. A discussion of the source of this injected pressure lies outside the scope of this work, but may be found in a paper by Childs (1954). It requires a consideration of the complete and continuous Gouy layer, both between the surfaces of finite plates and on the outer surfaces of the pair of micelles. A somewhat crude simile would be the effect produced by two persons with a sponge rubber cushion between them, both exerting pressure on the edges. The resulting pressure in the cushion would tend to separate them.

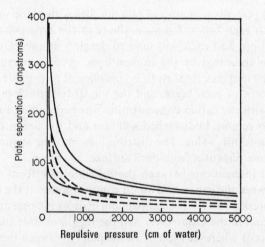

Figure 4.6 Repulsive pressure between plane parallel micelles as a function of the separation. Full lines show monovalent ions, broken lines show divalent ions. For a given valency the curves are respectively, reading from bottom to top N/10, N/100, N/1000.

It follows from what has been said that the separating pressure decreases as the anion and cation concentrations at the medial plane become more nearly equal, and, other things being equal, this occurs as the surfaces become more widely separated. Thus $\Delta P$, to use the nomenclature of Section 4.9, decreases as $d$ increases. This is shown by the curves of Figure (4.6), which have been constructed from calculations which are presented in detail in Note 4. Again, it has been shown in Section 4.9, and in particular by Figure (4.4), that the extent of the freely developed Gouy layer on a single charged surface is reduced both by increasing the concentration $n_0$ of ions in the remote solution and by increasing the valency $v$. Hence a pair of such surfaces may approach more closely for the same degree of interference or inter-penetration if the salt solution in which they are suspended is more concentrated or of higher valency; or conversely the degree of interference, of inequality of cation and anion concentration at the medial plane, and of osmotic and hydrostatic pressure differences is reduced for a given separation $2d$ when the suspension solution is of higher concentration, of higher valency, or both. All these effects are illustrated in Figure (4.6).

### 4.11 Collisions between micelles; dispersion and flocculation

In a sufficiently dilute suspension of charged particles there is on average plenty of room for each to develop its appropriate Gouy layer without restrictions due to the proximity of neighbours. It must happen, however, in the ordinary course of thermal motion of the micelles, that from time to time one micelle will find itself moving towards another. The one will not influence the other as long as the intervening distance is sufficiently great to accommodate the Gouy layer of each, freely developed, but as soon as the situation shown in Figure (4.5) sets in, a repulsive hydrostatic pressure is introduced which retards the particles, which therefore approach with steadily decreasing speed. There must come a time when the particles are brought to a halt, their original kinetic energy having been entirely converted into potential energy. The repulsive pressure, which is a maximum at this distance of closest approach, then operates to move the particles apart again, provided that no other kinds of force arise to overcome the repulsion and keep the particles together. The end of the encounter sees the particles once again outside the range of each other's influence, moving with speeds equal to those they had initially, but in the opposite directions. The particles have, in fact, suffered an elastic collision. A suspension of such particles, each repelling any other with which it collides, has a persistent distribution of concentration, appropriate to the masses of the individual particles, throughout the height of the column, in accordance with Equation (4.14). The suspension is said to be fully dispersed.

If the Gouy layer repulsion were the only force effective during an encounter, all collisions would evidently be elastic and complete dispersion of suspension would be the invariable rule. The fact is that dispersion is not invariable, and two phenomena have been invoked to account for this. There may well be others. The first is the Van der Waals force which was discussed in Section 1.3. It is a force of attraction between uncharged bodies and comes into operation only when the bodies approach each other very closely. In the case of charged bodies such as micelles, the force of attraction must be great enough to overcome the Gouy repulsion if the net force is to be attractive, and hence the second phenomenon invoked is one tending to reduce this Gouy repulsion so that it may be the more easily overcome. This phenomenon is the hydration of the micelles. The water molecules, being dipoles (see Sections 3.1 and 3.5), are tightly bound to the micelles where they are in close proximity, and are a part of the moving particle rather than a part of the liquid through which the particle moves. Thus the moving particle may more properly be conceived as a solid core with a shell of tightly bound water, and this shell will contain the Stern layer and that part of the Gouy layer nearest to the micelle. It may be called a part of the Gouy layer for convenience, although the ion distribution will not necessarily be that depicted in Figure (4.4) as calculated in Note 3, since that analysis was based on the concept of ions moving freely amongst freely mobile water molecules. However this may be, the Gouy layer which is responsible for the mutual repulsion of neighbouring moving complex particles consists only of that part outside the shell of bound water. The collision between particles is a collision between moving water shells, and the repulsion is the repulsion between particles, the charge on each of which is the charge on the particle together with the charge, of opposite sign, in the bound part of the Gouy layer. The net charge is reduced much below the magnitude of the particle charge, and close approach in the collision is possible. This close approach enhances the chances of the Van der Waals forces overcoming the repulsion and thereby producing a net attraction. The colliding particles cannot, of course, remain in contact without an apparent total loss of the initial kinetic energy and of the potential energy associated with the Van der Waals attraction. This energy may be imparted to third bodies in the collision, such as the water molecules. The conditions which favour the combination of a pair of particles, similarly favour such inelastic collisions generally throughout the suspension, and partnerships collide with other partnerships successively to produce visible large floccules. The process is called flocculation. The floccules settle out of suspension in accordance with Stokes law (Stokes, 1845) at a rate appropriate to the mass and density of the compound floccules and not of the individual particles. That is to say, the settlement is comparatively rapid and the supernatant liquid quickly clears.

An alternative suggestion, put forward by van Olphen (1950) and by Schofield and Samson (1952) is that hydroxyl groups located at the edges of clay mineral plates, as described in Section 4.4, may adsorb protons, particularly in acid solutions, just as neutral water molecules may capture additional protons to generate oxonium, or so-called hydrogen, ions. The resulting positive edge charge is then available to attract negative surface charge of nearby micelles. Electron-micrographical evidence for the existence of positive edge charges had been obtained earlier by Thiessen (1942). The postulation of such a mechanism avoids the difficulty of accepting the possibility of long range Van der Waals forces such as would be demanded in the absence of hydration.

The conditions which favour flocculation are clearly those which favour the formation of a thin Gouy layer, so that very little of it spreads outside the bound water layer. It has been seen that, for a particle charge density of given magnitude, reduction of the Gouy layer thickness is achieved (see Section 4.9) by increasing the concentration of salts in the solution in which the particles are suspended, by choosing salts of high valency, and in certain cases, by adjusting the acidity of the solution to depress the development of those charges on the particle which may be due to the dissociation or absorption of protons by certain ion groups at the surfaces of the particles (see Section 4.4). Since adjustment of the acidity can only be achieved by an addition of an acid or an alkali to the solution, and such an acid or alkali is at the same time a salt which tends to reduce the thickness of the Gouy layer for a given charge, the reverse process of enhancing dispersion by adjusting the acidity to favour the development of extra charges is less certain of success than is adjustment for flocculation. The two functions of increasing the charge and decreasing the Gouy layer thickness are then in opposition, and an optimum compromise must be sought. The dispersion of clay, for example, is usually achieved by the addition of hydroxides of monovalent elements or radicals (sodium and ammonium are common) to enhance the dissociation of protons from —OH groups, but an incautious excess straightway flocculates the suspension.

## NOTES

### Note 1. The distribution of particles in a vertical column when diffusion and segregation forces are in equilibrium

Consider a vertical column of unit cross-sectional area containing $n$ particles per unit volume, each of mass $m$ and true density $\rho$, moving freely in space as gas molecules with the velocities of thermal agitation. If the mass of the hydrogen atom is $a$, then $m/a$ is the molecular weight $M$. If $\sigma$ is the density of the gas, that is to say, the mass of matter per unit volume occupied, then

$$\sigma = nm \qquad\qquad (N1.1)$$

The thermal agitation of the particles is experienced by containing surfaces as the gas pressure $P$ which obeys the gas law,

$$PV = RT \qquad\qquad (N1.2)$$

where $V$ is the volume occupied by one gram molecule, namely $M$ grams, of the gas at the prevailing temperature $T$ on the Kelvin scale, which exceed the centigrade temperature by 273°, and $R$ is the gas constant. Because of the presumption that $n$ varies up the column, it follows that $P$ also will vary in sympathy.

Consider an element of the column between the heights $z$ and $z+\delta z$, as shown in Figure (N1.1). This has volume $\delta z$ and therefore weight $g\sigma\,\delta z$, where $g$ is the

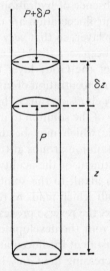

Figure N1.1 The pressure forces on an element of a vertical column of gas molecules.

gravitational acceleration of free fall. This weight is the segregating force which, if unopposed, must cause all the particles to settle at the base of the column. However, the agitated particles below the element exert a pressure $P$ on it tending to move it upward while the particles above it exert the downward pressure $P+\delta P$. Hence the net downward force, which must vanish if the element is to maintain its position in equilibrium, is

$$P+\delta P+g\sigma\,\delta z-P = 0$$

or

$$\delta P+g\sigma\,\delta z = 0 \qquad\qquad (N1.3)$$

There are three variables in this equation, since by Equation (N1.1) the density $\sigma$ varies due to the variation of $n$, which is responsible also for the variation of $P$.

However, since in Equation (N1.2) the volume $V$ is the volume occupied by $M$ grams, the density $\sigma$ is simply $M/V$ so that Equation (N1.2) may be written:

$$M/V = \sigma = MP/RT \qquad (N1.4)$$

where $M$, $R$ and $T$ are all either known constants or measurable parameters. Substitution in Equation (N1.3) results in

$$\delta P + (gMP/RT)\delta z = 0$$

or, in the limit when the increments $\delta P$ and $\delta z$ become vanishingly small and are written respectively $dP$ and $dz$,

$$dP/P = -(gM/RT)dz \qquad (N1.5)$$

Starting with a known pressure $P_0$ at the base where $z$ vanishes, the increment $dP$ may be calculated from this equation for a chosen increment $dz$ and the pressure $P$ may be plotted as a function of $z$ by summing the changes of $P$ for a continuous succession of increments of $z$. This is, of course, what is expressed by the integration of Equation (N1.5), which is an elementary form, the result being:

$$\ln(P_0/P) = Mgz/RT \qquad \begin{cases} (N1.6) \\ (4.13) \end{cases}$$

where $P$ is the pressure at the height $z$ and $\ln(P_0/P)$ is the natural logarithm of the ratio $(P_0/P)$.

Alternatively the volume $V$ in Equation (N1.2) may be regarded as the volume occupied by $N$ molecules, where $N$ is Avogadro's number or the number of molecules in a gram molecule. Then from Equations (N1.1) and (N1.4),

$$nm = MP/RT$$

or, since $M/m$ is Avogadro's number $N$

$$n = NP/RT \qquad (N1.7)$$

where all except $n$ and $P$ are constants of the system. Thus,

$$P_0/P = n_0/n \qquad (N1.8)$$

where $n_0$ and $n$ are respectively the numbers of particles per unit volume at the base of the column and at height $z$. Thus an alternative form of Equation (N1.6) is

$$\ln(n_0/n) = Mgz/RT \qquad (N1.9)$$

Now consider particles of solid suspended in liquid of density $\rho'$. Since the assumption to be tested is that such particles obey the gas laws, the analysis is the same as before except for the circumstance that the segregating weight of the particles is reduced by the buoyancy in accordance with Archimedes's principle. Thus each particle, of volume $m/\rho$, displaces the same volume of liquid of density $\rho'$, so that the weight $mg$ is reduced by the weight of displaced liquid $g(m/\rho)\rho'$. Hence the effective weight of $n$ particles in a unit volume of suspension is no longer $gnm$, or $g\sigma$ from Equation (N1.1), but $gnm - gn(m/\rho)\rho'$ or $g\sigma(1-\rho'/\rho)$. Hence Equation (N1.3) must now be written:

$$\delta P + g\sigma(1-\rho'/\rho)\delta z = 0 \qquad (N1.10)$$

and the solution now takes the form, instead of Equations (N1.6) and (N1.9),

$$\ln(P_0/P) = \ln(n_0/n) = (Mgz/RT)(1-\rho'/\rho) \qquad (\text{N1.11})$$

Where the liquid is water or very dilute solution, the density $\rho'$ becomes equal to unity, and Equation (N1.11) becomes:

$$\ln(P_0/P) = \ln(n_0/n) = (Mgz/RT)(1-1/\rho) \qquad \left\{ \begin{array}{l} (\text{N1.12}) \\ (4.14) \end{array} \right.$$

In the case of columns of gas, it is relatively easy to measure $P$ and impossible to measure $n$ directly, while in columns of suspended solid particles, it is easier to measure $n$ than $P$, but in any case, a measurement of either suffices to test the relationship. In the case of suspended solids the particles are not molecules and the ratio of the measured value of $m$ to the value of $a$ given in tables is a pseudo molecular weight calculated for the purposes of applying the formula.

### Note 2. The hydrostatic pressure due to particles in suspension

In the case of particulate matter in suspension, a unit volume of suspension contains a mass $\sigma$ of solid particles with a total volume of $\sigma/\rho$, while the remaining volume, $1-\sigma/\rho$, is of liquid with mass $\rho'(1-\sigma/\rho)$. The total mass of the unit volume of suspension, that is to say, the density of the suspension, is therefore $\rho''$, where

$$\rho'' = \sigma + \rho'(1-\sigma/\rho)$$
$$= \rho' + \sigma(1-\rho'/\rho) \qquad (\text{N2.1})$$

In Figure (N1.1) the element of height $\delta z$ has a hydrostatic pressure at the upper end, measured in cm head of water, which is in excess of that at the lower end by an amount $\delta h$, where

$$\delta h = -g\rho'' z = -g\delta z\{\rho' + \sigma(1-\rho'/\rho)\}$$

This is evidently in excess of the pressure difference which would have been experienced in the absence of suspended particles, when $\sigma$ takes the value zero, by the amount $\delta h_\sigma$ where

$$\delta h_\sigma = -(g\rho''\delta z - g\rho'\delta z) = -g\delta z\sigma(1-\rho'/\rho) \qquad (\text{N2.2})$$

Reference to Equation (N1.10) shows that

$$\delta P = \delta h_\sigma \qquad (\text{N2.3})$$

where $P$ is here the partial pressure which the particles in suspension would have if they were gas particles, that is to say, it is the osmotic pressure of the solution. Since both the excess hydrostatic pressure and the osmotic pressure are known to vanish together, the integrated form of Equation (N2.3) is simply

$$P = h_\sigma \qquad (\text{N2.4})$$

In the state of equilibrium, therefore, a difference of osmotic pressure is accompanied by a measurable and equal difference of hydrostatic pressure.

### Note 3. The distribution of ions in the Gouy layer

The analysis followed here is that of Verwey and Niessen (1939) rather than that of Gouy (1910), but the physical basis is the same. Equivalent solutions have been derived by several other authors.

Consider a solution of salt molecules of one kind only, which dissociate into cations of valency $v$ and anions of the same valency. In undisturbed solution the concentration of each of the two kinds of ions is the same, say $n_0$. If now there is an addition of clay particles, each of which dissociates to provide cations of the same kind as those of the salt molecules, itself remaining as a multianionic micelle, there will be a disturbance of the ionic concentrations in the neighbourhood of the charged clay surfaces. It will be assumed that the clay surface is typically plane and of large extent compared with the distance from it at which its disturbing influence is felt, so that it is necessary to consider only the variation of the ionic concentration in the direction, say $x$, perpendicular to the surface. Let the distributed charge density over the surface be $s$.

The charged plate will have an electric potential $\phi_s$ at its surface, relative to the zero potential in distant solution, and as the potential changes with distance from the surface there will be gradients of potential, $d\phi/dx$. Since $d\phi$ is, by definition, the work done on unit charge in moving it the distance $dx$, i.e. $d\phi$ is the product $-E\,dx$, where $E$ is the force of the electric field opposing the movement, it follows that $E$ is given by $-d\phi/dx$. The negative sign is introduced because the direction of $E$ is opposite to that in which $\phi$ increases, since it opposes the movement if work is done against it, as is required by definition.

Now consider an element of a column of unit cross-sectional area, with its axis perpendicular to the clay surface, at a distance $x$ from the surface and with length $\delta x$. At this distance, let the concentration of cations be $n_c$, so that if the electronic charge is written $\varepsilon$, the total charge on the cations in the volume element is $n_c v \varepsilon \delta x$ and the segregating force on this charge in the electric field $E$ is $F$ where

$$F = -n_c v \varepsilon \delta x (d\phi/dx) \qquad (N3.1)$$

The cations exert an osmotic pressure $P$ which obeys the gas law,

$$PV = RT$$

where $V$ is the volume of solution occupied by one gram ion of the cations, or, in other words, by $N$ cations where $N$ is Avogadro's number. Thus $n_c$, the number per unit volume, is related to $V$ by the equation,

$$n_c = N/V$$

Hence the gas law may be written more suitably for present purposes in the form,

$$P = n_c(R/N)T$$
$$= n_c kT \qquad (N3.2)$$

where $k$ is the Boltzmann form of the gas constant, to be found listed in standard tables.

The partial pressure of the cations outside the element exerted at $x+\delta x$ exceeds that at $x$ by the amount $(dP/dx)\delta x$, so that the net force on the element, $F'$, due to the thermal diffusive forces, is

$$F' = -(dP/dx)\delta x$$

where the negative sign is used because the force is in the opposite direction to that of increasing $x$. By the use of Equation (N3.2) this may be written:

$$F' = -kT(dn_c/dx)\delta x \qquad (N3.3)$$

The net force on the element is thus the sum of Equations (N3.1) and (N3.3), and since the element is in equilibrium, this net force must vanish. Hence

$$n_c v\varepsilon(d\phi/dx) + kT(dn_c/dx) = 0$$

Again, since $\phi$ varies with $x$ coincidentally with $n_c$, $dn_c/dx$ may be expanded in the form $(dn_c/d\phi)(d\phi/dx)$ so that the equilibrium equation reads

$$n_c v\varepsilon + kT(dn_c/d\phi) = 0 \tag{N3.4}$$

In the form,

$$dn_c/n_c = -(v\varepsilon/kT)d\phi$$

this is the equivalent of the gravitational Equation (N1.5), and on integration between the limits $n_0$ at remote distances where $\phi$ vanishes and the distance $x$ where the values $n_c$ and $\phi$ prevail, one obtains the equation,

$$\ln(n_c/n_0) = -v\varepsilon\phi/kT \tag{N3.5}$$

This is the equivalent of Equation (N1.6) of the gravitational case. It does not serve to provide the distribution of $n_c$ directly because $\phi$ varies with $x$ in a way which is not known at this stage. Equation (N3.5) leads directly to the form,

$$n_c/n_0 = e^{-v\varepsilon\phi/kT} \tag{N3.6}$$

where e is the well-known constant, 2·718, which is the base of the natural logarithms.

The calculation of the anion distribution follows the same course, except that the direction of the segregating electric force is opposite to that which results in Equation (N3.1) and the force is thus of opposite sign, so that the final result is

$$n_a/n_0 = e^{+v\varepsilon\phi/kT} \tag{N3.7}$$

The reason that the electric field force varies with distance from the clay surface is that the distributed space charge in the Gouy layer, which is on balance of opposite sign to the surface charge on the clay, increasingly screens the surface charge as distance from the surface increases. This is expressed by the Poisson equation which, for the unidirectional case appropriate to the present discussion, is

$$d^2\phi/dx^2 = -4\pi v\varepsilon(n_c - n_a)/\gamma \tag{N3.8}$$

where the product $v\varepsilon(n_c - n_a)$ is the net charge density in the Gouy layer at the distance $x$ and $\gamma$ is the dielectric constant of the suspending liquid, usually water. Using values of $n_c$ and $n_a$ from Equations (N3.6) and (N3.7), one arrives at the result:

$$d^2\phi/dx^2 = (4\pi n_0 v\varepsilon/\gamma)(e^{v\varepsilon\phi/kT} - e^{-v\varepsilon\phi/kT}) \tag{N3.9}$$

This may be put into simplified notation after the introduction of the contractions,

$$\psi = v\varepsilon\phi/kT \tag{N3.10}$$

$$\psi_s = v\varepsilon\phi_s/kT \tag{N3.11}$$

$$\xi = xv\varepsilon(8\pi n_0/\gamma kT)^{\frac{1}{2}} \tag{N3.12}$$

where the subscript $s$ refers to values at the clay surface.

From these it follows that

$$d\phi/dx = (d\phi/d\psi)(d\psi/d\xi)(d\xi/dx)$$
$$= (8\pi n_0 kT/\gamma)^{\frac{1}{2}} d\psi/d\xi \tag{N3.13}$$

and

$$d^2\phi/dx^2 = (8\pi n_0 kT/\gamma)^{\frac{1}{2}}(d^2\psi/d\xi^2)d\xi/dx$$
$$= (8\pi n_0 v\varepsilon/\gamma)d^2\psi/d\xi^2$$

Substitution of this result in Equation (N3.9), together with substitution of Equation (N3.10), yields

$$2\,d^2\psi/d\xi^2 = e^\psi - e^{-\psi} \tag{N3.14}$$

This may be written:

$$d(d\psi/d\xi)^2/d\psi = e^\psi - e^{-\psi}$$

At an infinite value of $x$, and therefore of $\xi$, the values of both $\phi$ and $d\phi/dx$ vanish and so, therefore, do the values of $\psi$ and $d\psi/d\xi$. Hence integration of Equation (N3.13) between this limit and the chosen value of $x$ and therefore $\xi$ gives the result,

$$(d\psi/d\xi)^2 = e^\psi + e^{-\psi} - 2$$

Since for positive potential on the clay, the potential decreases to zero at great distances, the gradient of potential is negative, so that the square root of the above solution must be taken with the negative sign thus,

$$d\psi/d\xi = -(e^\psi + e^{-\psi} - 2)^{\frac{1}{2}}$$
$$= -(e^{\frac{1}{2}\psi} - e^{-\frac{1}{2}\psi}) \tag{N3.15}$$

For integration this may be written in the form

$$d\psi/(e^{\frac{1}{2}\psi} - e^{-\frac{1}{2}\psi}) = -d\xi$$

Integration proceeds by the method of factorisation. Thus Equation (N3.15) is written

$$-d\xi = e^{\frac{1}{2}\psi}d\psi/\{(e^{\frac{1}{2}\psi})^2 - 1\}$$
$$= e^{\frac{1}{2}\psi}d\psi/\{(e^{\frac{1}{2}\psi} - 1)(e^{\frac{1}{2}\psi} + 1)\}$$
$$= \tfrac{1}{2}\,e^{\frac{1}{2}\psi}d\psi\{1/(e^{\frac{1}{2}\psi} - 1) - 1/(e^{\frac{1}{2}\psi} + 1)\} \tag{N3.16}$$

At the clay surface where $x$ and therefore $\xi$ vanish, $\psi$ has the value $\psi_s$, so that upon integrating Equation (N3.16) between the limits 0 and $\xi$, corresponding to $\psi_s$ and $\psi$, one arrives at

$$-\xi = \ln\{(e^{\frac{1}{2}\psi} - 1)(e^{\frac{1}{2}\psi_s} + 1)/(e^{\frac{1}{2}\psi} + 1)(e^{\frac{1}{2}\psi_s} - 1)\}$$

which, upon rearrangement to give $\psi$ explicitly yields

$$e^{\frac{1}{2}\psi} = \{e^{\frac{1}{2}\psi_s} + 1 + e^{-\xi}(e^{\frac{1}{2}\psi_s} - 1)\}/\{e^{\frac{1}{2}\psi_s} + 1 - e^{-\xi}(e^{\frac{1}{2}\psi_s} - 1)\} \tag{N3.17}$$

Thus if the surface potential is known, so is the potential at any distance $x$ by an application of Equation (N3.17), together with Equations (N3.10) to (N3.12); then from Equations (N3.6) and (N3.7) the ion distribution is revealed. However, the surface potential is not directly known, but must be expressed in terms of the known density of surface charge on the clay surface. Since the suspension is electrically neutral as a whole, the charge on the clay surface must be equal in magnitude and opposite in sign to the net charge in the complete extent of the Gouy layer.

Thus,

$$s = -\int_0^\infty v\varepsilon\,(n_c - n_a)\,dx$$

or, after combing this with Equation (N3.8),

$$s = (\gamma/4\pi)\int_0^\infty d^2\phi/dx^2\,dx$$

The potential gradient at infinite distance vanishes, so that this integral becomes

$$s = -(\gamma/4\pi)(d\phi/dx)_0 \tag{N3.18}$$

where the subscript indicates the potential gradient at the clay surface. From Equation (N3.15), where the potential gradient is expressed in terms of the transformed variables $d\psi/d\xi$, one has, with the help of Equation (N3.13),

$$d\phi/dx = -(8\pi n_0 kT/\gamma)^{\frac{1}{2}}(e^{\frac{1}{2}\psi} - e^{-\frac{1}{2}\psi})$$

and consequently, at the clay surface where $\psi$ has the value $\psi_s$,

$$(d\phi/dx)_0 = -(8\pi n_0 kT/\gamma)^{\frac{1}{2}}(e^{\frac{1}{2}\psi_s} - e^{-\frac{1}{2}\psi_s}) \tag{N3.19}$$

Finally the potential at the surface is expressed in terms of the known surface charge by combining Equations (N3.18) and (N3.19) thus:

$$s = (\gamma n_0 kT/2\pi)^{\frac{1}{2}}(e^{\frac{1}{2}\psi_s} - e^{-\frac{1}{2}\psi_s}) \tag{N3.20}$$

For clay minerals the surface charge density is known in terms of the chemically measured cation exchange capacity. It may be changed into units of electric charge per unit area of surface since the density $\rho$ and the thickness of the clay plates are known, or alternatively there are available chemical methods of measuring the surface area of the clay. Thus if the cation exchange capacity is $C$ milliequivalents per 100 g of dry clay mineral, the electrical charge carried is $-NC\varepsilon/1000$ units per hundred grams, where $N$ is Avogadro's number; and the area over which it is distributed is $200/\rho d$ cm$^2$ where $d$ is the thickness of a single clay plate as revealed by the structure. It must be remembered that each plate has two surfaces. Thus,

$$s = -5NC\varepsilon\rho d/10^6 \tag{N3.21}$$

The negative sign is due to the fact that the dissociation of cations leaves the surface negatively charged.

The distribution of ions in the Gouy layer is now calculable from Equations (N3.21), (N3.20), (N3.17), (N3.6), (N3.7) and (N3.12). Thus $s$ is first calculated from Equation (N3.21) and then the surface potential, or rather $\psi_s$, from Equation (N3.20). This is simpler than appears, since $\frac{1}{2}(e^n - e^{-n})$ is the function sinh $n$ tabulated in standard tables. Next one chooses the distance, $x$, at which the ion concentration is to be found, and calculates the value of $\xi$ from Equation (N3.12), and from this and the value of $\psi_s$ one next finds the value of $\psi$ from Equation (N3.17). The cation concentration follows at once from Equation (N3.6) and the anion concentration from Equation (N3.7), with a known value of undisturbed concentration $n_0$.

The various constants $N$, $k$, $\varepsilon$ and $\gamma$ are to be found in standard tables of physical constants, the density $\rho$ of clay minerals may be taken to be $2 \cdot 5$ g/cm³, and a convenient room temperature of 27°c may be assumed, to give a temperature $T$ of 300° on the Kelvin scale. Also the value of the concentration of undisturbed ions is usually known in terms of normality, say $\bar{n}_0$, so that $n_0$ may be replaced where it occurs by $\bar{n}_0 N/1000v$. These numerical values are substituted in the various equations for the sake of ready numerical calculation, to provide the following equivalent forms:

$$s = -3 \cdot 62 \; Cd \; (10^9) \text{e.s.u.} \tag{N3.21a}$$

where, for montmorillonite type minerals, $d$ has the value of about $9 \cdot 5 \times 10^{-8}$ cm.

$$\sinh \tfrac{1}{2}\psi_s = 2 \cdot 8s(v/\bar{n}_0)^{\frac{1}{2}}(10^{-5}) \tag{N3.20a}$$

$$\xi = 3 \cdot 24(\bar{n}_0 v)^{\frac{1}{2}}(10^7)x \tag{N3.12a}$$

$$\chi = e^{\frac{1}{2}\psi} = \{e^{\frac{1}{2}\psi_s}+1+e^{-\xi}(e^{\frac{1}{2}\psi_s}-1)\}/\{e^{\frac{1}{2}\psi_s}+1-e^{-\xi}(e^{\frac{1}{2}\psi_s}-1)\} \tag{N3.17a}$$

$$n_c/n_0 = \bar{n}_c/\bar{n}_0 = e^{-\psi} = 1/\chi^2 \tag{N3.6a}$$

$$n_a/n_0 = \bar{n}_a/\bar{n}_0 = e^{\psi} = \chi^2 \tag{N3.7a}$$

These formulas were used for the calculation of Figure (4.4) assuming montmorillonite type mineral with plate thickness of $9 \cdot 5 \times 10^{-8}$ and a cation exchange capacity of 100 mg equivalents per 100 g. The various values of valency $v$ and solution concentration $\bar{n}_0$ are indicated on the diagram.

**Note 4. The equilibrium in the Gouy layer between two parallel plane micelles**

The circumstances of the nature of the clay particles, the surface charge density, and the constitution of the solution in which the clay is suspended are taken to be the same as in Section 4.9 and Note 3, with the exception that there are now two micelles with their surfaces parallel and facing each other, the separation between them being $2d$. The equations of the equilibrium of the ions and of the distribution of potential, as also of the simplifying contractions, Equations (N3.10) to (N3.12), apply equally to the present case down to the differential Equation (N3.14), namely

$$2 \, d^2\psi/d\xi^2 = e^{\psi}-e^{-\psi} \tag{N4.1}$$

The boundary conditions are, however, different, since at the medial plane the gradient of potential, $d\phi/dx$, vanishes as therefore does the transformed gradient $d\psi/d\xi$, while the potential itself has the value $\phi_d$ and the quantity $\psi$ which represents it takes the value $\psi_d$. Hence integrating between the limits of $\psi_d$ and $\psi$ on the one hand and the corresponding limits of zero and $d\psi/d\xi$ on the other, one arrives by the same process as was used for treating Equation (N3.14) at the result, corresponding to Equation (N3.15),

$$d\psi/d\xi = -(e^{\psi}+e^{-\psi}-e^{\psi_d}-e^{-\psi_d})^{\frac{1}{2}} \tag{N4.2}$$

The second stage of the integration to give $\xi$ as a function of $\psi$ involves the elliptic integral $F(\alpha,\beta)$ defined as

$$F(\alpha,\beta) = \int_0^{\beta} d\omega/(1-\alpha^2\sin^2\omega)^{\frac{1}{2}} \tag{N4.3}$$

Numerical values of this integral are tabulated in the tables of Jahnke and Emde (1938) for combinations of ranges of $\alpha$ and $\beta$.

To show that Equation (N4.2) does in fact lead to an elliptic integral, put

$$e^{-\psi_d} = \alpha \tag{N4.4}$$

$$e^{-\psi} = \alpha \sin^2\omega \tag{N4.5}$$

so that from Equation (N4.5) one has

$$-e^{-\psi}d\psi/d\omega = 2\,\alpha\,\sin\,\omega\,\cos\,\omega$$

or, using Equation (N4.5) again to substitute for $e^{-\psi}$,

$$d\psi/d\omega = -2/\tan\,\omega \tag{N4.6}$$

Thus by the use of Equations (N4.3) to (N4.6), the Equation (N4.2) may be written in terms of the variable $\omega$ in the form,

$$d\xi = 2\,d\omega/\{\tan^2\omega(\alpha\sin^2\omega + 1/\alpha\,\sin^2\omega - \alpha - 1/\alpha)\}^{\frac{1}{2}}$$

By the use of the elementary relationships between trigonometrical ratios this may readily be rearranged to take the simple form,

$$d\xi = 2\,d\omega/\{\tan^2\omega[(1/\alpha)\cot^2\omega - \alpha\cos^2\omega]\}^{\frac{1}{2}}$$
$$= 2\,d\omega/\{1/\alpha - \alpha\sin^2\omega\}^{\frac{1}{2}}$$
$$= 2\,\alpha^{\frac{1}{2}}\,d\omega/\{1 - \alpha^2\,\sin^2\omega\}^{\frac{1}{2}} \tag{N4.7}$$

In integrating between limiting values of $x$, namely zero and $d$, the corresponding limits of $\xi$ and $\omega$ must be established first. The corresponding limits of $\xi$ are respectively zero and $\xi_d$ as given by Equation (N3.12) The corresponding limits of $\omega$ are given by Equations (N4.4) and (N4.5), from

$$e^{-(\psi-\psi_d)/2} = \sin\,\omega \tag{N4.8}$$

Thus at the medial plane where $\psi$ takes the value $\psi_a$

$$\omega_d = \pi/2 \tag{N4.9}$$

while at the clay surface where $\psi$ has the value $\psi_s$

$$\omega_s = \sin^{-1}e^{-(\psi_s-\psi_d)/2}$$

Thus upon integrating Equation (N4.7) between the stated limits one has the result, after substituting for $\alpha$ outside the integral from Equation (N4.4),

$$\xi_d = 2e^{-\psi_d/2}\int_{\sin^{-1}e^{-(\psi_s-\psi_d)/2}}^{\pi/2} d\omega\Big/\Big(1 - e^{-2\psi_d}\sin^2\omega\Big)^{\frac{1}{2}}$$

$$= 2e^{-\psi_d/2}\left[\int_0^{\pi/2} d\omega\Big/\Big(1 - e^{-2\psi_d}\sin^2\omega\Big)^{\frac{1}{2}} + \right.$$

$$\left. -\int_0^{\sin^{-1}e^{-(\psi_s-\psi_d)/2}} d\omega\Big/\Big(1 - e^{-2\psi_d}\sin^2\omega\Big)^{\frac{1}{2}}\right]$$

Comparison with the definition of the elliptical integral, Equation (N4.3), shows that this expression for $\xi_d$ is equivalent to the elliptical integral form

$$\xi_d = 2\,e^{-\psi_d/2}\{F(e^{-\psi_d},\pi/2) - F(e^{-\psi_d}, \sin^{-1}e^{-(\psi_s-\psi_d)/2})\} \quad (N4.10)$$

Verwey and Overbeek (1948) present a square table derived from tables of elliptic integrals in which $\xi_d$ is given for combinations of a range of values of $\psi_s$ and $\psi_d$. Thus with $\psi_s$ given, the value of $\psi_d$ may be determined for any chosen $\xi_d$, that is to say for any given separation of the micelles, and the anion and cation concentrations at the medial plane follow from an application of Equations (N3.6) and (N3.7). The osmotic and hydrostatic repulsion pressures then follow. However, again the surface potential is unknown except in terms of the surface charge density of the micelle, and must be determined.

Equation (N3.18) applies equally to the present case, giving

$$s = -(\gamma/4\pi)(d\phi/dx)_0 \quad (N4.11)$$

where the potential gradient at the clay surface, $(d\phi/dx)_0$, follows from Equation (N4.2) in conjunction with the transformation Equation (N3.13). Thus,

$$d\phi/dx = (8\pi n_0 kT/\gamma)^{\frac{1}{2}} d\psi/d\xi \quad (N4.12)$$

so that, from Equation (N4.2) with the $\psi$ value of $\psi_s$ appropriate to the micelle surface,

$$(d\phi/dx)_0 = -(8\pi n_0 kT/\gamma)^{\frac{1}{2}}(e^{\psi_s} + e^{-\psi_s} - e^{\psi_d} - e^{-\psi_d})^{\frac{1}{2}}$$

and, from Equation (N4.11),

$$s = (n_0 kT\gamma/2\pi)^{\frac{1}{2}}(e^{\psi_s} + e^{-\psi_s} - e^{\psi_d} - e^{-\psi_d})^{\frac{1}{2}} \quad (N4.13)$$

This is the equation which corresponds to Equation (N3.20) for the single Gouy layer.

The repulsion pressure is calculated from the ion density at the medial plane as compared with that in remote solution. Thus by an appropriate modification of Equation (N3.2) to sum the osmotic pressure contributions of all the molecular and ionic particles present, the pressure at the medial plane is

$$P_d = kT(n_m + {}_d n_c + {}_d n_a)$$

while in remote solution it is

$$P_0 = kT(n_m + 2n_0)$$

Hence the difference, $P,\Delta$ is

$$\Delta P = kT({}_d n_c + {}_d n_a - 2n_0) \quad (N4.14)$$

and this is the magnitude of the excess of pressure on the confronting surfaces of the micelles over that on the outer surfaces, where the Gouy layers develop freely. Thus the equivalent accompanying hydrostatic pressure is the swelling or repulsion pressure.

Equations (N3.6) and (N3.7), together with the contraction (N3.10) combine with Equation (N4.14) to yield

$$\Delta P = kTn_0(e^{-\psi_d} + e^{\psi_d} - 2) \quad (N4.15)$$

The procedure for calculating the repulsion pressure for a given separation of the micelles is now clear. It is simplest to assume a value of $\psi_d$ from which the repulsion pressure follows at once from Equation (N4.15), and to calculate the separation $2d$, appropriate to this repulsion. Thus with $s$ known from the cation exchange capacity and surface area of the micelle, in accordance with Equation (N3.21), $\psi_s$ is determined from Equation (N4.13). Again this is simpler than appears, due to the fact that $e^{\psi_s} + e^{-\psi_s}$ is 2 cosh $\psi_s$, which is a tabulated function. With both $\psi_d$ and $\psi_s$ now known, the quantity $\xi_d$ may now be calculated from the elliptic integrals of Equation (N4.10) or by interpolation in the square table derived from these integrals and presented by Verwey and Overbeek, as already described. Then the value of $d$ follows from the transformation (N3.12) with $d$ substituted for $x$.

For the purpose of ready computation it is convenient again to rewrite the relevant equations after substitution of the known values of the physical constants and parameters, as was done also at the conclusion of Note 3, with the concentration of the ions in the suspending solution expressed in terms of normality, $\bar{n}_0$. For a room temperature of $27°C$, the equations are

$$P = 25(\bar{n}_0/v)(e^{-\psi_d} + e^{\psi_d} - 2)(10^6)$$

$$= 50(\bar{n}_0/v)(\cosh \psi_d - 1)(10^6) \text{ dynes/cm}^2 \qquad \text{(N4.15a)}$$

$$\cosh \psi_s = \cosh \psi_d + 1 \cdot 57 s^2 (v/\bar{n}_0)/10^9 \qquad \text{(N4.13a)}$$

$$\xi_d = 2e^{-\frac{1}{2}\psi_d}\{F(e^{-\psi_d}, \pi/2) - F(e^{-\psi_d}, \sin^{-1} e^{-\frac{1}{2}(\psi_s - \psi_d)})\} \qquad \text{(N4.10a)}$$

$$d = 3 \cdot 1 \xi_d/10^8 (v\bar{n}_0)^{\frac{1}{2}} \qquad \text{(N3.12a)}$$

# CHAPTER 5
# The sizes of soil particles

## 5.1 The nature and purpose of mechanical analysis

THE size distribution of the particles which make up a soil is called the mechanical composition of that soil, and any process for determining that property is called mechanical analysis. This is not the place to describe in detail the methods of mechanical analysis, which may be referred to in many practical textbooks. It is enough to say here that all the usual techniques depend upon the relationship between the size of a particle and the speed with which it falls through a column of viscous liquid, commonly water. Sedimentation is certainly a form of interaction between soil particles and water, and therefore an analysis of this phenomenon lies within the scope of this book. In passing, it may be said that at one time mechanical composition was regarded as a topic of importance because it showed promise of providing an objective basis of assessment of soil texture, but in recent years this aspect of the subject has tended to recede into the background since it has been recognized that the mineralogical identity of the clay is at least as important as the amount of clay in affecting soil texture. The methods of mechanical analysis have, nevertheless, remained of importance in separating soil fractions for further mineralogical study.

The obvious method of separating particles into size groups is by the use of graded sieves, and this is indeed the method used where the sieve required is not too fine meshed. To separate particles of finer grain, one uses the fact that small particles fall more slowly through a column of water than do larger particles. The shape of the particle also has some influence on the speed of fall, and when interpreting the speed of settling, one assumes conventionally that the particles are spherical. Hence the basis of the estimate of size is different in the different groups. In the larger groups a particle of size between $r_1$ and $r_2$ is one which passes a sieve of mesh size $r_1$ but is retained on one of mesh $r_2$. Whether it is retained or allowed to pass, may well depend upon whether the sieves have square or round apertures and whether the particle is round or rod-shaped. In the groups of smaller particles the size range $r_3$ to $r_4$ is defined by a measured settlement

77

velocity range $v_3$ to $v_4$, and the size is calculated from the velocity by an application of Stoke's law which was developed for spherical particles. The size so derived is, of course, the size which the particle would have had if it had been spherical and falling with the same velocity, and it is usual to refer to it in some such hedging terms as "effective size". This chapter presents a discussion of the three basic ways in which the observation of the settlement of suspended particles can provide an estimate of mechanical composition.

## 5.2 Stokes's law

If a particle is held motionless in a column of water and suddenly released, it will begin to fall with the acceleration of gravity, $g$, since there are no opposing forces. As soon as the acceleration, acting for a finite time, results in the particle gaining a finite velocity, the water exerts a frictional retarding force and the acceleration consequently decreases. So long as there is any acceleration at all the velocity continues to increase and the frictional retardation continues to increase with it, until a point is reached at which the force of gravity is just balanced by the opposing frictional force. From this point onwards the acceleration is zero, and the body continues to fall at constant speed. If the particle is a big one, it may well have reached the bottom of the water column before this constancy of speed is attained, and with such particles we are not concerned. They can readily be sorted out by sieves. Again, if the particle is sufficiently minute, the thermal motions may complicate the issue so that the final state is one of equilibrium with a distribution of particles in the column, as described in Section 4.7 and Note 1, and not one of constant speed of fall until the bottom of the column is reached. Such small particles are not the subject of this chapter. Between these two extremes there are particles of a great range of sizes which may be regarded as achieving their steady speed of settlement practically at the moment of release. This speed, $v$, is related to the radius $a$ of the particle (which is assumed to have specific gravity $\rho$) and to the viscosity $\eta$ of the water (of unit specific gravity) by the equation, derived in Note 5,

$$v = (2a^2g/9\eta)(\rho - 1) \qquad (5.1)$$

Substituting values for the viscosity of water and the gravitational constant from tables, and assuming for $\rho$ the value 2·6, which for most soil minerals is approximately true, Equation (5.1) takes the approximate form,

$$v = 34700 \, a^2 \text{ cm/sec}$$

This is the equation commonly used to derive $a$ from a knowledge of $v$. For accurate work, the viscosity appropriate to the prevailing temperature must be used.

The law that speed of settlement is determined by particle size may be applied in three different ways as follows. In the first, one divides the particles into two separate groups, one of which contains all particles settling faster than a preselected speed $v$ and no particles at all with slower settlements, the other containing all particles settling at speeds slower than $v$ and no other particles. This is called a complete separation of fractions. In the second application, one samples the column of sedimenting particles at a measured depth $d$ after a known time interval $t$, and from the result one calculates the proportions of particles settling with greater and smaller velocities respectively than $v$, where $v$ has the value $d/t$. In the third application, one weighs the sediment as it settles on the bottom of the sedimentation vessel, and deduces the mechanical composition from this rate of precipitation.

### 5.3 The complete separation of fractions

Attention may first be focused upon those particles present which fall at exactly the speed $v$. It will be supposed that the suspension occupies a height $h$ in a cylindrical vessel, and that all particles, of whatever size, are initially uniformly dispersed throughout this height by vigorous shaking. From the moment that settlement is allowed to proceed, all particles of the exact selected velocity of fall descend without overtaking each other or being overtaken. Those which started at the top will constitute a "ceiling" for that group, and as the ceiling descends at velocity $v$, the base of the column of those particles collapses as sediment on the bottom of the vessel. At the moment that the ceiling arrives at the "floor", the time $t$ will have elapsed where $t$ has the value $h/v$, and at this moment the sediment will contain all the particles with settling velocity $v$, no matter at what height they started. The ceilings of particles having other settling velocities will have arrived at the floor already if the particles are larger and fall faster, or will not yet have arrived if the particles fall more slowly than $v$. Hence at the time $t$, the sediment will contain not only all particles of settling velocity $v$, but in addition all those of faster fall and some particles with slower rates of fall. Hence if at the time $t$, the supernatant liquor be decanted from the sediment, it will contain *only* particles with rates of fall slower than $v$, but not *all* of those particles. If the sediment be now redispersed in fresh water to the height $h$ and the shaking and sedimentation be repeated, again the supernatant liquor will contain only, but not all of, the finer particles which were contained in the first sediment. After a sufficient number of repetitions, usually three or four, it will be found that the decanted liquor is clear water, so that the remaining sediment contains particles with a velocity of fall greater than $v$ and no other, while the total collected liquors contain particles with velocity of fall less than $v$ and no other.

Complete separation of fractions is time consuming, especially when the division between fractions is set at a rather low velocity, and when it is required to divide the particles into a number of fractions at various size boundaries. In the latter case, one naturally divides first at the slowest settling velocity, so that it is the sediment from the first fractionation which is further divided, rather than the supernatant liquor. Elutriation has been developed as a means of making the separation into fractions a continuous process, and in this modification the dispersed suspension is made to flow vertically upward at the velocity chosen to define the division between the fractions. Particles with a faster settling velocity then fall against the stream, whilst those with a slower settling velocity are carried with the stream. Thus the elutriation vessel ultimately contains a sediment consisting of all particles, and only particles, of settling velocity greater than $v$. Passage through a number of elutriation chambers in series, of different diameters and consequently through which the suspension travels with different speeds, effects division into a number of fractions in one continuous process. The fastest flow, of course, has to occur in the first chamber in the train, for in the reverse order the suspension in subsequent chambers would have none of the particles which those chambers were designed to separate.

### 5.4 Mechanical analysis by sampling the suspension

It was remarked in Section 5.3 that the "ceiling" of those particles which settle with velocity $v$ forms a boundary below which such particles are present in their original concentration, since all fall together without changing their relative position. Above the ceiling there are no particles at all of that size. Hence if one extracts a sample by means of, say, a pipette, that sample must contain either the initial concentration of particles or no particles at all, depending upon whether the sample is taken from above or below the ceiling. Now suppose that a number of groups of particles of different settling velocities are shaken up together, and that after a time interval $t$ a sample of suspension is taken from a depth $d$. The ceiling of those particles whose settling velocity is equal to $d/t$ will just have arrived at the depth $d$; the ceilings of all faster settling particles will have fallen below $d$; and the ceilings of all more slowly settling particles will still be above the depth $d$. Hence the sample will contain particles of the latter groups in their original concentrations, and none at all of the faster settling particles, since the sample is from below the ceilings of the one and above the ceilings of the other.

Consider a sample of volume $V_s$ taken from the total volume $V_T$ of suspension, which sample is found to contain $m_s$ gm of solid particles. Then $m_s$ is the mass of particles whose settling velocity is less than $d/t$, and the mass of such particles in the whole suspension is, by proportion, $m_T$, where

$$m_T = m_s(V_T/V_s) \qquad (5.2)$$

Thus sampling attains the result got by the more laborious complete separation of fractions.

Direct sampling by pipette is attended by some disadvantages. The finite volume extracted inevitably includes suspension from both below and above the sampling depth $d$, so that the sample must contain some particles having a faster rate of fall than $d/t$, and must lack some particles with a slower rate of fall than this. However, provided the sample is small, these two errors approximately neutralize each other. A more serious objection is that even the most careful manipulation of the pipette must disturb the settling of the column to some extent, and in an attempt to avoid this various indirect sampling methods have been proposed from time to time, without achieving much popularity.

One such indirect method employs manometers to measure hydrostatic pressure differences between points in the suspension. The hydrostatic pressure difference depends upon the density of the suspension and that, in turn, is related to the percentage of suspended solids. Let the manometers measure the pressures at two points separated by the vertical distance $\Delta d$ at the mean depth $d$, and let the mean density of the suspension in this neighbourhood be $\sigma$. The water columns in the manometers, measured from the respective points of communication with the suspension, have lengths differing by the amount $\Delta h$. This gives

$$\Delta h = \sigma \Delta d \qquad (5.3)$$

from which one can estimate $\sigma$. A sample of 1 cm$^3$ of suspension containing $m_s$ gm of solids of specific gravity $\rho$ consists of $m_s/\rho$ cm$^3$ of solids and $1-(m_s/\rho)$ cm$^3$ of water, with a total mass of $1+m_s(1-1/\rho)$ gm, i.e.

$$\sigma = 1+m_s(1-1/\rho)$$

or

$$m_s = \rho(\sigma-1)/(\rho-1) \qquad (5.4)$$

$\sigma$ having been found from the experiment (see Equation 5.3), one can at once infer the value of $m_s$ and proceed as in the direct sampling method.

In yet another variation, the suspension density $\sigma$ is measured by hydrometer. For adequate sensitivity the bulb of the hydrometer must be large compared with the stem. On the other hand, because the density varies with depth, the bulb ought to be small in order that it may define a depth at which the density is being measured. With the exception of Bouyoucos's (1927) version, the hydrometer has not been much used. This version employs a hydrometer bulb of considerable length, so that the density measured is the mean density of a long column over which density variation is considerable. A precise analysis and interpretation of the results is not possible, and therefore appeal is made to empirical calibration.

None of these methods entirely avoids disturbance of the column of settling particles, but the most recent method is unobjectionable on these grounds. A beam of light is allowed to traverse the column of suspension at a chosen depth and time, and the intensity of the emergent beam is measured by a photoelectric cell. The degree of absorption of light in the column is a measure of the turbidity, i.e. of the concentration of solid particles in the suspension. Again the concentration is derived by appeal to an initial empirical calibration of the apparatus, using standard suspensions of known concentrations.

### 5.5 The rate of deposit of sediment

The analysis of the rate of sedimentation as a means of mechanical analysis is, after Odén (1915), due to Svedberg and his collaborators (1923a). Again it is noted that all the particles of a given size fall at the same rate, so that the rate of increase of the mass of sediment due to these particles is constant as long as the "ceiling" is still falling. As soon as the ceiling of these particles arrives at the bottom of the vessel, increase of mass of sediment abruptly ceases. If the velocity of settling of this group is $v$, and the height of the column of suspension is $h$, then the ceiling arrives at the bottom after the lapse of time $t$, where $t$ is equal to $h/v$. Thus the rate of increase of sediment is constant until this time has elapsed, and is thereafter zero. Now, in a suspension containing particles of many different settling velocities, each group makes its contribution to the increase of the sediment at its own constant rate, and ceases to make that contribution at the moment that the ceiling of that group arrives at the bottom of the vessel. Since ceilings are arriving at the bottom continuously, beginning with that of the group with the fastest rate of descent and continuing with those of the groups of slower and slower descent, the rate of growth of the sediment is a maximum at the beginning of the period of settling and decreases steadily with lapse of time as more and more groups of particles cease to make their contribution. The curve obtained by plotting the mass of sediment against elapsed time is therefore of the form shown in Figure (5.1).

Now consider the point $B$ on the curve $ABC$ of Figure (5.1). If the total height of the settling column is $h$, then at the time $t$ corresponding to the

c

point $B$, the ceiling of that group of particles whose velocity is $h/t$ has arrived at the bottom and the group ceases to contribute further. The slope of the curve therefore decreases beyond $B$. Other groups of faster descent have already ceased to contribute at times earlier than $t$, and the slope of the curve is therefore steeper between $A$ and $B$. Had these faster groups been absent, the ceiling of the group with velocity of descent $h/t$ would have been the first to arrive, all the groups present would have contributed

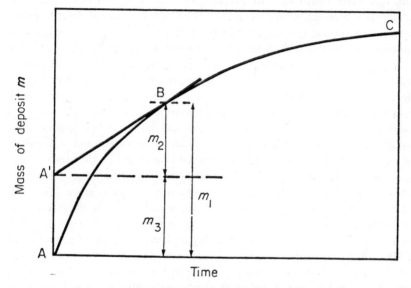

Figure 5.1 The rate of deposit of sedimenting particles in suspension.

to the increase of sediment at a constant rate, and the curve would have been a straight line up to time $t$, with a slope equal to that of the curve $ABC$ at the point $B$. This curve is shown as $A'BC$, $A'B$ being the tangent to $ABC$ at $B$. The origin of the curve is now $A'$. Had the faster particles been absent, therefore, the mass of sediment at time $t$ would have been $m_2$. In fact it is $m_1$. The difference must therefore be due to those particles which had ceased to make their contribution by the time $t$, i.e. to those particles with a velocity of descent greater than $h/t$. This contribution is evidently $m_3$, the difference between $m_1$ and $m_2$, and is given by the intercept of the tangent on the axis representing the mass of sediment. Hence at any point on the curve at which the elapsed time is $t$, the intercept of the tangent on the mass axis gives the mass of those particles with velocity of descent in excess of $h/t$.

The experimental arrangements necessary to obtain a record such as

D

that of Figure (5.1) involve the introduction of a plate into the settling suspension. This plate is hung from a balance arm and collects the sediment. This necessarily interferes with the freedom of settling, and errors are thereby introduced, as in other methods.

## 5.6 The limits of validity of Stokes's law

It has been mentioned in Section 5.2 that large particles fall with acceleration through an appreciable fraction of their descent. Furthermore, their descent is not governed by Stokes's law even if they attain constant velocity, due to the turbulence set up by their rapid passage. Again, the total time of descent would be small and incapable of accurate timing. Fractionation of large particles by sedimentation is therefore ruled out, but the easy method of sifting is available. At the other extreme, Brownian movement may prevent complete sedimentation, and furthermore, there is some evidence that particles which are too large to be suspended in equilibrium on this account may nevertheless be suspended indefinitely through the suspension forming a weak gel in certain circumstances. Then again, it must be admitted that slowness of settling of very fine particles, if not an invalidation of the method, is certainly a disadvantage. The conventional settling speed adopted to divide the clay from the silt fractions is the equivalent of 10 cm in 8 hours, and to subdivide the clay at a size one-tenth of the upper conventional clay limit would need an elapsed time of 800 hours, i.e. some 33 days. It is true that Russell (1943) has circumvented this disadvantage by using a micropipette technique with a sampling depth of only a few mm, but niceties of manipulation of this kind are hardly likely to be widely adopted. On all counts some method other than gravitational sedimentation needs to be sought for further analysis of the conventional clay fraction. The centrifuge provides the answer.

## 5.7 Sedimentation in the centrifuge

The acceleration in a centrifuge bucket may be many thousand times that of free fall under gravity, and the effective "weight" of a particle in the centrifuge may thus be increased in that ratio. This artificial increase of weight solves the difficulties outlined in Section 5.6. When there is equilibrium between diffusion and the segregating weight of particles, Equation (4.14) shows that an increase of "$g$" (which is what the centrifuge accomplishes) concentrates the suspended particles near the bottom of the vessel, i.e. tends to produce a true sediment. The increase of weight also breaks down the resistance to descent which may be due to the formation of weak gels, and at the same time reduces the time of the whole process.

Set off against this there is the disadvantage that the centrifugal force which, in the centrifuge, plays the part performed by weight in gravitational settling, increases as the particle travels farther from the axis of rotation, so that the effective "weight" is not constant, but increases as the settling proceeds. Since there is no constancy of centrifugal force, there is no constancy of velocity at equilibrium with the viscous retarding forces, and the simple Equation (5.1) must be replaced by the more complicated expression,

$$\ln(r_2/r_1) = \{8\pi^2 a^2 n^2(\rho - 1)/9\eta\}t \tag{5.5}$$

where the particle, of radius $a$, starts at a distance $r_1$ from the axis of rotation of the centrifuge and reaches a distance $r_2$ in time $t$, the centrifuge turning at the rate of $n$ revolutions per second. The derivation is given in Note 6.

## 5.8 Complete separation of the fractions in the centrifuge

Let the distance $r_1$ be the distance from the axis of rotation to the surface of the suspension in the centrifuge bucket, and let $r_2$ be the distance to the bottom of the bucket. In the first place let it be again supposed that there are particles of one size only in the suspension. After shaking vigorously, the particles will be uniformly dispersed at the beginning of centrifugation. From Equation (N6.1) it follows that particles which are farther from the axis move faster than those nearer the axis, so that particles which start from the axis steadily increase their distance from those starting nearer. Hence those starting at the surface of the suspension always fetch up the rear, and provide the "ceiling" of that particular group of particles. The argument on which separation is based in the centrifuge thus follows identically that for separation under gravity. As the ceiling travels out to the bottom of the bucket, clear fluid is left behind. For a group of particles of size $a$ the ceiling reaches the bottom of the bucket at time $t$ given by Equation (5.5), and all these particles are included in the sediment. If, instead, the suspension contains particles of a range of sizes, then at this time $t$ all particles of size larger than $a$ are included in the sediment, together with some particles of smaller size which started sufficiently near the bottom of the bucket. The case is exactly similar to the complete separation of fractions by gravitational settling (Section 5.3), and the procedure is the same. The supernatant liquor is poured from the sediment after the time $t$, the sediment is shaken up again with fresh water, and centrifugation proceeds a second time. After a sufficient number of repetitions, the supernatant liquor is found to be almost clear water, and at this stage the final sediment contains all and only particles of size larger than $a$, whilst the bulk liquor contains all and only particles smaller than this.

It should be noted that particles move radially in the bucket, whilst the bucket itself is almost invariably cylindrical. It therefore happens that some particles are thrown on the wall instead of settling to the bottom, but this phenomenon is usually ignored. Secondly, if the total time of centrifugation is short, an appreciable proportion is occupied by the initial period of acceleration up to running speed and the final period of deceleration to a stop. Account must be taken of this.

### 5.9 The two-layer method

Marshall (1930) has introduced a modification of the method of complete separation of fractions, known as the two-layer method, which is designed to reduce the tedium due to repetition of centrifugation. This is achieved by locating all the particles near the surface at the beginning of centrifugation, instead of uniformly distributing them throughout the volume contained in the bucket. The technique consists of floating a thin layer of aqueous suspension onto the main bulk of particle-free fluid in the bucket. To facilitate this operation, this fluid is of a somewhat higher specific gravity than the aqueous suspension. It may be, for example, a dilute solution of glycerol in water. Marshall's papers may be referred to for details and for the necessary modification of Equation (5.5) appropriate to the case. Since all particles start from the same mark, namely the surface, complete separation of the fractions is attained with a single centrifugation. The method may, of course, only be used if the solute used to increase the specific gravity of the fluid has no flocculating effect on the particles.

### 5.10 The method of sampling in the centrifuge

The method of sampling a column of suspension settling under gravity was shown in Section 5.4 to depend upon the fact that the concentration of particles of a given size remains constant underneath the ceiling of that group. This condition does not hold in the centrifuge. It is shown, however, in Note 7, that sampling is still capable of interpretation, although in a somewhat complicated way. It is there shown that as the ceiling of a particular size travels towards the bottom of the centrifuge bucket during centrifugation, the concentration of particles below the ceiling decreases as time goes on, but decreases in the same proportion everywhere, so that if the concentration is initially uniform, it remains uniform at a decreasing value under the ceiling. At any given instant of time during the centrifugation, the variation of particle concentration along the axis of the bucket, when particles of many sizes are present, is therefore due solely to the distribution of the ceilings of the various size groups, and not to any variation of concentration with axial distance for any one size group. By sampling at a particular distance from the axis of rotation at a given time, and repeating at a slightly different distance, one can say that the difference of concentration at the two points is due to particles in that size range whose ceilings have fallen at least to the one point but not beyond the other. The concentration of that group at the time of sampling can therefore be found, and the size range can also be found in terms of the time and distance at which the samples were taken, by application of Equation (5.5). The decrease in concentration of particles of this size range due to centrifugation is calculable, and consequently the initial concentration of particles of the size group, in the uniformly dispersed suspension before centrifugation, can be found. The information required to compute the mechanical composition as described in Section 5.4 is then to hand.

As developed by Svedberg and his collaborators (1923b, 1924), this method employs indirect estimation of the density of the suspension in the centrifuge bucket by a measurement of the reduction of the intensity of a beam of light passing through it. The intensity is recorded photographically whilst the centrifuge is in motion. As described in Section 5.4, the mass of solid matter suspended per unit volume is derived from the absorption of the light beam by appeal to an initial empirical calibration.

The equation used for interpreting the variation of density of suspension in the centrifuge bucket is

$$\delta M/\delta r = (\delta m/\delta a)(r_1{}^2/2r^3)/\{\alpha t \ \ln(r/r_1)\}^{\frac{1}{2}} \tag{5.6}$$

where

$$\alpha = 8\pi^2 n^2(\rho - 1)/9\eta$$

In this expression, $\delta M$ is the difference between the masses of solid particles per unit volume of suspension at the distances $r$ and $r + \delta r$ from the centrifuge axis after centrifugation time $t$, the surface of the liquid in the bucket being at the distance $r_1$; whilst $\delta m$ is the mass of particles per unit volume in the size range $a$ to $a + \delta a$ in the well-shaken suspension before centrifugation. The size $a$ is related to $t$, to $r$ and to the centrifuge constant $\alpha$ by Equation (5.5). The ranges $\delta r$ and $\delta a$ must be small, and in practice one plots the curve of $M$ against $r$ and takes tangents at selected values of $r$. The slope of each tangent gives the value of $dM/dr$ (i.e. $\delta M/\delta r$ when $\delta r$ becomes very small indeed) at the corresponding value of $r$. These values of $dM/dr$ and $r$ are used for insertion in the Equation (5.6). The quantity $\delta m/\delta a$ derived from Equation (5.6), plotted against the particle size $a$, is the required expression of the mechanical composition of the particular material.

## NOTES

### Note 5. The velocity of settlement of a particle in water

The mass of a spherical particle of radius $a$ and specific gravity $\rho$ is $4\pi a^3 \rho/3$, and the mass of water which it displaces is $4\pi a^3/3$. Hence the weight $W$ of the particle in water is

$$W = (4\pi a^3 g/3)(\rho - 1)$$

The resistance $F$ to movement when the particle moves with velocity $v$ in water of viscosity $\eta$ has been shown by Stokes to be

$$F = 6\pi a\eta v$$

The state of steady fall is attained at such a velocity that $W$ is equal to $F$, when the resultant force, and therefore the acceleration, are zero; i.e.

$$6\pi a\eta v = (4\pi a^3 g/3)(\rho - 1)$$

whence
$$v = (2a^2 g/9\eta)(\rho - 1) \qquad \left\{ \begin{array}{l} \text{(N5.1)} \\ \text{(5.1)} \end{array} \right.$$

### Note 6. The application of Stokes's law to sedimentation in the centrifuge

A plan view of the centrifuge head is shown sketched in Figure (N6.1). The

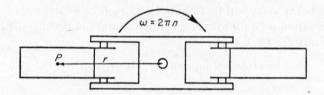

Figure N6.1 A particle $P$ sedimenting in a centrifuge.

bucket containing the suspension of particles hangs vertically until the centrifuge is put into motion, but with rotation it flies out into the horizontal position shown. The same force which affects this acts also on the particles in the suspension. Consider the particle $P$, of radius $a$ and specific gravity $\rho$, at a distance $r$ from the axis of rotation. If the centrifuge performs $n$ revolutions per second, so that the angular velocity $\omega$ is $2\pi n$ radians per second, the centrifugal acceleration of a free particle at $P$ is $4\pi^2 n^2 r$ and this takes the place of $g$ in Equation (5.1) to give

$$v = (2a^2/9\eta)(4\pi^2 n^2 r)(\rho - 1) \qquad \text{(N6.1)}$$

At any given instant the particle is in equilibrium, the centrifugal force just balancing the viscous retarding force at the velocity $v$, but as time goes on $r$ increases and $v$ with it, as shown by Equation (N6.1). Hence, writing $dr/dt$ for $v$, one has

$$v = dr/dt = (2a^2/9\eta)(4\pi^2 n^2 r)(\rho - 1)$$

or
$$dr/r = \{8\pi^2 a^2 n^2(\rho - 1)/9\eta\}dt$$

Integrating between the limits 0 and $t$, at which times $r$ takes the values $r_1$ and $r_2$, one has

$$\ln(r_2/r_1) = \{8\pi^2 a^2 n^2(\rho - 1)/9\eta\}t \qquad \left\{ \begin{array}{l} \text{(N6.2)} \\ \text{(5.5)} \end{array} \right.$$

**Note 7. The variation of concentration of particles in suspension in the centrifuge**

Instead of a centrifuge bucket, consider a cylindrical mass of suspension rotating about its axis at $n$ revolutions per second, as illustrated in Figure (N7.1). During centrifugation, the particles travel outwards along radii, and all particles of the same size and at the same distance from the axis travel at the same speed. Hence particles of the size $a$ lying on a cylinder of radius $R_1$ at a given instant,

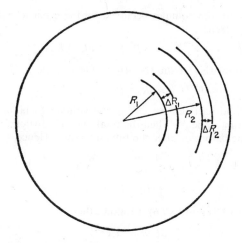

Figure N7.1 Sedimentation in the centrifuge of an element of volume of suspension. The element with boundaries at $R_1$ and $R_1 + \Delta R_1$ moves to the boundaries $R_2$ and $\Delta R_2$.

move outwards all at the same speed, and therefore always lie on a cylinder with ever increasing radius. After a time $t$ this radius will be $R_2$, given by Equation (5.5), i.e.

$$\ln(R_2/R_1) = 8\pi^2 a^2 n^2(\rho - 1)t/9\eta$$
$$= \alpha a^2 t \qquad \text{(N7.1)}$$

where $\alpha$ is a constant, used for convenience, characteristic of the centrifuge, the specific gravity of the particles, and the viscosity of the water, i.e. it collects up all the constant terms in Equation (5.5):

$$\alpha = 8\pi^2 n^2(\rho - 1)/9\eta$$

Similarly, those particles starting at the slightly greater distance $R_1 + \Delta R_1$, and lying on the cylinder of that radius, will after time $t$, lie on the cylinder of radius $R_2 + \Delta R_2$ where

$$\ln\{(R_2 + \Delta R_2)/(R_1 + \Delta R_1)\} = \alpha a^2 t \qquad \text{(N7.2)}$$

From Equations (N7.1) and (N7.2) it follows that

$$R_2/R_1 = (R_2+\Delta R_2)/(R_1+\Delta R_1)$$

and therefore

$$R_2/R_1 = \Delta R_2/\Delta R_1 = e^{\alpha a^2 t} \tag{N7.3}$$

All the particles which initially lay between the cylinders of radii $R_1$ and $R_1+\Delta R_1$ lie, after time $t$, between the cylinders of radii $R_2$ and $R_2+\Delta R_2$. Hence, for unit length of cylinder,

$$V_1 = 2\pi R_1 \Delta R_1$$
$$V_2 = 2\pi R_2 \Delta R_2$$

where $V_1$ and $V_2$ are the volumes occupied by these particles before and after time $t$ respectively. Hence,

$$V_1/V_2 = (R_1 \Delta R_1)/(R_2 \Delta R_2) \tag{N7.4}$$

Combining Equations (N7.3) and (N7.4) one has

$$V_1/V_2 = R_1^2/R_2^2 = 1/e^{2\alpha a^2 t} \tag{N7.5}$$

Let the mass of solid particles per cm³ in the volume $V_1$ be $m$, and that in the volume $V_2$, after time $t$, be $M$. Then the total mass of particles in the element of volume is the same before or after the time interval $t$. Hence,

$$V_1 m = V_2 M$$

or

$$V_1/V_2 = M/m \tag{N7.6}$$

From Equations (N7.5) and (N7.6) it follows that

$$M/m = 1/e^{2\alpha a^2 t} \tag{N7.7}$$

It will be observed that this depends only upon the size of particle $a$ and the elapsed time $t$, since $\alpha$ is constant. No matter at what radius one chooses the element of volume to begin with, and to which distance in consequence it travels in the time $t$, one arrives at the same value of $M/m$. Since the suspension is well shaken before centrifugation, the particle concentration is uniform to begin with, i.e. $m$ is the same everywhere. Hence, from Equation (N7.7), $M$ is also the same everywhere, for a given group of particles of size $a$, i.e. the particle concentration is uniform also after time $t$.

The next step is to consider a suspension, the surface of which is at distance $r_1$ from the axis of rotation. Particles cannot now start nearer the axis than this; there is a discontinuity of particle concentration at the beginning and this must persist. After time $t$, the ceiling of these particles will have reached the distance $r$, where, by Equation (5.5),

$$\ln(r/r_1) = \alpha a^2 t \tag{N7.8}$$

At this time there are no particles of size $a$ nearer the axis than $r$, and at all points farther from the axis the particles are present in uniform concentration given by Equation (N7.7). There is no variation of concentration along a radius other than that due to the presence of the ceiling at $r$.

Now consider the practical case where the particles of size $a$ form but one group of many different sizes. At the time $t$, the ceilings of the various size groups will be distributed outwards from the inner surface, and there will be a variation of particle concentation with distance from the axis due to this cause alone. Consider the group of particles of size $a+\delta a$; its ceiling will be at the distance $r+\delta r$, and the relationship between $\delta a$ and $\delta r$ is most simply derived from Equation (N7.8) by differentiation thus:

$$(1/r)\delta r = 2\,\alpha\,at\,\delta a \qquad \text{(N7.9)}$$

provided $\delta r$ and $\delta a$ are sufficiently small. Now let $\delta M$ be the mass per $cm^3$ of suspension, contributed by all particles in the size range $a$ to $a+\delta a$ at the time $t$. The ceiling of the largest particles of the range has not travelled farther from the axis than $r+\delta r$, hence at this and greater distances all the particles are present at the concentration $\delta M$. The ceiling of the smallest particles having penetrated to the distance $r$, none of the particles are present at distances closer to the axis. Since the whole change of concentration of suspension in the distance $\delta r$ at this time is due to this size group making its full contribution at $r+\delta r$ and no contribution at all at $r$, then $\delta M$ is the difference between the total concentration of solids at $r+\delta r$ and that at $r$; this can be measured by experiment. One may now determine the mass per $cm^3$ of this group of particles in the original suspension before centrifugation. Let this be $\delta m$. Then, from Equation (N7.7),

$$\delta M/\delta m = 1/e^{2\alpha a^2 t} \qquad \text{(N7.10)}$$

Since $\delta r$ has to be specified before $\delta M$ can be measured, and since this specification determines the size range of the group in accordance with Equation (N7.9), one may combine Equations (N7.9) and (N7.10) to give

$$\delta M/\delta r = (\delta m/\delta a)/2\alpha atr\ e^{2\alpha a^2 t} \qquad \text{(N7.11)}$$

Since $a$ is not directly measured, but inferred from $r$ and $t$, Equation (N7.8) is used to turn Equation (N7.11) into the form,

$$\delta M/\delta r = (\delta m/\delta a)(r_1/r)^2/2r\{\alpha t\ \ln(r/r_1)\}^{\frac{1}{2}}$$
$$= (\delta m/\delta a)(r_1{}^2/2r^3)/\{\alpha t\ \ln(r/r_1)\}^{\frac{1}{2}} \qquad \left\{\begin{array}{l}\text{(N7.12)}\\ \text{(5.6)}\end{array}\right.$$

# CHAPTER 6

# The soil pore space

## 6.1 The nature of the pore space

BETWEEN the packed soil particles is an interstitial space occupied partly by soil solution (loosely referred to as soil water) and partly by the soil atmosphere. The most important single physical attribute of the pore space is that it is continuous; a path can be drawn from any point in it to any other point without leaving the space. This is a very different state of affairs from the discrete bubble nature of the air spaces in such materials as pumice. This is not to say that the portions occupied by water and by air are separately continuous, since there may be isolated air bubbles with intervening water, and bodies of water which are, for all practical purposes, isolated from each other. Whilst the pore space is continuous, it is also essentially cellular in nature, consisting of "caverns" interconnected by narrower channels. It is this element of discreteness which has led to the use of such words as soil "pores", "pore space", "porosity" and, of course, "porous" itself, as well as to the more recent innovations "pore size" and "pore size distribution". Some workers, notably in the United States of America, have held this usage to be regrettable in that the word "pore" implies an isolated bubble; they have introduced the words "voids", "percentage voids", and so on. It is possible to concede the validity of the criticism without admitting the impeccability of the alternative; the word "porous" is itself so universally in use as to be unavoidable, and in practice no misunderstanding does in fact ever arise. The older nomenclature will be adopted here.

The more elementary properties of the pore space are most simply dealt with by listing some quantitative definitions. The volume, $V$, of a mass of soil enclosed by its outer envelope is called the "apparent volume". Of this volume, the amount $v$ is devoted to interstitial space, the remainder being the volume of the solid component. The ratio $v/V$ is called the "porosity", and $100v/V$ the "percentage porosity". For comparison, the ratio of pore volume to volume of solids, $v/(V-v)$, is referred to as the "voids ratio" by those who prefer this terminology. It must be emphasized at the outset that these are properties of the soil as encountered, and may change with circumstances. If the total mass of the volume $V$ is $M$, of which $m$ gm is dry matter of specific gravity $\rho$ and $M-m$ is water, then $M/V$ is called the "apparent density" (or "volume weight" in United States usage) and $\rho$ is called the "true specific gravity". Although it is not proposed to discourse on laboratory techniques, an elementary account of the measurement of the space relationships in soils may help to focus the subject.

## 6.2 Laboratory determinations of the relative volumes of solids, water and air in soils

The true specific gravity of the solids may be measured by means of a specific gravity bottle. The only special precaution is that the dry soil needs to be boiled for a short time in a little of the fluid to be used in the determination in order to expel trapped air. It is to be noted that the specific gravity of the fluid enters into the determination, and that the specific gravity of water is locally increased to an indeterminable value by attraction to solid surfaces and to clay surfaces in particular. The use of water as the fluid in the determination is therefore attended by error, and it is better to use some non-polar liquid such as xylol. In that case, since the liquid is inflammable, the usual precautions must be taken in boiling to expel air. If the mass of dry soil be $m$, if the mass of liquid of density $d$ required to fill the bottle to the mark be $M$, and if the total mass of dry soil plus liquid to the mark when both are in the bottle be $\mu$, then the specific gravity of the dry solid, $\rho$, is

$$\rho = md/(M+m-\mu)$$

The choice of method for determining the apparent volume of soil is governed by the condition of the soil. If the soil is friable and loose, it may be sampled with a sampler of known volume designed to extract the sample without modifying its condition. If the soil is strongly coherent, a natural clod may be selected and its volume found by an application of Archimedes' principle. The clod is weighed, given a thin coating of wax by quick dipping, and weighed again. From the gain in weight and the known specific gravity of the wax, the volume of the coating may be determined. The coated clod is then weighed whilst suspended in water, the difference between the weight in air and that in water giving a measure of the total volume of the coated clod. Subtraction of the known volume of the wax results in a value of the volume of the clod itself. The mass and apparent volume of the clod being now known, the apparent density is at once calculable. The mass, and therefore volume, of the solid particles alone cannot be found by drying this clod, since this is forbidden by the presence of the wax. A second clod is therefore taken, which is presumed to be in a similar condition as regards the relative proportions of solids, water and air space. This also is weighed before and after drying. The percentage loss of weight on drying being known, the actual loss of weight (i.e. the water content) which would have been suffered by the first clod is calculated, and hence the weight of the solids and water are separately known. The volumes of solids and water follow at once since the respective densities are known, and the only part of the total volume unaccounted for clearly represents the air space.

### 6.3 Pore size distribution; the "textural" pore space

Although the pore space is continuous, from its cavernous nature it can be regarded as having a pore size distribution, just as soil particles have a size distribution. In modern soil physics much of the emphasis which was at one time laid upon mechanical composition of soil has been transferred, with profit, to pore size distribution, and an exploitation of this concept will loom large in later chapters.

The particle size distribution is commonly of the type known to statisticians as "normal"; that is to say, a certain size occurs more frequently than any other, and the frequency of occurence decreases steadily as the size departs in either direction from this favoured group. A random packing of this kind commonly results in a "normal" kind of pore size distribution. It is readily seen that a soil of predominantly coarse particles will have pores which are large relative to those between finely divided material. A small proportion of fine particles, by occupying the pores between the larger proportion of coarse particles, can profoundly modify the pore size distribution, and, indeed, the total porosity. Since the mechanical composition of soil (i.e. the size distribution of the ultimate particles) is usually closely correlated with soil "texture" (i.e. the assessment of agricultural heaviness or lightness), the related pore size distribution may conveniently be referred to as the "textural pore size distribution", and the pore space defined by the spatial distribution of the ultimate particles may be referred to as the "textural pore space".

### 6.4 The structural pore space

The colloidal fraction of soil, and the organic matter in particular, confers the power of aggregation into clods and crumbs. The phenomena associated with aggregation are by no means well understood, but the facts are fairly well known. In a well-aggregated soil the ultimate particles are to be found bound together to form clods and crumbs; the binding between the particles in any individual aggregate is much stronger than the binding between different aggregates. Although the aggregates are by no means permanent, and may become greatly modified during the seasonal cycle of weathering, nevertheless, over short periods they persist as individuals in spite of disruptive influences. Indeed, the stability of the aggregates in the face of disruptive forces is of greater significance than the mere presence of aggregates at a particular moment. Aggregates of a sort may be produced by cracking in the course of drying of dispersed colloidal materials, being no more than manifestations of shrinkage; such aggregates will not survive a remoistening and redispersion. Since, as we shall see, aggregation is of greatest significance in wet soils, conferring favourable soil aeration and mobility of soil water just when those properties are most needed, the mere presence of aggregates consequent upon shrinkage phenomena in dry soils may be of only minor importance. The requirement of major importance is that the aggregate should withstand rewetting and should persist in the saturated soil.

Given the existence of aggregates, then clearly their positions and orientations with respect to one another define an inter-aggregate pore space. Just as the ultimate particles define a textural pore size distribution, so do the aggregates define a characteristic pore size distribution which may conveniently be referred to as the "structural pore size distribution". Soil structure has hitherto been discussed almost solely in terms of the size and shape of aggregates, but future developments, arising out of the current preoccupation with the quantitative aspects of structure, will almost certainly be based upon the structural pore space and upon those physical properties of soil, capable of precise definition, which the structural pore space controls. If aggregates fit closely together to form a three-dimensional mosaic, as is commonly observed in deep clay subsoils, the structural pore space will be of a very different kind from that which exists between those same aggregates when disturbed, as for example during tillage. The tilth produced by tillage thus takes its place as a form of structure, being that evanescent surface structure of randomly arranged aggregates due to man's interference. In such a tilth it is possible to demonstrate, by means to be described in Chapter 7, the presence of a "normal" type of pore size distribution, characteristic of the aggregates, in addition to another "normal" distribution, of smaller pores, characteristic of the ultimate particles; i.e. distinct groups of structural and textural pores are evident. In deep subsoils in wet winter conditions, on the other hand, the structural pore space may consist of no more than potential planes of weakness between well-fitting aggregates, and in that case distinct groups of textural and structural pores are not to be found. It is worth emphasizing that modern techniques are able to demonstrate the presence of structural pore space where it exists, and in the absence of such evidence of its existence it is indefensible to divide the pore space arbitrarily into structural and textural components. The subject has advanced beyond the stage of arbitrary division of the pore space into the macro and micro pore space, or into the equivalent non-capillary and capillary pore space, defined qualitatively, or arbitrarily, by the ability to retain water against the force of gravity which tends to effect drainage. The introduction of the words "structural pore space" and "textural pore space" as mere descriptive alternatives to the older terms has no value. The value lies in the ability to specify and to assess quantitatively the amounts of the respective pore spaces and thereby to invest the description with meaning.

# CHAPTER 7

# The measurement of soil water pressure and suction

## 7.1 The purpose and nature of pressure measurements

THE different kinds of equilibrium of soil water, the dependence of the pore water content on the prevailing hydrostatic pressure which as often as not is less than atmospheric pressure and therefore commonly called a suction, and the dependence of the rate of movement of pore water on differences of pore water pressure, will all be discussed in subsequent chapters. It is therefore necessary to describe the way in which the soil water pressure is measured.

It must first be noted that the pore space is inaccessible to measuring probes, and therefore the internal hydrostatic pressure is inferred from the measurable pressure of an external body of water which is, either directly or indirectly, in hydrostatic equilibrium with the soil water. This measured pressure is in fact defined to be the soil water pressure. Where the soil is unusually simple in composition, as is the case when it is a coarse sand mass, one may infer the internal pore water pressure without much ambiguity. In more normal circumstances where the soil has an appreciable content of colloidal material such as clay minerals, the internal pore hydrostatic pressure has no unambiguous meaning, since, as has been described in Section 4.10, such material is characterized by the formation of Gouy layers in which the hydrostatic pressure varies with distance from the surface of the colloidal micelle. The pressure in the external water body in equilibrium with it, however, retains its precise and unambiguous significance in these circumstances, and specifies a state of the soil water. It is invariably referred to as the soil water pressure, and this terminology will be adopted here.

In the case of colloidal materials the measurement of the pore water pressure must usually be accompanied by a disturbance of the state of the equilibrium and ultimately of the pressure being measured, because it will usually be the case that the external body of water will not have the same concentration of salts and their ions as the pore water. Slow interdiffusion will occur to dilute the latter and, in accordance with the Gouy layer phenomena elucidated in Section 4.10, the particle repulsion pressure will be altered, and with it the prevailing pore pressure itself. It may often be assumed without much error, however, that the pressure measurement is completed before such interference has had time to develop seriously.

Methods of measurement are divisible into three main categories. Firstly, one may measure directly, by means of a monometer, the hydrostatic pressure in the external water body in equilibrium with the soil. Alternatively, the external body is allowed to come into equilibrium via the vapour phase, and the soil water pressure (invariably a high suction in these circumstances) is inferred from a knowledge of the state (usually a knowledge of the osmotic pressure) of the external water body. Thirdly, one may infer the soil water pressure from measurements of the freezing point of the soil water itself. These methods will be discussed in turn.

### 7.2 Manometers for insertion in soil; the porous membrane

In order to understand the particular difficulty attending the measurement of soil water pressure by manometric methods, it is necessary to anticipate some of the findings of Chapter 8, but this is perhaps preferable to dealing with the matter of that chapter without first having indicated the nature of the measurements there taken for granted. Briefly, when the soil water pressure is positive (i.e. when there is a head of water in addition to the prevailing atmospheric pressure), the soil is in general saturated. There is no air present other than isolated trapped bubbles. This state prevails until the pressure is reduced first to zero (relative to atmospheric) and then to some small negative value (suction). Upon further reduction of pressure (increase of suction), water leaves the soil and air enters a portion of the pore space, this air being in continuity with the external atmosphere. At positive pressures, therefore, a manometer inserted into the pore space is bound to be in contact with soil water and will indicate the soil water pressure. On the other hand, when the soil water is under suction the manometer aperture may or may not open into the soil water. If it does not, but debouches into air-filled pore space, then the manometer merely records the atmospheric pressure, i.e. air enters the probing limb and the level in both limbs becomes the same.

The solution to this problem lies in constructing the limb of the manometer which is in contact with the soil in such a manner that there are a large number of minute orifices instead of one aperture which is the full diameter of the tube. Some of these orifices will then open into soil water and some into air-filled pores. The latter situation is shown in Figure (7.1), where the suction in the water at the orifices is measured by the distance *h* of the water meniscus in the open limb of the manometer below the level of the orifices. It will be shown in Note 11 to Chapter 8 that the suction on the water side of an air-water interface, such as occurs in some of the orifices shown, results in a curvature of that surface, equilibrium being maintained if the necessary degree of curvature is allowed by the geometry of the system. If the manometer orifice is cylindrical, the water surface meets the surface of the solid forming the orifice tangentially, and the curvature of the interface, as it becomes drawn under suction into the orifice, is spherical with radius of curvature *R*. This curvature supports a suction in the water equal to $2T/R$, where *T* is the surface tension. As the suction increases, the radius of curvature therefore decreases, the interface being drawn farther down into the orifice and taking on sharper curvature.

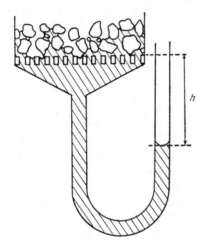

Figure 7.1 The continuity through a porous membrane of water in a reservoir under suction *h* and in an unsaturated porous material.

The limit is reached when the interface becomes a hemispherical surface of radius *r* where *r* is the radius of the orifice. Since the interface must retain contact with the solid surface, no smaller radius is possible and therefore no higher suction than $2T/r$ can be supported. When this suction is reached, the hemispherical interface is withdrawn down the orifice, followed by the air from the soil pore space and outer atmosphere. This free entrance of air into the manometer allows the water to fall away from the remaining orifices and the manometer indicates only atmospheric pressure, which is to say that the water in both limbs settles to the same level. Provided, however, that the suction does not exceed the maximum value which the *largest* orifice can support, air does not enter the manometer, the water in which remains in continuity with soil water via those orifices which open into water-filled pore space. In such circumstances the manometer indicates soil-water suction.

The necessary large number of small orifices is commonly provided by a sheet of porous material, the grade of material being selected in accordance with the magnitude of the suctions to be measured. Higher suctions demand, as has been shown, pores of smaller size than those suited to lower suctions, but pores of unnecessarily small size are avoided since water moves but slowly through them and equilibrium is attained only after an unnecessarily long time. Thus for suctions up to about 150 or 200 cm of water, filter paper or sintered glass discs may be used; for suctions up to about half or three-quarters of an atmosphere, unglazed earthenware or porcelain are suitable; whilst for the highest suctions recourse must be had to membranes of cellophane or fabricated sausage casing.

### 7.3 Forms of tensiometer

A combination of manometer and porous membrane for measuring soil water suction is commonly called a tensiometer, particularly when it is designed for installation in soil in the field. It seems to have been described first by Kornev (1921-23). The precise form of tensiometer chosen depends upon circumstances. If the soil is an extracted sample in the laboratory, it is usual to support the membrane on the platform of a Buchner funnel or, if the membrane is a sintered glass disc, the disc is itself the platform of such a funnel. The apparatus is completed by connecting a burette by rubber tubing, or by some similar means, to the funnel to form a flexible U-tube. The apparatus is filled with water to the under side of the porous platform, on top of which is placed the soil sample. The suction at the platform, which is communicated to the soil sample, is measured by the distance, $h$, of the meniscus in the burette below the platform. The suction may be varied either by raising or lowering the burette, if the U-tube is flexible, or by adding water to or removing it from the burette if the U-tube is rigid. The amounts of water admitted to or extracted from the soil as the suction is changed may be read on the graduated scale of the burette. This form of apparatus is often called a tension table.

Where it is required to measure the suction of soil water in the field, at a given site and depth, the membrane is commonly in the form of a cylindrical or slightly tapering conical pot of unglazed ceramic material inserted in a borehole previously driven to the position of interest. The tool used for boring the hole is shaped so as to leave a cavity into which the pot fits snugly, so that contact between the pot and the soil (i.e. continuity between the soil water and the manometer water) is as good as can be contrived.

Since the pot is below the surface, while the manometer meniscus must be read above ground level, it is necessary to achieve a suction in the pot by employing mercury as part of the manometer column. The water-filled riser from the buried pot is bent over at about a metre above ground level and is allowed to dip under the surface of a pot of mercury at ground level. This is illustrated in Figure (7.2). As suction develops, mercury rises in the

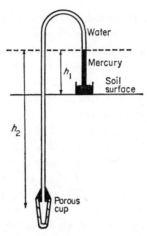

Figure 7.2 A form of tensiometer for use in the field.

manometer tube. At equilibrium with the junction between the mercury and water at a height $h_1$ above the free surface of the mercury in the pot, the suction at this level is $h_1\rho_m$ cm of water, where $\rho_m$ is the density of mercury. If $h_2$ is the height of this junction above the buried porous pot, then the pressure at the pot exceeds that at the junction by the amount $h_2$ cm of water. Hence the net suction at the porous pot is $h_1\rho_m - h_2$ cm of water, and because $\rho_m$ is large the net result may be the maintenance of suction in spite of the fact that $h_2$ is necessarily larger than $h_1$.

When discussing the measurement of soil water suction in the field, it must be remembered that the act of installing the buried pot may so alter the soil structure in the immediate neighbourhood, as for example by causing fissures which may permit the unduly easy penetration of water direct from the surface, that the measured suction may no longer fairly reflect the suction which would have prevailed had the soil not been disturbed, and which does in fact prevail at other similar but undisturbed sites to which the measurements purport to have application.

One practical point deserves mention. It is commonly found that air bubbles form in the tensiometer pot after a time. These do not cause the complete breakdown of the method, as is the case when air penetrates the porous membrane from outside and is continuous with the external atmosphere, but nevertheless they are a source of error. Recent developments of design have been concerned to provide means of removing this collected air from time to time without disturbing the installation.

### 7.4 Limitations of application of the tensiometer

A tensiometer ceases to function as soon as the suction applied is sufficient to draw air through the porous membrane at any point. A further, and more fundamental, limitation is due to the fact that no manometer can apply a suction in excess of one atmosphere. What is conventionally called a suction is a pressure less than the ubiquitous atmospheric pressure. Thus in Figure (7.1) the pressure at the membrane is less than atmospheric pressure by the amount $h$ cm of water, and there is said to be a suction of $h$ cm. But unless this height $h$ is as great as the height of the water barometer at the time, there will in fact be a residual positive absolute pressure. When the height $h$ is equal to the barometric height, i.e. when the whole pressure of the atmosphere on the surface of the water in the burette is required to support the column of water of height $h$ in the closed limb, then the absolute pressure at the top of that limb, i.e. at the membrane, is reduced to zero. If an attempt be made to increase the suction still further by lowering the water level in the burette, success can only be achieved if the upper part of the column in the closed limb, above that part which is supported by the atmospheric pressure on the surface in the open limb, is suspended from the under surface of the porous plate, i.e. if this upper part is under tensile stress. In certain circumstances water can be observed to have a tensile strength, but usually a water column breaks under tension, and in the tensiometer of Figure (7.1) the water falls away from under the porous plate or membrane. The apparatus becomes, in fact, a water barometer. Suction is, of course, still applied via the vapour phase at the top of the closed limb but at this stage the apparatus ceases to be a practicable means of applying suctions in excess of one atmosphere.

### 7.5 Indirect measurement of high suctions; relative humidity

Suction and pressure are measured relative to an arbitrary datum, and in deriving quantitative expressions the symbol $P$ will be adopted to represent the physical quantity of hydrostatic pressure. If this is above datum $P$ will be positive, if below datum $P$ will be negative and recognized as a suction. The datum level is commonly atmospheric pressure.

When the limitations of tensiometers at high suctions require the use of indirect methods, advantage may be taken of the dependence of the vapour pressure over a water surface on the hydrostatic pressure difference maintained between the two sides of the air-water interface. The vapour pressure is also affected by the osmotic pressure of the solution, if the water should contain dissolved matter, so that the method cannot by itself differentiate between hydrostatic suction and osmotic pressure. It is shown in Note 8 that a pressure $P$ on the water side of the interface relative to the air side causes a vapour pressure $p$ which differs from the value $p_0$ in the absence of a hydrostatic pressure difference, in accordance with the equation,

$$\ln(p/p_0) = \ln(\sigma/\sigma_0) = MP/RT\rho \qquad (7.1)$$

where $M$ is the molecular weight of water, the density of which is $\rho$, $R$ is the gas constant and $T$ is the prevailing temperature on the absolute scale. $\sigma$ and $\sigma_0$ are the water vapour densities corresponding to the vapour pressures $p$ and $p_0$ respectively. If the water experiences a suction, then $P$ is negative and $p$ is less than $p_0$. A measurement of the relative humidity $p/p_0$ thus suffices for a determination of $P$. It is shown in Note 10 that the vapour pressure $p$ over the surface of a solution whose osmotic pressure is $P_0$ is, in the absence of an interfacial hydrostatic pressure difference, related to the vapour pressure $p_0$ in the absence of osmotic pressure, by an expression which is almost identical to Equation (7.1), namely,

$$\ln(p/p_0) = \ln(\sigma/\sigma_0) = -(MP_0)/RT\rho \qquad (7.2)$$

When both hydrostatic and osmotic pressures operate at the same time, then

$$\ln(p/p_0) = \ln(\sigma/\sigma_0) = (P-P_0)M/RT\rho \qquad (7.3)$$

Evidently a measurement of the relative humidity in equilibrium with a soil sample does not by itself suffice to provide a measure of the hydrostatic pressure of the soil water. Ambiguity may be removed only by an independent assessment of the osmotic pressure, $P_0$, of the soil solution. There is no difficulty associated with the concept of the osmotic pressure of the soil solution in inert materials such as sand, for the content of dissolved ions is the same everywhere throughout the pore space, and provided that in some way a small sample of the solution can be extracted, its osmotic pressure can be estimated by a measurement of the vapour pressure over the extracted sample, free of hydrostatic suction. With surface active materials such as clays, however, the same difficulty of concept arises as was met when discussing the internal hydrostatic pressure. Like the latter, the osmotic pressure is known to vary throughout the Gouy layer, and, also as in the case of the hydrostatic pressure, the difficulty is removed by defining the osmotic pressure to be that of the external body of solution in equilibrium with the soil. A sufficiently small sample of soil solution may be expressed without seriously affecting the moisture content of the soil being examined and therefore without appreciably influencing the state of the solution which is being examined. At the moment of extraction this small sample is free of the soil, but in equilibrium with it ionically, since it constitutes the 'remote solution' referred to in the discussion of the Gouy layer theory. Hence it constitutes the external body of solution for the purpose of defining the osmotic pressure of the soil solution, and the osmotic pressure may be estimated in the usual way.

If it were possible to make up a bulk solution of the same chemical constitution as the expressed sample, and to use this bulk solution as the fluid in a tensiometer, then at hydrostatic equilibrium the tensiometer would be in complete hydrostatic and osmotic equilibrium with the soil solution, since there would be neither a tendency for bulk movement due to hydrostatic pressure imbalance nor diffusion of dissolved ions and molecules due to mismatching of the external and internal solutions. Thus measurements of the hydrostatic and osmotic pressures of this external body could be made independently with great ease. The preparation of such a bulk solution would be possible in principle, but is hardly necessary, since the osmotic pressure can be measured from the expressed sample which would be the basis of the preparation of the bulk solution, and the hydrostatic pressure, as has been described, can be measured by a water tensiometer if the measurement is concluded before ionic diffusion has gone very far.

The technique of measuring vapour pressures over soil samples depends upon whether one wishes to determine the moisture content corresponding to a preassigned suction or vice versa. In the former case, the sample is enclosed in an airtight vessel together with a large bulk of material of a kind which maintains a known vapour pressure. This is commonly a solution of sulphuric acid of known concentration; a table relating concentration to relative humidity, given by Wilson (1921), is presented in Table 2. If the sample is wetter than is appropriate to the prevailing vapour pressure, it loses water by evaporation and the sulphuric acid absorbs this water. The acid must therefore be present in sufficient bulk sensibly to maintain its concentration in spite of absorption. Conversely, if the sample is initially drier than is appropriate to the prevailing humidity, it absorbs water from the acid solution. The ultimate equilibrium is attained more rapidly if the enclosure contains no gas other than the water vapour, and a vacuum dessiccator is usually employed.

If the soil sample is at a preassigned moisture content and the vapour pressure is to be determined, then the sample must be present in such bulk as to dominate the enclosure and ensure that the prevailing relative humidity is that of the sample and not of the vapour pressure indicators. These indicators may conveniently be small pieces of filter paper soaked in

*Table 2*

| per cent $H_2SO_4$ | Relative humidity per cent at | | | |
|---|---|---|---|---|
| | 0°C | 25°C | 50°C | 75°C |
| 0 | 100·0 | 100·0 | 100·0 | 100·0 |
| 5 | 98·4 | 98·5 | 98·5 | 98·6 |
| 10 | 95·9 | 96·1 | 96·3 | 96·5 |
| 15 | 92·4 | 92·9 | 93·4 | 93·8 |
| 20 | 87·8 | 88·5 | 89·3 | 90·0 |
| 25 | 81·7 | 82·9 | 84·0 | 85·0 |
| 30 | 73·8 | 75·6 | 77·2 | 78·6 |
| 35 | 64·6 | 66·8 | 68·9 | 70·8 |
| 40 | 54·2 | 56·8 | 59·3 | 61·6 |
| 45 | 44·0 | 46·8 | 49·3 | 52·0 |
| 50 | 33·6 | 36·8 | 39·9 | 42·8 |
| 55 | 23·5 | 26·8 | 30·0 | 33·0 |
| 60 | 14·6 | 17·2 | 20·0 | 22·8 |
| 65 | 7·8 | 9·8 | 12·0 | 14·2 |
| 70 | 3·9 | 5·2 | 6·7 | 8·3 |
| 75 | 1·6 | 2·3 | 3·2 | 4·4 |
| 80 | 0·5 | 0·8 | 1·2 | 1·8 |

solutions of different known osmotic pressure, graded through the required range. All those indicators whose osmotic pressures correspond to vapour pressures in excess of that imposed by the soil sample, lose water by evaporation, and all those whose vapour pressures are less, absorb water. By interpolation, that osmotic pressure $P'_0$ which results neither in loss nor gain of water can be determined, and is a direct measure of the sum of suction and osmotic pressure of the soil sample. Since both the measuring standard and the soil sample are in equilibrium with the same vapour pressure and obey Equation (7.3), it follows that

$$(P - P_0) = -P'_0$$

where the left-hand side refers to the soil, and the right-hand side to the saturated filter paper standard in equilibrium with it.

Insertion of the numerical values of the constants in Equation (7.1) for a room temperature of 300°A shows that a relative humidity of 99% corresponds to a soil water suction of about fifteen atmospheres, lower suctions giving humidities between 99 % and 100 %. Since it is a matter of extreme experimental difficulty to maintain or to measure with the required precision relative humidities in this restricted range, it follows that this particular indirect method of measuring soil water suction is unsuited to suctions of less than about fifteen atmospheres.

### 7.6 Indirect measurement of high suctions; freezing point depression

It is a well-known result of thermodynamic analysis that the vapour pressure over a liquid is intimately connected with the freezing point of that liquid, so that the soil water suction which reduces the vapour pressure over the soil sample also affects the freezing point of the soil water. Here again osmotic pressure has its similar effect, so that a single measurement of the freezing point does not suffice to provide separate estimates of suction and osmotic pressure. Furthermore, it is necessary to make some assumptions about the way in which soil water freezes in order to relate the suction to the freezing point. To this end it is usually supposed, with a certain amount of justification from experimental evidence, that the ice formed upon freezing becomes relieved from the suction, just as it becomes relieved from the osmotic pressure. It is shown in Notes 9 and 10 that if this is the case, the incidence of hydrostatic pressure $P$ and osmotic pressure $P_0$ changes the freezing point from $T°$A, appropriate to zero values of $P$ and $P_0$, to $T + \Delta T$, where

$$\Delta T/(P - P_0) = V_w T /_i L_w \qquad (7.4)$$

where $_i L_w$ is the latent heat of fusion of ice and $V_w$ is the volume of 1 gm of water. In this equation, the latent heat must be expressed in mechanical units, e.g. ergs/gm.

Substitution of appropriate values for the physical quantities and constants in Equation (7.4) reveals the magnitude of the freezing point depression. The freezing point of pure free water is 273°A and this may be substituted for $T$. The latent heat of fusion of ice is about $336 \times 10^7$ ergs/gm of water at the freezing point. $V_w$ is, of course, very nearly unity. Equation (7.4) thus becomes,

$$\Delta T = (P - P_0) \times 0.813 \times 10^{-7} °A$$

Hence with zero osmotic pressure each 1° of freezing point *depression* (negative $\Delta T$) indicates a suction (negative $P$) of $1.23 \times 10^7$ ergs/cm$^3$, or about 12.5 atmospheres. Since a temperature difference of this magnitude may readily be measured with an error of as little as 0.01°, the method is suited to the range of suctions between 1 and 15 atmospheres, covering the gap between direct manometric measurements and the relative humidity method.

In order to outline the experimental techniques which have been proposed and used, it is necessary to anticipate some of the phenomena described in Chapter 8. Briefly, the magnitude of the soil water suction depends, for a given soil, on the moisture content; the lower the moisture content, the greater is the suction. If some of the soil water is frozen, then the content of liquid water is reduced, the suction prevailing in this residue is increased, and *its* freezing point decreased. Thus as the temperature is progressively reduced, more and more soil water is frozen. At any particular stage of freezing, the temperature is the freezing point appropriate to a moisture content defined by the remaining liquid portion of the soil water. Since any freezing point experiment involves the freezing of some water, the freezing point at the initial moisture content can only be found as an extrapolation of the curve which results from plotting the freezing point versus residual moisture content. Alternatively, one may be satisfied with a measurement of the freezing point at a moisture content only very little less than the given initial value, involving the freezing of only a little soil water.

A good list of references of work going back to 1916, much of it by Bouyoucos and his colleagues, is given in a paper by Anderson and Edlefsen (1942a). In one experimental technique, described by Schofield and Botelho da Costa (1935), a Beckmann thermometer is embedded in the soil sample contained in a boiling tube, and the whole is placed in a freezing mixture. Because soil water often is reluctant to begin freezing, some super-cooling takes place. The onset of freezing involves the release, as sensible heat, of the latent heat of fusion, and the temperature is observed to rise. When a final steady temperature is reached, the amount of ice formed may be calculated from the heat evolved during this stage of rising temperature. The final temperature is thus the freezing point corresponding to a moisture content less than the initial value by an amount equal to the mass of ice formed.

In another method, due to Alexander, Shaw and Muckenhirn (1936), the onset of freezing is indicated electrically. The dielectric constant of liquid water is many times as great as that of ice, so that if moist soil forms the dielectric of an electrical condenser, the capacitance of the latter falls progressively as more and more ice forms. If, therefore, one measures the capacitance and plots it against the prevailing temperature, one may find by interpolation the temperature at which the capacitance *just* begins to change. This temperature is the freezing point of the soil water at the given initial moisture content.

In yet another method, described by Anderson and Edlefsen (1942a), the formation of ice may be observed in a dilatometer, depending on the fact that the freezing of water is accompanied by an increase of volume.

The amount of ice formed at a measured temperature is calculated from the change of volume displaced by the soil with its accompanying water. This method is suited to the recording of the change of freezing point with moisture content, starting with saturated soil, rather than with the measurement of freezing point and suction of any given sample at any given moisture content less than saturation, since the presence of air in the dilatometer would prevent the interpretation of the observation.

## 7.7 Other indirect suction measurements

There are occasions when it is desired, not so much to measure the existing soil water suction, as to bring the soil to a moisture content appropriate to a certain suction. This may be done by expelling water by increasing the pore air pressure instead of by sucking it out, or alternatively by throwing it out, as in a centrifuge. From measurements of this kind, the suction corresponding to a given moisture content, in the absence of the expelling or throwing manipulations, may be inferred. The pressure plate and centrifuge methods briefly referred to here will be dealt with more fully and more appropriately in Sections 8.10 and 8.11.

## NOTES

### Note 8. The relative humidity of the atmosphere over soil water which experiences suction

Equation (7.1) may be derived by arguments identical in form with those of Note 1. It may be supposed that the suction is maintained by placing the soil sample on the sintered glass platform of a Buchner funnel apparatus as shown in Figure (7.2), the level of the free water surface in the open limb being $h$ cm below the sample. Taking the saturation vapour pressure at this open surface to be $p_0$, the vapour pressure at any greater height will be less by an amount equal to the pressure exerted by the column of vapour between this height and the free surface. At height $h$ the vapour pressure will be in equilibrium with the soil moisture, since otherwise there would be continuous distillation of water to or from the soil sample, i.e. perpetual motion. Hence if one calculates the vapour pressure at $h$ as that at a point on the vapour pressure profile above the free surface in the open limb, the result is the vapour pressure in equilibrium with the soil sample at the known suction, $h$ cm of water or $-P$ dyn/cm$^2$, where if $\rho$ is the density of water,

$$P = -g\rho h$$

Let the density of the vapour at height $h$ above the free surface, where the vapour pressure is $p$, be $\sigma$. Then at height $h+dh$, the vapour pressure is $p+dp$ where

$$dp = -g\sigma \, dh \qquad (N8.1)$$

At the corresponding heights in the column of water in the manometer the hydrostatic pressures are $P$ and $P+dP$ where

$$dP = -g\rho \, dh \qquad (N8.2)$$

Hence, combining Equations (N8.1) and (N8.2),

$$dp = (\sigma/\rho)dP \qquad (N8.3)$$

If it be assumed that the vapour obeys the gas laws, one may write:

$$p/\sigma = RT/M \qquad (N8.4)$$

where $R$ is the gas constant, $T$ is the prevailing temperature on the absolute scale and $M$ is the molecular weight of water. Hence, from Equations (N8.3) and (N8.4),

$$(1/p)dp = M \, dP/RT\rho$$

On integrating this between the limits of the free surface, where $P$ is zero and $p$ has the value $p_0$, and the soil surface in the Buchner funnel, where the hydrostatic pressure is $P$ as measured by the manometer and the vapour pressure is $p$, one has the result,

$$\ln(p/p_0) = MP/RT\rho \qquad (N8.5)$$

From arguments similar to those by which Equation (N1.8) was derived, it follows that

$$\ln(\sigma/\sigma_0) = \ln(p/p_0)$$

Hence finally,

$$\ln(\sigma/\sigma_0) = \ln(p/p_0) = MP/RT\rho \qquad \begin{cases} (N8.6) \\ (7.1) \end{cases}$$

**Note 9. The depression of the freezing point of water which experiences suction**

At the freezing point of water all the phases, solid, liquid and vapour, are in equilibrium. If the vapour pressure over the dry ice were not equal to that over the liquid water, there would be a steady transfer of water from the liquid to the solid phase via the vapour phase, or vice versa. One or other of the solid and liquid phases would thus grow at the expense of the other, which is to say that the prevailing temperature would *not* be the freezing point. If, at the freezing point, the suction of the water at the air-water interface were to be increased without any change of the ice condition, then in accordance with Equation (7.1), the vapour pressure over the water, but not over the ice which does not suffer the suction, would be decreased, and the consequent water vapour movement would transfer water from the ice to the liquid phase. That is to say, the ice would melt, and to counteract this effect the temperature would have to be reduced, so that the freezing point of the water at increased suction would be lowered. In order to deduce the change of temperature required to restore equilibrium between all three phases (i.e. in order to calculate the reduction of the freezing point), one must first examine the variation of vapour pressure with temperature both over ice and water.

An experiment on a given mass of vapour, designed to demonstrate how its volume depends upon the pressure exerted upon it at constant temperature $T°A$, provides some such result as is shown in the curve $ABCD$ of Figure (N9.1). In the early stages $AB$ the vapour obeys the gas laws, the volume being inversely proportional to the pressure. At the maximum pressure $p$ which the vapour can sustain, i.e. at the point $B$, the vapour begins to condense to liquid form, and along $BC$ the volume decreases without increase of pressure due, of course, to the fact that the liquid occupies less space than the vapour. When the whole of the vapour has liquefied, i.e. at the point $C$, further increases of pressure result in slight progressive decreases of volume in accordance with the bulk modulus of elasticity of the liquid. Heat is evolved at all stages of this process, due to the work done in compression and, during the stage $BC$, to the release of the latent heat of vapourization; hence provision must be made for the escape of the heat

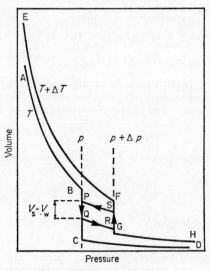

AB and EF, vapour phase
CD and GH, liquid phase
PQRS, Carnot cycle in condensing vapour

Figure N9.1 Curves at two different temperatures relating the volume to pressure of a condensing vapour.

in order to maintain constant temperature during the experiment. These changes of pressure and volume are said to be isothermal. Had the experiment been performed at some slightly higher temperature $T+\Delta T$, the curve would have been as shown at $EFGH$; at all pressures the volumes would have been larger, and vice versa. In particular, the saturation vapour pressure would have increased from $p$ to $p+\Delta p$, and it is necessary to deduce the relationship between $\Delta p$ and $\Delta T$. It should be noted that the relationship between the volume and the pressure of the gas and liquid at a given temperature is unique, i.e. the same curves would have resulted had the direction of the experiments been reversed, the pressure being steadily relaxed from the points $D$ or $H$.

Suppose that during the course of an experiment of this kind a point $P$ is reached on the $T$ isothermal starting at $A$. This is at the stage of liquefaction at the saturated vapour pressure $p$. Let a further mass of 1 gm of vapour, occupying the volume $V_s$, be liquefied to water occupying the smaller volume $V_w$, so that the reduction of volume, represented by the distance $PQ$, is $V_s - V_w$. The heat emitted during this stage is the latent heat of vapourization, $_wL_s$, and it must be dissipated to waste in order to keep the temperature down to $T$. At the point $Q$ let steps be taken, as for example, by heat insulation, to prevent further heat waste. Further liquefaction must therefore be accompanied by a rise of temperature and hence of vapour pressure, so that the next stage will be represented by some such curve as $QR$. At $R$, where the temperature has risen to $T + \Delta T$, let the adiabatic (i.e. the heat insulated) condition revert to the isothermal condition at this higher temperature, and let the direction of the experiment be reversed, heat being supplied from an external source to evaporate liquid at the higher vapour pressure $p + \Delta p$. Let one gram of liquid evaporate, bringing the mass of vapour back to that which existed at $P$. Provided that the increment of temperature $\Delta T$ is but small, the volumes $V_s$ and $V_w$ occupied by one gram of water in the vapour and liquid states respectively will not differ sensibly from those at the lower temperature $T$. The heat that must be *supplied* during this stage $RS$ is the latent heat of evaporation, $_wL_s$. Finally let a further adiabatic stage of evaporation, in which the necessary heat is supplied by the system itself with consequent reduction of temperature, bring the experiment to an end with a return to the point of departure $P$. The cycle of operations represented by $PQRS$ is known as a Carnot cycle.

The mechanical work done by the system during any small increase of volume $\delta V$ at a pressure $p$ is given by the product $p\delta V$, whence it is readily seen that the work performed by the system during the cycle, that is to say, the work done *by* the system during the stage of expansion minus the work done on it during the stage of contraction, is measured by the area of the figure $PQRS$ calculated in the units indicated on the axes of Figure (N9.1). Thus,

$$\text{Work done during the cycle} = \Delta p(V_s - V_w)$$

In order to perform this work, the system has had to absorb a quantity of heat $_wL_s$ and later to reject it to waste. Hence the mechanical efficiency of the cycle, $E$, is given by

$$E = \Delta p(V_s - V_w)/_wL_s \qquad (\text{N9.1})$$

From the definition of the absolute or Kelvin scale of temperature, it is known that

$$E = \Delta T/T \qquad (\text{N9.2})$$

Combining Equations (N9.1) and (N9.2) yields

$$\Delta p/\Delta T = {_wL_s}/\{T(V_s - V_w)\} \qquad (\text{N9.3})$$

This is the relationship sought.

A similar cycle carried out in circumstances such that vapour condenses directly to dry ice, or dry ice sublimes, leads to a similar relationship,

$$\Delta p/\Delta T = {_iL_s}/\{T(V_s - V_i)\} \qquad (\text{N9.4})$$

where $p$ is now the saturated vapour pressure over dry ice at temperature $T$, $_iL_s$ is the latent heat of sublimation per gram, and $V_s$ and $V_i$ are the respective volumes of one gram of vapour and of one gram of ice at the temperature $T$.

At the freezing point the vapour is in equilibrium simultaneously with ice and with liquid water, so that both Equations (N9.3) and (N9.4) hold with identical values of $T$ and therefore of $V_s$. Suppose that the water is soil water which can experience a suction. Then a change of suction upsets the equilibrium by changing the vapour pressure over the water without affecting the vapour pressure over the ice, which is plausibly assumed to be free of the suction. Equilibrium may be restored by a change of temperature of such a magnitude that the consequent change of vapour pressure over the ice is just equal to the sum of the changes of vapour pressure over the water due to the simultaneous change of suction and of temperature. Over the ice one has, from Equation (N9.4),

$$\Delta p = \Delta T \ _iL_s/[T(V_s - V_i)] \qquad (N9.5)$$

Because the freezing point depression method is not used when the relative humidity over the soil is less than about 99%, the range of variation of vapour pressure is so small that one may treat the vapour density, $\sigma$, as a constant and use Equation (N8.3) to relate suction to vapour pressure instead of the integrated form of Equation (7.1), so that

$$\Delta p_P = P(\sigma/\rho)$$

where $\Delta p_P$ is the increase of vapour pressure due to hydrostatic pressure $P$. Since $\rho$ is the mass of 1 cm$^3$ of water, it is the inverse of $V_w$, the volume of 1 gm; and similarly $\sigma$ is the inverse of $V_s$, hence,

$$\Delta p_P = PV_w/V_s \qquad (N9.6)$$

The increase of vapour pressure, $\Delta p_T$, due to an increase of temperature $\Delta T$ of the water, is given by Equation (N9.3) with $\Delta p_T$ written for $\Delta p$. The total change of vapour pressure, $\Delta p$, over the water, due to a change of hydrostatic pressure and a simultaneous change of temperature, is thus

$$\Delta p = \Delta p_P + \Delta p_T$$
$$= PV_w/V_s + \Delta T \ _wL_s/[T(V_s - V_w)] \qquad (N9.7)$$

If the new temperature $T + \Delta T$ is to be the new freezing point, the increase of vapour pressure over the water, given by Equation (N9.7), must equal the increase of vapour pressure over the ice, given by Equation (N9.5). Hence,

$$\Delta T \ _iL_s/[T(V_s - V_i)] = \Delta T \ _wL_s/[T(V_s - V_w)] + PV_w/V_s \qquad (N9.8)$$

This equation may be much simplified by introducing approximations which result in only small errors. The volumes of 1 gm of ice and of liquid water are negligibly small compared with the volume of 1 gm of water vapour, so that Equation (N9.8) may be written, with only negligible error,

$$\Delta T(_iL_s - _wL_s)/(TV_s) = P \ V_w/V_s$$

Remembering that the heat required to convert 1 gm of ice directly into 1 gm of vapour is the sum of the heat contributions required to convert the ice first into liquid water and then into vapour, one may finally write Equation (N9.8) in the form,

$$\Delta T \ _iL_w/T = PV_w \qquad (N9.9)$$

where $_iL_w$ is the latent heat of fusion of ice.

This gives the dependence of the change of freezing point on hydrostatic pressure alone. When $P$ is negative, i.e. a suction, $\Delta T$ also is negative, and the freezing point falls.

**Note 10. The contribution of osmotic pressure to relative humidity and freezing point depression**

Figure (N10.1) represents an osmometer in which the solution, whose osmotic pressure is $P_0$, is separated from pure solvent by a semi-permeable membrane, as referred to in Section 4.8. Preferential diffusion of solvent takes place from the pure solvent side to the solution side of the membrane until equilibrium is attained,

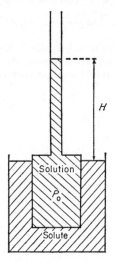

Figure N10.1 Osmometer with solution separated from pure solute by a semi-permeable membrane.

when the hydrostatic pressure built up, as measured by the height $H$ of the column in the standpipe, is a measure of the osmotic pressure. The vapour pressure over the surface of the solution is in equilibrium with and equals the vapour pressure at the height $H$ above the free surface of the pure solvent, and the calculation of the vapour pressure at height $H$ follows exactly the same course as in Note 8. It makes no difference whether the height $H$ is due to suction at the air-water interface or to osmotic pressure. The only difference arises in expressing the vapour pressure at the height $H$ in terms of the cause. The hydrostatic pressure on the water side of the interface in Note 8 is given by

$$P = -g\rho H$$

since hydrostatic pressure decreases with height above the free surface. The osmotic pressure, on the other hand, is given by

$$P_0 = g\rho H$$

Hence the result of the analysis in the present case will give an equation which differs from Equation (7.1) only in the replacement of $P$ by $-P_0$, thus

$$\ln(\sigma/\sigma_0) = \ln(p/p_0) = -(MP_0)/RT\rho \qquad \begin{cases} \text{(N10.1)} \\ \text{(7.2)} \end{cases}$$

Similarly, it may be shown that if hydrostatic pressure $P$ prevails on that side of an interface where there is a solution of osmotic pressure $P_0$, as for example when the standpipe of the osmometer in Figure (N10.1) is a capillary tube accommodating a strongly-curved meniscus, the total height $H$ of the column in the standpipe is given by

$$P_0 - P = g\rho H$$

whence the analysis of the vapour profile now gives

$$\ln(\sigma/\sigma_0) = \ln(p/p_0) = M(P-P_0)/RT\rho \qquad \begin{cases} \text{(N10.2)} \\ \text{(7.3)} \end{cases}$$

Where both $P$ and $P_0$ operate together, the factor $(P-P_0)$ takes the place of $P$ or $-P_0$ in any vapour pressure relationship derived from those factors separately, so that the general form of Equation (N9.9) in Note 9 is

$$\Delta T \, _iL_w/T = (P-P_0)V_w \qquad \begin{cases} \text{(N10.3)} \\ \text{(7.4)} \end{cases}$$

CHAPTER 8

# The hydrostatic equilibrium of soil water

## 8.1 The alternative mechanisms of withdrawal of water from soil

THE suction necessary to remove water from soil is required to overcome soil water forces of different kinds. In very light soils, whose particles are predominantly of very coarse grain in physical contact with each other at all moisture contents, there can be but very little diminution of overall volume as water is withdrawn. Such very slight shrinkage as may occur can only be due to rearrangement of the particles to a form of somewhat closer packing. Hence as the initially saturated soil loses water, air must enter to replace the water in the pore space. Since the unsaturated soil is permeated by both air and water, it must clearly be permeated also by an interface between these fluids, and in that surface will act the force of surface tension.

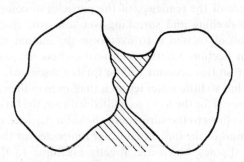

Figure 8.1 The curved meniscus of the water between
soil particles.

As is indicated in Figure (8.1), the solid particle surfaces which meet the interface are not, except fortuitously, parallel; and since water is a wetting fluid, the surface of which meets solid bounding surfaces tangentially, different parts of the continuous interface must slope in different directions. The interface must therefore be curved. Surface tension acting in a liquid surface tends to keep that surface stretched flat, so that curvature can only

E                                      115

be maintained if there is a difference of pressure on the two sides, the side from which the surface appears concave naturally having the higher pressure. The surface has to be "blown" from the concave side or sucked from the convex side. It will be seen from Figure (8.1) that water occupies the side from which the interface appears convex; and since the air in the pore space is continuous with the external atmosphere and is therefore at atmospheric pressure, to which hydrostatic pressures are commonly referred as datum, it follows that the circumstances depicted can only comprise a state of static equilibrium if the water experiences a suction. The function of suction in water in non-shrinking wet soils is thus to maintain curvature of the air-water interface, and Section 8.2 will deal with the relationship between the magnitude of the suction and the nature of the curvature. In Sections 8.3 and 8.4, the consequent relationship between the suction and the soil moisture content will be developed.

If the soil is of heavy texture, consisting largely of electrically charged and fine-grained particles such as those of the clay minerals, then as has been seen in Section 4.10, such particles repel each other and, provided that there is sufficient water in which they can swim, they take up positions remote from each other. If the water content of the soil is reduced, then the particles are forced into closer proximity than would otherwise be the case. Withdrawal of water is accompanied by an equivalent reduction of overall volume (i.e. by shrinkage) without the entry of air into the interstitial pore space, and the suction is required to maintain this smaller volume in the face of the tendency of the particles to compel more living space. Hence in swelling and shrinking wet soils (the so-called colloidal soils) the function of suction is to overcome the mutual repulsion of the particles, and in Section 8.6 the relationship between suction and soil moisture content on this account will be further discussed.

When the soil has so little water left in it that, even in the case of colloidal soils, the pore space is for the most part filled with air, the little water present is just that which clings to the surfaces of the solid particles by virtue of the electrical attractions of the individual water molecules for those surfaces or for other water dipoles which are already anchored to the surfaces, as described in Sections 3.2 and 3.5. The suction necessary to remove further water is then required to overcome these molecular attractions, and is rather high. Water may be held in yet other forms, as, for example, in the form of hydroxyl ions in the structure of soil minerals. By forced alteration of the structure, two such ions may combine to form a water molecule, leaving an oxygen ion to occupy a place in the residual structure. Such radical changes as this require such large suctions for the extraction of water that they are commonly brought about only at high temperatures of some hundreds of degrees centigrade. Since studies in this field throw light on the chemical properties of soil constituents rather than on the physical properties of the soil itself, phenomena in this range of high suctions will not be discussed here.

## 8.2 The shapes of air-water interfaces in soil

Even when a porous material consists of particles of very regular shapes and sizes, as for example when it is an array of spheres of uniform size, the shape of the air-water interface which those particles support is far from simple. In a mass of heterogeneous sand particles, and even more so in a real soil, the interface shape almost defeats the imagination. Nevertheless,

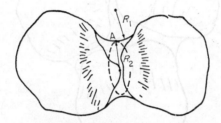

Figure 8.2 The shape of the water surface round the
contact point between two particles.

the most complicated interface shape is made up of certain typical elements which one may discuss quantitatively. Thus it may be noted in Figure (8.2) how a water surface is suspended between a particular pair of particles. A section in a plane through both particles reveals a surface curved so that the convex side is presented to the water, whilst in a plane between the two particles (perpendicular to the first plane) the concave curvature is presented to the water. The water surface is saddle-shaped. The surface suspended between a group of three particles, as in Figure (8.3), presents three such "saddles", which merge together at the centre of the group to form a bowl-shaped interface, concave as viewed from the air side. In order to relate the pressure difference on the two sides of the surface to the curvature, one must know how to specify the curvature. The diagrams referred to in this section suggest how this is to be done. Through the point in the surface at which the curvature is to be specified, draw two planes perpendicular to each other and to the surface. The intersection of each of these planes with the surface results in a curved line. Only exceptionally will these curves be circles, but a sufficiently small portion of *any* curved line may be regarded as a part of a circle without appreciable error, and the radius of this circle specifies the radius of curvature of the particular small element. Hence a

statement of the radii of curvature of the two curves of intersection, together with an indication as to which side of the interface accommodates
the respective centres of curvature, provides a complete specification of the
curvature of the surface at the chosen point. For example, in Figure (8.2)
the curvature at the point $A$ consists of two components, one with radius
of curvature $R_1$ and with centre of curvature on the air side of the interface

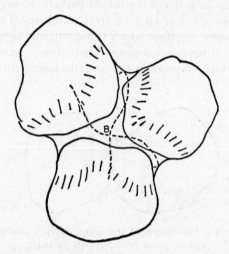

Figure 8.3 The shape of the water surface contained
between three soil particles.

and one with radius of curvature $R_2$ and with centre of curvature on the
water side of the interface. In Figure (8.3) the curvature at point $B$ is such
that the two centres of curvature are on the air side of the interface. In the
special case of a spherical surface, such as is presented by the interface
between a small drop of water and the surrounding air, or a small bubble of
air in water, the two radii of curvature are equal and have the same value
at all points, and the centres of curvature are one and the same point,
namely the centre of the sphere.

A curved interface can only be maintained by excess pressure on the
concave side or suction on the convex side, as was mentioned in Section
8.1. Where the two curvature components are in opposition (i.e. with
centres of curvature on different sides), as in Figure (8.2), the net pressure
difference is a result of one curvature component outweighing the other.
Let the two components of curvature have radii $R_1$ and $R_2$ respectively,
and let the surface tension acting in the surface be $T$. Then it is shown in
Note 11 that the pressure $P$ on that side of the interface which does *not*
contain the centre of curvature from which $R_1$ radiates is less than that, $A$,

on the side which does contain that centre by an amount which is given by the formula,

$$P = A - T(1/R_1 - 1/R_2) \qquad (8.1)$$

if the centres of curvature are on opposite sides, as illustrated by the saddle-shaped surface of Figure (8.2). If, however, the radii of curvature, now written $R_3$ and $R_4$, extend from centres of curvature on the same side of the interface, namely that on which the pressure $A$ operates, as shown in Figure (8.3), the relationship becomes

$$P = A - T(1/R_3 + 1/R_4) \qquad (8.2)$$

If the interface is spherical, as in the case of a small drop of water or a small bubble of air in water, both radii have the same value, $R$, and the pressure $P$ outside the sphere is related to that, $A$, inside it by the appropriate modification of Equation (8.2), namely,

$$P = A - 2T/R \qquad (8.3)$$

In the case of soil water it is almost invariably the case that one side of the interface is in free communication with the atmosphere and is therefore at constant pressure which is taken to be the datum for measurements of $P$. Thus a conventional value of zero being assigned to the constant $A$ results in the following formulas appropriate to the respective cases:

$$P = -T(1/R_1 - 1/R_2) \qquad (8.1a)$$
$$P = -T(1/R_3 + 1/R_4) \qquad (8.2a)$$
$$P = -2T/R \qquad (8.3a)$$

where $P$ is the pressure in the water. In Equations (8.2a) and (8.3a) this pressure is evidently negative and therefore represents a suction, as is commonly the case, while in Equation (8.1a) the relative magnitudes of $R_1$ and $R_2$ determine the sign of $P$. In fact the geometry of the system of granular material determines that $R_2$ will always exceed $R_1$ in the interface which pervades the pore space, so that again $P$ represents a suction.

It must be borne in mind during subsequent discussions that the reference to soil water suction in absolute terms is a convention, and what is meant is always the amount by which the pressure on the water side of the interface is less than that on the air side. Absolute values of pressure on the two sides are irrelevant to the equilibrium, which, as Note 11 shows, depends only on the difference. It is sometimes convenient to alter the air pressure by experimental manipulations, but the equilibrium is not upset if the water pressure is changed by the same amount.

It is to be observed from Equations (8.1a) and (8.2a) that very different shapes and degrees of curvature may yet be in equilibrium at the same suction. Thus if $R_1$ and $R_2$ are small but not very different, the terms in brackets in Equation (8.1a) will be large but not very different, so that the suction determined by the difference between these terms may yet be small. If $R_3$ and $R_4$ are large, the terms in brackets in Equation (8.2a) will be small and the suction determined by the sum of those terms will be small. It is therefore possible to choose small values of $R_1$ and $R_2$ and large values of $R_3$ and $R_4$ which provide the same values of suction; that is to say, saddle-shaped interfaces of sharp curvature may be in equilibrium at just the same suction as for cup-shaped interfaces of more gentle curvature. Indeed, if Figure (8.3) represents a state of equilibrium, it can only be for this reason, since one and the same interface, necessarily maintained by the same suction everywhere since there is no migration of water from one part to another, is in some places saddle-shaped and in others cup-shaped. It follows that, whilst the shape of an interface determines the suction uniquely, the converse is not true; one and the same suction may support a variety of differently shaped interfaces. If one happens to know in a particular case, or may be permitted to assume, that the interface is supported in a capillary tube of circular cross-section, then the meniscus is known to be hemispherical and Equation (8.3a) applies. Then, from a knowledge of suction and surface tension, the radius of curvature at once follows. Equation (8.3a) is sometimes applied even where it is known that the surface can hardly be spherical, and the radius of curvature calculated in such circumstances may be called the *effective radius*; it is the radius which the curvature would have had, had it been spherical. The situation is analogous to the specification of particle size in mechanical analysis on a basis of the velocity of settlement in water. In that case Stokes's law is applied boldly as though the particles were spherical, although it is known very well that they are very irregular in shape. The size which emerges is then described as the effective radius of the particles.

## 8.3 The withdrawal of water from non-shrinking soil

In Figure (8.4) is shown a section through a hypothetical soil specimen, showing particles and pore space near the surface. Suppose that initially the soil is saturated, with water standing above the surface as indicated by the air-water interface marked "stage 1". This interface is the plane surface of free water in bulk, and therefore the water suction just under the surface is of zero magnitude, in accordance with Equations (8.1a) and (8.2a). The water at the soil surface is therefore at the hydrostatic pressure $g\rho H$, where $H$ is the height of the head of standing water. It will be assumed that the soil sample is sufficiently small for differences of "head" at different parts of it to be negligible; strictly speaking one cannot speak of the pressure or suction prevailing in a soil sample, but only of that at a point in the sample. With that proviso, it may be said that the soil water is at positive pressure $g\rho H$ and is saturated at that pressure. If now the sample is supposed to be supported in an apparatus of the kind shown in Figure (7.2), water may be withdrawn from it via the porous plate of the Buchner funnel by lowering the burette. At all times until stage 2 is reached, the falling surface of the water above the soil remains plane and the suction just under the surface remains zero. The head of standing water decreases and therefore the soil water pressure falls until it takes the value zero at stage 2. The

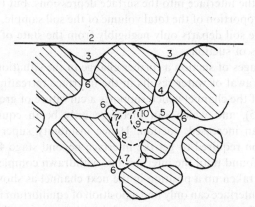

Figure 8.4 Stages of the withdrawal of water from and its reentry into the pore space of a porous material. Numbers represent the order of stages. Full lines represent retreating stages and broken lines advancing stages.

soil remains saturated the whole time, the water emerging into the burette being merely due to the reduction of standing water. Hence it is seen that for as long as the soil is saturated, there is a positive (or zero) hydrostatic pressure prevailing in the soil water.

As soon as surface particles cut the air-water interface, the situation changes. Curvature of the interface, of the kind shown in Figure (8.3) is imposed; stage 3 of Figure (8.4) may be regarded as a section through the cup-shaped depression of Figure (8.3). The magnitude of the suction experienced by the soil water is that which conforms to the interface curvature. As the interface is drawn farther down, the curvature becomes sharper, and the suction, in accord with Equation (8.2a), increases. At the same time water is, of course, withdrawn from the saddle-shaped masses at the points of contact between neighbouring particles (see Figures 8.2

and 8.3), both curvature components becoming increasingly sharp but the net result being an increase of suction in accordance with Equation (8.1a). The greatest suction which can be maintained by the interface corresponds to the sharpest curvature which can be accommodated in the channel through which the interface is being withdrawn, and the sharpest curvature occurs at the narrowest part, as shown at stage 4. As an illustration of the order of magnitude of the suction at this stage, a coarse sand with particle sizes of about 1mm would have channels which could support an interface with curvature of effective radius of the order of 0·25 mm, at a water suction of about 6 cm of water, as derived from Equation (8.3a). During the stage of increasing suction up to this magnitude, water is lost by the soil due to the sinking of the interface into the surface depressions, but this represents only a small proportion of the total volume of the soil sample. Hence it may be said that the soil departs only negligibly from the state of saturation in the early stages of suction.

At higher stages of suction a further change in the situation takes place. Further withdrawal of water results in the interface retreating still further into a region of the channel which supports a curvature of greater, not less, radius (stage 5), and which can therefore only be in equilibrium at a reduced, not an increased, suction. Hence instability supervenes; at the imposed suction required to get the interface beyond stage 4, no position of rest can be found until the interface has withdrawn completely from the "cell" and has taken up a position in the next channel as shown at stage 6. Even then the interface can only find a position of equilibrium if this channel is narrower and can support a surface of sharper curvature than is the case at stage 4. If the soil were to consist of a mass of particles enclosing channels all of exactly the same geometry as that shown at stage 4, there would be no resting place at all for the interface at suctions greater than that corresponding to stage 4; all pores would suddenly empty at that suction. Thus at this point one would observe a sudden and almost complete withdrawal of the soil water.

In fact some water is left behind; the saddle-shaped masses are reduced by water withdrawal as suction is increased, but at the moment of instability their connection with the withdrawing main body is ruptured, and they spring back as isolated water rings round the points of contact between the particles. The fact that they do so spring back means, of course, that the suction corresponding to the final curvature of the isolated ring is less than that of the connected saddle-shaped mass at the moment of rupture, so that one cannot infer from the suction at the stage of sudden withdrawal the precise value of the suction in the isolated ring *after* the rupture which isolated it. It is of course obvious that any increasing suction which may be imposed on the main body of soil water after its separation from the isolated rings cannot be transmitted to those rings. One must therefore exercise some caution when speaking of the suction in soil water as indicated by the observed suction in external bodies of water in equilibrium with the soil. It is, of course, true that the water rings round points of contact between particles are not to be regarded as isolated from the main body of water if we accept the vapour phase as a connecting link. Differences of suction must be accompanied by differences of vapour pressure in accordance with Equation (7.1), so that transference of water from a site at low suction to one at high suction can take place by distillation. Furthermore a movement in the liquid phase may take place via the very thin films left on the wet particles because of molecular attraction, but all such means of water transport are extremely slow compared with movement through water-filled pores. States of complete equilibrium of soil water are probably rarely achieved either in natural soils in the field or experimental manipulations in the laboratory, and what is referred to as soil water suction is really the suction on the main body of continuous soil water, which is also in liquid continuity with such external water masses (e.g. in manometers) as are used to indicate that suction.

In general, a porous body will not contain pores of uniform size and shape, all emptying at the same suction, but, as described in Section 6.3, will be characterized by a pore size distribution. It follows that not all pores will be emptied at the same suction. Those with large channels of entry, in

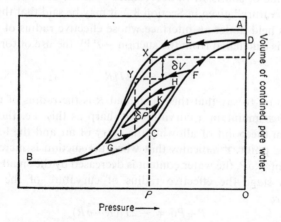

*AB*, drying from saturation
*BCD*, rewetting with air entrapment
*FGHJK*, internal branches due to successive reversals

Figure 8.5 Typical moisture characteristics of a non-shrinking granular material.

E*

which only gentle interface curvatures can be maintained, will empty at low suctions, while those with narrow channels of entry, supporting interfaces of sharp curvature, will not empty until larger suctions are imposed. Hence, as the soil water suction is progressively increased, the soil moisture content is progressively reduced, the larger pores (which may plausibly be supposed to have the larger entry channels) emptying at the lower suctions and the smaller pores at the higher suctions. The curve obtained by plotting the moisture content against the suction will be of some such form as is shown at *AB* in Figure (8.5). Curves of this kind, it will be found, are of fundamental importance and form the starting point of the quantitative discussion of some physical soil properties of great practical significance. Because they are being increasingly referred to in soil physics literature, some convenient short name is desirable. It is too early to propose any nomenclature for conventional adoption, but the name "soil moisture characteristic" has been suggested (Childs, 1940), and has achieved some degree of general usage. It will be adopted here.

## 8.4 The quantitative interpretation of the soil moisture characteristic of a non-shrinking soil

Consider the situation represented by the point *X* on the curve *AB* in Figure (8.5). This indicates that at a suction $-P$, the volume of water held in unit volume of soil is *V*. In accordance with the definition of "effective radius of curvature" given in Section 8.2, it may be said that this volume of soil water is held under an interface whose effective radius of curvature is *R*, where *R* is calculated from the suction $-P$ by the use of formula (8.3a), namely,

$$P = -2T/R$$

One may go on to say that this calculated *R* is the radius of the channels which can *just* maintain a curvature as sharp as this, i.e. those channels which are on the point of allowing the entry of air and the loss of water. Similarly the point *Y* indicates that when the suction is increased by the small amount $-\delta P$, the water content is decreased by the small amount $\delta V$, and at this stage the effective radius of curvature of the interface is $R - \delta R$ where

$$P + P\delta = -2T/(R - \delta R) \tag{8.4}$$

The retreat of the interface is halted in channels of size $R - \delta R$. The increase of suction from $-P$ to $-(P + \delta P)$ causes the emptying of those pores whose channels of air entry are smaller than *R*, but larger than $R - \delta R$, and the total volume occupied by those pores is $\delta V$. The whole of the curve *AB* may be divided into a number of such short intervals, to each of which the above argument applies. The shape of the curve thus provides evidence of a quantitative kind as to the distribution of the pore space among pores of different sizes of channels of entry and therefore, by inference, among pores of different sizes.

In applying this interpretation one must recognize that the nature of the material precludes precision. One can imagine, for example, a pore of a larger size group to be surrounded by pores of a much smaller size group with characteristically narrow channels of entry. The large pore cannot empty until at least one of the surrounding pores does so; that is to say, the larger pore makes its contribution to the water loss, together with the smaller size group, instead of with that group to which it properly belongs. Again, it must sometimes happen that a small pore is entirely surrounded by pores of a larger size group, all of which empty at a suction much below that at which the small pore is due to empty. As a consequence this small pore becomes isolated from the main body of retreating water, and it has no channels of connection through which it can empty when the appropriate suction is applied to the main body. This small pore therefore fails to make its contribution at any stage of water withdrawal. Such circumstances may often cancel each other out, and it is noteworthy that Swanson and Peterson (1942) obtained reasonable agreement between pore size distributions observed directly under the microscope and those inferred from the appropriate moisture characteristics.

## 8.5 The relaxation of suction and the uptake of water; hysteresis

Suppose that the emptying of the cell shown in Figure (8.4) has reached stage 6, and that, instead of increasing the suction in order to empty more and smaller cells, one relaxes the suction in an attempt to refill the cell. In order that the cell may refill, the interface must climb back through the widest part, that is to say, through stages such as those shown at stages 7 and 8. The interface curvature becomes progressively less sharp, demanding the relaxation of suction to progressively smaller values. It will be observed that because of the shape of the interface no very great proportion of the cell is filled by the time stage 8 is reached. Beyond stage 8, as for example at stage 9, the interface curvature becomes sharper again as the cavity walls become more nearly parallel. This sharper curvature would require for its equilibrium a greater suction than that which allowed the interface to reach stage 8. The interface is therefore unstable and the suction is insufficient to prevent the cell refilling completely until a stage 10 is reached in a neighbouring cell where equilibrium is again possible. Further uptake of water then requires a further relaxation of suction. Thus the suction required to empty a cell is the relatively high value corresponding to the sharp curvature at stage 4 of Figure (8.4), whilst the suction for refilling is the relatively low value corresponding to the gentler curvature of the interface at stage 8 of Figure (8.4), the suctions being related to curvature in accordance with Equation (8.3a). A further contribution to the difference of suction between the emptying and the filling stages is made possible by a difference of angle of contact between the solid and water surfaces. This angle is zero only when the most meticulous attention is paid to the scrupulous cleanliness of the solid surfaces, which is clearly impossible when dealing with soil; for surfaces which are other than clean, the contact angle tends to be smaller when the interface is retreating than when it is advancing. Other things being equal, this implies that the radius of curvature in a given channel is greater for a greater angle of contact, i.e. is greater when the interface is advancing and the cell refilling than when the cell is emptying. The suctions on this account are therefore less in the refilling than in the corresponding emptying stages.

If now one considers the whole range of pore sizes and the whole course of water entry from the dry to the saturated states, it is clear that the smaller pores, with higher suctions at reentry of water, will fill before the larger pores, which must await further relaxation of suction. The suction at which each size group refills is, however, less than that at which it empties. Hence the curve obtained by plotting water content against suction is similar in shape to that obtained during water removal, but is displaced in the direction of lower suction. In Figure (8.5), the curve *AB* is appropriate to the condition of decreasing moisture content, as indicated by the arrow. If the trend is reversed at *B* then the curve for increasing moisture content is some such curve as *BCD*. This irreversibility of the moisture characteristic is called hysteresis.

In order that a pore may fill with water, there must be a free passage of escape for the air, and this free passage is not always available. If a large pore is entirely surrounded by smaller pores, these latter will fill first at relatively high suctions so that the air in the large pore becomes an isolated bubble. When the suction is relaxed to the value at which this large pore should refill, there will be no entry of water because the air cannot escape. Thus when the suction is relaxed to zero, at which stage the soil should be saturated, complete saturation will usually fail to occur because of such air imprisonment. It is for this reason that the point *D* of Figure (8.5) does not coincide with *A*. Such a failure of the first hysteresis cycle, starting from saturation, to close is very commonly observed. This first cycle entraps all the air that the pore geometry renders capable of entrapment, and subsequent cycles such as *DEBCD*, if carefully performed, are both closed and reproducible. Closed loops of this kind will be implied in the following discussion.

Hysteresis of the moisture characteristic depends upon the irregularity of shape of the pore space, comprising larger caverns connected by narrower channels. The greater the disparity between the size of the cavern and the size of the channel, the more marked is the difference between the suctions of emptying and refilling. There is clearly no reason to suppose that in such a material as a random array of sand particles all the pores should be of the same shape, differing only in size. If this were the case, the ratio of cavern size to size of the channel of entry or exit would be constant, and all caverns of a given size would have a common channel size. This being so, all pores with a given emptying suction would have a common refilling suction, and the ratio of emptying to refilling suction would be constant. Thus if the wetting were to be interrupted at some intermediate point $F$ on $BCD$, and the suction then increased, the last pores to be filled would all be the first to empty again and would all empty at a common increased suction, after which smaller pores would empty in reversed order of filling. Hence the suction would be increased without any re-emptying of water until it corresponded to the suction on the drying curve $DEB$ appropriate to the prevailing moisture content; i.e. the hysteresis loop would be initiated by a section parallel to the suction axis as far as the intersection with the drying portion of the boundary loop, $DEB$, which would there-after be followed. At all moisture contents the ratio of suctions at the wetting and drying arms would be constant, so that the width of the loop would be proportional to the emptying suction (or to the refilling suction) and would therefore be greater at the higher than at the lower suctions. In fact it is commonly observed that the secondary hysteresis curves, such as $FG$, show some immediate loss of water with the change from wetting to drying, as indicated, and moreover the main hysteresis loop is very commonly wider at the low suction end than at high suctions. The evidence is, therefore, that some pores have a greater discrepancy between cavern and channel sizes, and therefore between emptying and refilling suctions, than do others, and the greater discrepancy seems to be more common in the case of large pores than of small. This is to be expected, since the smallest channels can conceivably connect with the largest caverns.

A complete moisture characteristic between saturation at vanishing suction as the initial state and the maximum suction attainable is called the boundary drying curve. In the reverse direction, the characteristic between the maximum suction and vanishing suction is the boundary wetting curve, and the pair of such curves is the boundary loop. A characteristic beginning at an intermediate point on the boundary wetting curve and carried to maximum suction is a primary drying curve, while similarly a characteristic from an intermediate point on a boundary drying curve and carried to saturation at vanishing suction is a primary wetting curve. A characteristic beginning at an intermediate point on a primary curve is called a scanning curve. Such curves are illustrated in Figure (8.6).

It is sometimes said that the boundary wetting curve *BCD* of Figure (8.5) is a better indication of pore size distribution than is the drying curve *DEB*, on the grounds that the former is governed by the size of the cavern, and the latter by the size of the channel of entry. This is an important matter, for the pore size distribution, unlike the mechanical composition of the particles, is a fundamental soil property which, as will be seen in

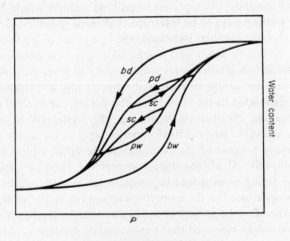

bd, boundary drying curve
bw, boundary wetting curve
pd, primary drying curve
pw, primary wetting curve
sc, scanning curve

Figure 8.6 Hysteresis of the soil moisture characteristic.

Chapter 10, forms a basis for deriving quantitatively some other soil properties of great importance in the study of soil water movement. However, the matter is not as simple as this, as the foregoing discussion makes quite clear, and moreover the experimental difficulties associated with the wetting part of the cycle are greater than those experienced with the drying arm, so that errors must be offset against possible theoretical superiority. In what follows, if the term "moisture characteristic" is used without special reference to the wetting process, the boundary drying arm will be implied.

### 8.6 The independent domain concept

In the preceding section, an explanation of some soil water phenomena has been offered in physical terms on the basis of a postulated pore geometry. In fact the pore geometry cannot be directly observed and specified and in many respects it is more satisfactory to build a self-consistent theory of hysteresis, based only on the observed properties of moisture content and prevailing pore water suction. To do this, the assumption is made that the whole body of pore water may be divided into elements of volume, each of which is defined unambiguously by two properties, namely, the pressure range $P_e$ to $P_e + \delta P_e$ at which it is withdrawn from the pore space, and the pressure range $P_i$ to $P_i + \delta P_i$ at which it re-enters. All pressures are negative (i.e. suctions) and are so indicated in Figure (8.7).

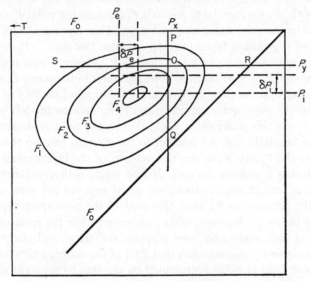

Figure 8.7 The specification of elements of pore water
in the independent domain diagram.

This treatment amounts to a particular application of the general concept of independent domains developed by Everett and his colleagues (1952 to 1955) and by Enderby (1955, 1956). The assumption is not, of course, self-evidently acceptable, for one may readily postulate a kind of pore geometry for which the suction values at entry and exit of water depend upon the history of wetting and drying of the sample. However, it will emerge that one result of the analysis is the design of an experiment to test the validity of the concept in any given case.

According to the independent domain concept, one may indicate on a diagram, as in Figure (8.7), the specification of any given element of pore water. The pore emptying pressure $P_e$ is plotted as abscissa and the filling pressure $P_i$ as ordinate. Thus the elementary rectangle shown specifies the element of pore space which loses water as the emptying pressure changes from $P_e + \delta P_e$ to $P_e$ and refills as the filling pressure changes from $P_i$ to $P_i + \delta P_i$. The whole area of the quadrant clearly covers all the possible combinations of ranges of entry and exit pressures. Furthermore, since the filling pressure corresponds to a suction which exceeds the emptying suction, only those elements have an existence which are represented by rectangles lying above the diagonal bisector of the quadrant which passes through the origin.

The changes of state of the porous body that are recorded by the various paths shown in Figure (8.5) are indicated in a different way in Figure (8.7). If a line $PQ$ is drawn parallel to the axis $P_i$ through the point $P_x$ on the $P_e$ axis, it divides the pore space into two parts. That represented by the area to the left of $PQ$ cannot be emptied by suctions less than $-P_x$, while that to the right empties at suction less than $-P_x$. Hence if one begins with saturated material at zero suction, corresponding to the point $D$ in Figure (8.5), and decreases the pressure (increases the suction) steadily to $P_x$, then that part of the pore space represented by the area to the left of $PQ$ in Figure (8.7) remains water-filled while that to the right represents empty pore space. Similarly line $RS$ parallel to the $P_e$ axis, passing through the point $P_y$ on the $P_i$ axis separates two regions of the independent domain diagram during a wetting process. If one starts with completely empty pore space at infinite suction (infinitely great negative pressure) and then increases the pressure to $P_y$, then that part of the pore space represented by the area below $RS$ becomes filled with water, while the remainder stays empty. If at this stage one now reverses the trend and decreases the pressure again to $P_x$, one removes that part of the water, which had previously entered, held in pores represented by the area to the right of $PQ$, so that at the end of this stage the bent line $QOS$ is the boundary between the area representing water-filled pores and that representing empty pores. The filled pore space is always the part represented by the area on that side of the boundary which is the more remote from the origin.

In a similar way one may trace on diagrams such as Figure (8.7) the course of suction changes shown in Figure (8.5), starting at $D$ and passing through $B$, $F$, $G$, $H$, $J$ and $K$. The result is shown in Figure (8.8). In this example the suction ranges converge steadily, but this is not necessarily always the case. If during any stage of water extraction the suction is increased beyond the limit reached in some previous drying stage, or if during any wetting stage the suction is relaxed to a value less than that reached in some earlier wetting stage, then, by following through the changes on a diagram such as Figure (8.8) it may be shown that the effects of all stages from and including the exceeded stage to the current stage are erased. The prevailing boundary between filled and empty pore space in such diagrams as Figure (8.8) is at all times a step-shaped curve as shown.

So far nothing has been said about the indication on Figure (8.7) of the amounts of pore space devoted to elements specified by their position in the diagram. These volumes may be indicated by plotting a third variable $F(P_e, P_i)$, a unique function of $P_e$ and $P_i$, in the direction perpendicular to the plane of the diagram, thus producing a surface in a three-dimensional system of Cartesian coordinates. Because of difficulty in representing

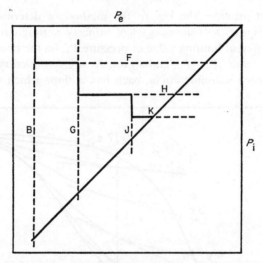

Figure 8.8 The boundary between water-filled and air-filled pore spaces in the independent domain diagram after a series of reversals from wetting to drying and vice versa. The boundaries *BFGHJK* correspond to the reversals shown in Figure (8.5).

three-dimensional figures quantitatively, this surface is indicated in Figure (8.7) by the $F$ contours $F_0$, $F_1$, $F_2$, and so on. The value of $F$ is defined in the following manner. Consider the element $\delta P_e$, $\delta P_i$ at the point $P_e$, $P_i$ lying at the base of a right prism of height $F$ whose upper boundary is the $F$ surface. The volume of this prism is defined to be that part of the pore space, expressed as a fraction $\delta N$ of the porosity $f$, which is accounted for by the specified element; and provided that $\delta P_e$ and $\delta P_i$ are so small that $F$ does not vary appreciably over the element, it follows that

$$F\delta P_e \delta P_i = f\delta N \tag{8.5}$$

Referring to Figure (8.7) it is seen that that part of the pore space, again expressed as the fraction $N$ of the pore space $f$, which retains water when the pressure is decreased to $P_e$, after having been relaxed from minus infinity to $P_i$, is obtained by integrating the volume elements under the $F$ surface on the side of the boundary $QOS$ remote from the origin. That is to say, in the limit when $\delta P_e$ and $\delta P_i$ become infinitesimally small,

$$\int_{P_e}^{-\infty} \int_{P_i}^{-\infty} F \partial P_e \partial P = fN \tag{8.6}$$

The function $F$ may from this be defined explicitly in the form

$$F = f[\partial(\partial N/\partial P_i)/\partial P_e] \tag{8.7}$$

This definition provides the key to the method of determining $F$ from experiments. Figure (8.9) shows a set of primary wetting curves initiating each from a different limiting value of pressure $P_e$, on the boundary drying curve, such as from $_1P_e$, $_2P_e$, and so on. At a given re-entry pressure $P_i$, the same for each scanning curve, each has a slope which measures the

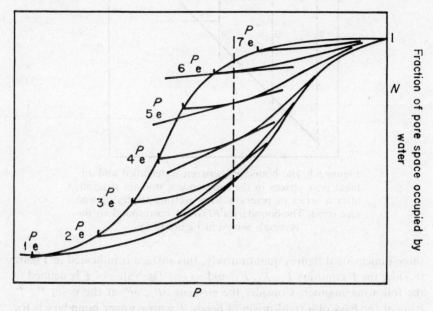

Figure 8.9 A sequence of primary wetting curves from which the distribution function $F$ may be determined.

value $f\partial N/\partial P_i$ for that particular curve, that is to say, for the appropriate value of $P_e$, at that particular value of $P_i$. One may now plot $f\partial N/\partial P_i$ as a function of $P_e$, and the slope of this curve at a given value of $P_e$ is the value of $F$, in accordance with Equation (8.7), at the particular values of $P_i$ and $P_e$. By choosing a sufficient number of values of $P_e$ and $P_i$ a survey of $F$ may be made over the whole area of Figure (8.7) and the $F$ contours plotted.

With the shape of the $F$ surface known, one can evidently predict the moisture content after any given sequence of suction changes by constructing the boundary between filled and empty pores in an independent domain diagram as described above in producing Figure (8.8), and then calculating the volume under the $F$ surface on the side of the boundary remote from the origin by graphical or numerical integration.

In particular, one may calculate the course of the primary drying curves, the $F$ contours having been revealed from analysis of the primary wetting curves as described. Conversely, from a knowledge of the primary drying curves one may compute the $F$ function and contours from the alternative expression of the differentation of Equation (8.6), namely

$$F = f[\partial(\delta N/\delta P_e)/\partial P_i] \qquad (8.8)$$

and thence calculate the course of the primary wetting curves. Agreement with observed results provides a test of the validity of the assumed concept of the independence of the domains. Such agreement has been found in a porous material examined by Poulovassilis (1962). Topp and Miller (1966) failed to confirm this finding for more highly mono-disperse systems, unlikely to occur in nature.

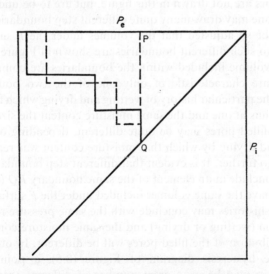

Figure 8.10 Illustrating the different states of pore water at the same moisture content and pressure after different histories of wetting and drying.

A little remains to be said about the size distribution of the water-filled pores in an unsaturated material. In the independent domain diagram of Figure (8.7) let the line $PQ$ represent the boundary between empty and filled pores when the pressure is reduced to $P_e$ from zero. The water content is the volume under the $F$ surface to the left of $PQ$. Equally one could have reached the same moisture content by wetting, allowing the pressure to increase from a very large negative value to $P_i$. The boundary between empty and filled pores in the independent domain diagram would now be $RS$. Equality of moisture content indicates equality of the volumes under the $F$ surface to the left of $PQ$ and below $RS$. The two boundaries intersect at $O$, and the volume under the $F$ surface within the boundary $SOQ$ is common to both the final wetting and final drying states. This volume represents identical pores which are filled with water whether the final moisture content was reached by wetting or drying. On the other hand, the pores represented by $SOPT$, where both $S$ and $T$ are at infinity, are filled in the drying state but empty in the wetting state, while those represented by $QOR$ are filled in the wetting state but empty in the drying state. The argument may be generalized. It is evident that the step-shaped boundary in Figure (8.8) which describes the state of the soil water after a specified number of reversals of wetting and drying, that is to say, after a sequence of different scanning curves, corresponds to a water content which is indicated by the volume under the $F$ surface and within the step boundary. The $F$ contours are not drawn in this figure, but are to be understood. It is evident that one may draw many quite different step boundaries which may nevertheless be so adjusted that the volumes under the $F$ surfaces are all the same. Two such different boundaries are shown in Figure (8.10). Some parts of the volume included within the boundaries are common to both, while some are characteristic of only one of the two boundaries, and therefore of the particular history of wetting and drying which this boundary represents. Thus at one and the same moisture content the size distribution of the water-filled pores may be quite different, depending on the history of wetting and drying by which that moisture content was reached.

One may go further. It is evident that different step boundaries in Figure (8.10) may conclude at an element of the same boundary $PQ$ (or, of course, $RS$) and yet have the same volumes included under the $F$ surface. That is to say, different histories may conclude with the same pressure reached in the same direction (wetting or drying) and the same moisture content, and yet the size distribution of the filled pores will be different. In other words, in terms of the hysteresis diagram of Figure (8.5), a point within the hysteresis loop may lie on a great number of different scanning curves, which intersect at that point, and the pore size distribution of the filled pores will differ with each curve.

## 8.7 The withdrawal of water from a shrinking soil

The subject of this section is an idealized soil consisting solely of particles of colloidal clay material such as is discussed in Chapter 4, particularly in Sections 4.10 and 4.11. It was shown there that two neighbouring particles of colloidal material in suspension in a solution repel each other with a force whose magnitude depends upon the surface density of electric charge on the solid surface, on the distance between their opposing surfaces, and on the concentration and valency of those ions in solution which have charge of sign opposite to that of the particle surfaces. Examples illustrating the relationships between these variables were given in Figure (4.6). If no other force is present to oppose the repulsion, the particles separate until the repulsion becomes negligible. Consider an idealized case of a mass of material consisting of a large number of parallel plane plates of the kind discussed in Chapter 4, as shown in Figure (8.11).

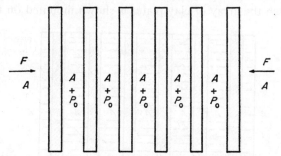

Figure 8.11 A mass of parallel surface-active plates in suspension experiencing swelling pressure.

The repulsion between the plates is due to the hydrostatic pressure in each of the spaces between a pair of plates being higher than in the region outside the mass. Each of the two end plates of the array experiences a higher pressure from the side facing a neighbouring plate than from the side presented to the outside solution, and thus moves away from its immediate neighbour unless prevented in some way from doing so. This movement increases the separation from that neighbour, which reduces the hydrostatic pressure in the intervening space and thus upsets the equilibrium of that neighbour, which now has a lower hydrostatic pressure on the face nearer to the outer solution than on that presented to the next plate in order. The second plate thus separates from the third, and so on; the whole pile of parallel plates expands and the mass is said to swell. The swelling may be prevented in two different ways. If a mechanical pressure is applied to the two end plates of the pile, of just sufficient magnitude to equal the amount by which the hydrostatic pressure between two plates exceeds the pressure in the outer solution, then these two plates are in equilibrium and, as has been seen, this is sufficient to maintain the equilibrium of the pile. In practice the external mechanical pressure may be applied by squeezing the clay between porous plates, the pores of which are small enough to retain the clay, but permit the continuity of the water between the plates and that outside. If the mechanical pressure is increased, the equilibrium is disturbed, and the clay plates move closer together until the internal pressure in accordance with Gouy's theory increases to equal the higher mechanical pressure. Similarly, if the pressure is relaxed, the clay mass will swell until equilibrium is again attained between the lower hydrostatic and mechanical pressures. In the latter case it is, of course, assumed that the clay mass between its porous plates is in contact with an external supply of solution. This mechanical pressure which maintains equilibrium is called the swelling pressure.

The second mode of prevention of free swelling is illustrated in Figure (8.12) in which the array of clay plates is shown mounted on the sintered

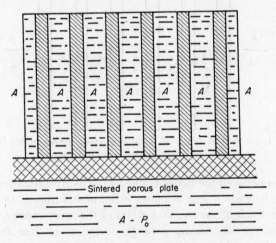

Figure 8.12 As Figure (8.11), but the plates are maintained in equilibrium by pore water suction.

glass platform of a Buchner funnel type of tensiometer. The solution below the platform is the external solution in equilibrium with the clay solution, and it will be supposed that it is kept at a pressure which is less than atmospheric by an amount, $P_0$, by which the hydrostatic pressure between the clay plates exceeds the pressure in the external solution, in accordance with the Gouy hypothesis, at the prevailing plate separation. If the atmospheric pressure is $A$, then the pressure under the funnel platform is $A - P_0$ and the hydrostatic pressure between the clay plates is $A - P_0 + P_0$, that is to say simply atmospheric pressure $A$. Since the external pressure acting on the end plates of the array is also atmospheric pressure $A$ there is no difference of pressure on these plates and consequently the whole pile is in equilibrium. The suction indicated by the manometer attached to the funnel is what we have defined to be the soil water suction. The equilibrium of the clay mass is thus maintained by a suction applied to the interstitial water just as effectively as by a mechanical pressure applied to the mass. If the manometer is raised, thus reducing the suction, the equilibrium is disturbed, the net hydrostatic pressure between the clay plates thereby being increased. The plates thus separate, drawing in water from the supply in the funnel. The clay is said to imbibe water and to swell in the process. Similarly an increase of suction draws the plates together and the clay mass shrinks. This kind of apparatus has been chosen for simplicity of illustration, but the argument holds good for any alternative methods of applying suction. If the suction is applied by controlling the vapour pressure as in Section 4, the external body of solution in equilibrium with the clay is that body which maintains the vapour pressure at the set value.

Figure (4.6) illustrates the relationship between the separation of a pair of clay plates and the hydrostatic pressure between them according to the Gouy hypothesis. It is now seen that the same curves result from plotting the separation against the mechanical pressure or water suction required to maintain the separation as an equilibrium state. Furthermore, since the separation of the plates is proportional to the moisture content of a specified number of plates, i.e. to a given mass of clay, it would be expected that curves of the same kind represent the relationship between moisture content and soil water suction, i.e. the moisture characteristics of such colloidal materials. One cannot in general expect a satisfactory quantitative agreement with observed relationships because in practice masses of clay are not composed of arrays of precisely parallel clay plates, but the observed curve for kaolin, shown in Figure (8.13), is at least of the required monotonic shape in contrast with the moisture characteristic of a non-shrinking material. Warkentin, Bolt and Miller (1957) have described experiments in which agreement between observed and calculated swelling pressures was obtained after a preparatory treatment of the clay sample.

An explanation of hysteresis based upon the retreat and advance of an air-water interface in the pore space can have no relevance to colloidal material which remains saturated at all moisture contents. Inasmuch as the Gouy analysis provides a unique relationship between suction and moisture content, there can be no accounting on this basis for any hysteresis which may be observed. Assuming that an observed hysteresis effect is genuine and not due to a failure to wait for a sufficient length of time for equilibrium to be reached at each stage (and the necessary time intervals may be great in the case of clays), the following mechanism has been held to account plausibly for it, although no incontrovertible evidence in support seems to have been adduced. If the clay mass consists of plates which are not parallel, the cavities between them will be larger than in the array of parallel plates hitherto considered, and the potential energy will be higher.

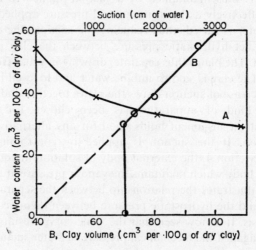

Figure 8.13 The moisture characteristic and swelling
curve of a sample of kaolin clay.

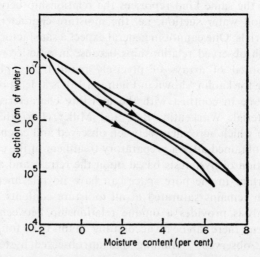

Figure 8.14 "Hysteresis" of the moisture characteristic
of clay (after **R. K. Schofield**).

An increase of suction may not only draw the plates closer together, but may also reorientate them into more nearly parallel positions of lower energy, water being lost in the process. At the same stage of moisture content during rewetting, a relaxation of suction will allow the plates to separate, but will not necessarily cause them to return to their original orientations, so that less water is absorbed than was lost during drying. On this hypothesis only one cycle of drying and wetting should show a hysteresis loop, for once the plates are orientated on the first drying process, further cycles should show reversible changes of moisture content. In fact, curves of the kind shown in Figure (8.14) have been observed (Schofield, 1938). One may refer also to preparatory cycles presented by Warkentin, Bolt and Miller (1959).

## 8.8 Water loss by different mechanisms in the same material

In certain circumstances water may be lost with equivalent air entry over a part of the suction range and with equivalent shrinkage over the remainder of the range. In yet other cases the loss of water may be accompanied in part by air entry and in part by shrinkage. Consider a structured clay soil consisting of aggregates within which the textural pore space and the water which it holds are governed by the equilibrium of the Gouy layers, but between which the structural pore space is governed by the geometry of the array of aggregates. Beginning with a state of saturation at zero suction, an increase of suction will cause some shrinkage of the crumbs and some withdrawal of the air-water interface into the structural pore space. If the aggregates are of commonly occuring sizes, of the order of millimetres up to centimetres, and are in random array as in surface tilths, the structural pores will be large and will empty at low suctions of the order of a few centimetres of water, in accordance with Equation (8.3a). At such suctions, relatively little water may be lost by the crumbs with the accompaniment of shrinkage, so that in the early stages of drying there will be air entry equivalent to the loss of water, and the moisture characteristic is to be interpreted as reflecting the structural pore size distribution in accordance with Section 8.4. In illustration of this, Figure (8.15) shows the low suction stages of the moisture characteristic of a clay crumb fraction separated between sieves of 1mm and 2mm mesh size compared with the moisture characteristic of a similar size fraction of sand grains. The only significant difference between the two curves is that the former is displaced from the latter in the direction of higher moisture content, the difference of moisture contents being due, of course, to water held in the crumbs which would be lost at higher suctions to the accompaniment of crumb shrinkage. Figure (8.16) shows a comparison between the moisture characteristics of different

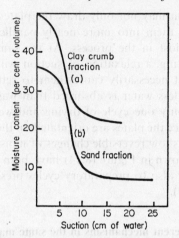

Figure 8.15 A comparison between the moisture characteristics at low suctions of (a) a clay crumb fraction and (b) a similar size fraction of sand.

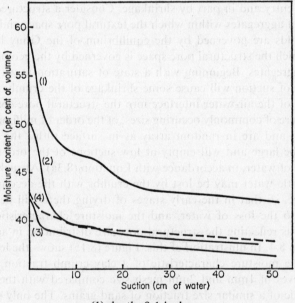

(1) depth, 0-1 inch
(2) depth, 2-3 inches
(3) depth, 4-6 inches
(4) depth, 15-16 inches

Figure 8.16 The moisture characteristics of undisturbed soil samples at different depths in the profile.

horizons of the same profile in natural undisturbed structure. The site of sampling was on grassland on clay. The difference of structural porosity is clearly indicated. Again, Figure (8.17) shows differential moisture characteristics of a clay crumb fraction (i.e. pore size distributions), before and after disturbance due to drying and quick wetting. One sample has obviously had its structural porosity modified to a much greater extent than has the other. The use of moisture characteristics is thus a powerful tool in the study of structure and the stability of aggregation.

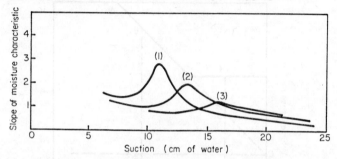

Figure 8.17 The differential moisture characteristics before and after the breakdown of an unstable soil. The numbers indicate stages of breakdown.

At the opposite extreme, one may consider a clay which entirely lacks structural porosity. In the unlikely event of the particles being all plates of equal thickness arranged parallel to each other, increasing suction draws the plates closer together until the separation is negligible, so that loss of water is accompanied by equivalent shrinkage until the water content is negligibly small. In the more likely event of the plates being of uneven thickness, arranged with more or less random orientation, then there will be points at which physical contact is made between neighbouring plates while there is still appreciable separation at others. Shrinkage is thus inhibited while there is still an appreciable amount of water held in the pore space, and further increases of suction must cause curvature of the air-water interface and loss of water with equivalent air entry. An indication of the mechanism of a particular case of water loss is provided by the shrinkage curve. Such a curve is shown for a hypothetical case in Figure (8.18). Starting with a given overall volume of saturated soil, one measures at each stage of suction the volume of water extracted and the volume of shrunken soil. The result is some curve, such as *ABD*. Over that part of the range which corresponds to perfect equivalence of water lost and shrinkage, the curve is a straight line *AB*, whose slope is 45° if the scales of the two axes are the same, for each cm³ of water lost is accompanied by 1 cm³ of shrinkage. If equivalence of shrinkage and water loss were to continue to the end, the curve would be a straight line, such as *ABC*, in which the final volume, *OC*, is clearly the volume of solids since the soil at this stage is dry. If, however, shrinkage ends at the intermediate stage *B*, then the portion *BD* is parallel to *OG* since the volume over this range is constant. At any intermediate stage *H*, one may say that for a moisture content *OL*, the volume would have been *LJ* had there been equivalent

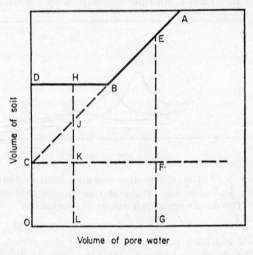

Figure 8.18 Shrinkage during the removal of water
from soil.

shrinkage, but in fact is *LH*. Hence that part *JH* which is due to the fact that no shrinkage occurred subsequently to the stage *B* must be the volume of air which has entered since that stage. Thus if one extrapolates the part *AB* to *C*, one can interpret the curve in the following way. At any point *E* on *AB*, the ordinate *EG* is made up of a part *GF* (equal to *OC*) which is the volume of solid material, and a remainder *FE* which is the volume of water held. At any point *H* on the part *BD*, the ordinate *LH* is made up of a part *LK* (equal to *OC*) which is the volume of solids, a part *KJ* (where *J* is the intercept on *AC*) which is the volume of held water, and a remainder *JH* which is the volume of soil air. The moisture characteristic which is observed simultaneously with the shrinkage curve may then be interpreted either in terms of Gouy layer compression or of pore size distribution, according to the evidence of mechanism provided by the shrinkage curve at the appropriate suction. Thus in Figure (8.13) which shows not only the moisture characteristic of a kaolin mass but also the shrinkage curve, it is seen that the whole of the former over the range of the experiment corresponds to a regime of equivalent shrinkage, and is to be interpreted in the light of the Gouy hypothesis. If the moisture characteristic had been continued to higher suctions, it would have been interpretable in terms of interface penetration and pore size distribution in the shrunk clay. In this example there is a sharp discontinuity between the regimes of shrinkage and of interface penetration, but experiments with soil clays frequently exhibit a smooth gradation from the one to the other. Over the region of partial shrinkage (i.e. shrinkage of smaller volume than that of the water removed), both mechanisms of water withdrawal operate at once.

### 8.9 The equilibrium moisture profile in the field

The moisture characteristic shows the relationship between the soil moisture content and the prevailing suction, in whatever way one interprets it. In those circumstances in the field which give rise to a water table, defined below, one can assign values to the hydrostatic pressure at known depths below, or to the suction at known heights above the water table if the soil water is in hydrostatic equilibrium. Such a state of equilibrium is fortuitous in nature and no doubt rare, but provides a convenient starting point for subsequent discussion, in Chapter 12, of vertical movements of soil water and the development of soil moisture profiles.

The water table is defined to be that level in the soil at which the hydrostatic pressure of soil water is zero. If a well is dug, the water table is the level at which water stands in it, for at that level the hydrostatic pressure in the well water is clearly zero, and is at the same time in equilibrium with the pressure in the adjoining soil at that level. It follows that the pressure at a depth $d$ below the water table, both in the well and in the soil when the two are in equilibrium, corresponds to a head $d$ of water; i.e. in c.g.s. units the pressure is $g\rho d$ where $\rho$ is the density of water. Similarly, at a height $h$ above the water table the pressure is less than that at the water table by a head $h$, i.e. there is a suction of magnitude measured by the height $h$. If one inserts a tensiometer of the form described in Sections 7.2 and 7.3 in the soil at height $h$, with the manometer tube dipping into the well water as shown in Figure (8.19), the water in the tensiometer cell must be in equilibrium with the water in the adjacent soil as otherwise there would be a perpetual circulation of water through the system. The suction inside the tensiometer is clearly $h$ and so, therefore, is the soil water suction at the same height. A measure of the height above the water table is thus also a measure of the prevailing suction; and since the suction is related to the soil moisture content, a plot of moisture content against height above the water table is the same as a plot against suction. The former is called the soil moisture profile, and one sees therefore that the equilibrium soil moisture profile is the same curve as the soil moisture characteristic. Water movements require a change of the pressure gradients, and so what has been said above has no relevance to profiles in which water is moving.

In those cases where the moisture characteristic shows a well-defined suction at which desaturation recognizably sets in, the moisture profile will exhibit a similar feature, namely the maintenance of sensible saturation over a well-defined height above the water table, above which height the

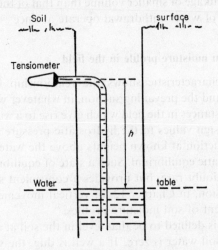

Figure 8.19 The suction in the soil profile above a water table equals the height *h*.

soil is suddenly and markedly unsaturated. It is convenient to refer to this saturated zone above the water table, within which suction prevails, as the capillary fringe. Although its upper limit is not a precisely defined surface, and in fact in some soils cannot be even approximately located, it is often found convenient to refer to the zone within which sudden unsaturation is recognizable as the upper boundary of the fringe.

## 8.10 Equilibrium moisture profiles in the centrifuge

An object whirled in the centrifuge experiences a centripetal acceleration which may be many thousands of times as great as the gravitational acceleration *g*, as described in Section 5.7 and Note 6. Let it be supposed that the object is a soil sample in a container with a perforated base to permit the passage of water, this perforated base lying against the inner surface of the centrifuge head as shown in plan view in Figure (8.20). If the sample is initially saturated, rotation of the centrifuge head causes the centrifugal force to drive water radially outward to escape at the base. The absence of curvature of the surface of the escaping water indicates that the hydrostatic pressure at the base of the sample is vanishingly small. The situation is similar to gravity causing water to drain from a vertical sample to escape at zero pressure at a water table. When equilibrium is eventually attained there will be a moisture profile in the sample as though the water were draining to a water table in a world where gravity is many times as great as that on the earth. Suppose that the centripetal acceleration is $Ng$ where $N$ is a number which is controlled by the speed of rotation and the

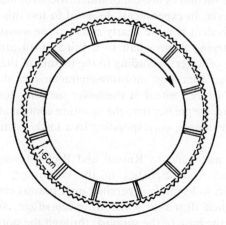

Figure 8.20 The moisture-equivalent centrifuge. Each cell has a perforated base resting against the serrated rim.

radius of the head of the centrifuge. Then at a distance $h$ from the base of the sample, or the water table, the suction will be $(Ng)h\rho$, i.e. that of a head $h$ in a gravitational field of strength $(Ng)$. This suction may clearly just as well be expressed as $(Nh)g\rho$, that is to say, it is equal to the suction which would be experienced at a height $Nh$ above the water table in the terrestrial gravitational field $g$. In effect the moisture profile in the sample of thickness $H$ is the compressed profile in a thickness $NH$ draining to a water table under gravity. In particular, it should be noted that centrifugation does *not* produce a sample at known uniform suction and moisture content, except approximately in the case of a special piece of apparatus which is described below.

Two applications of the centrifuge may be noted here. A centrifuge of the pattern shown in Figure (8.20), containing cells designed to hold samples 1·6 cm in thickness and spun at a speed calculated to produce a centripetal acceleration of 1000 $g$ at the rim, produces a measured moisture content (i.e. an average over the sample) which is known as the moisture equivalent. The suction varies linearly from zero at the base of the sample to 1600 cm of water (i.e. $1000 \times 1\cdot6$ cm) at the exposed surface, and the moisture content varies accordingly in keeping with the moisture characteristic of the soil. No more can be known with precision, but it is sometimes assumed that the mean moisture content cannot be greatly different from the moisture content at the mean height (i.e. at $0\cdot8$ cm) where the suction is known to be 800 cm of water. If this be granted, then the measured mean moisture content is approximately equal to the moisture content of this particular soil at a suction of 800 cm of water. Richards and Weaver (1943), have found, however, in experiments designed to test this conclusion, that the moisture equivalent is more nearly equal to the moisture content at a lower suction, varying from a half to a third of an atmosphere (about 500 cm to 350 cm of water) according to the texture of the soil. This agrees with the known fact that most moisture characteristics show marked concentrations of moisture content at the lower suctions, the mean moisture content being therefore greater than the moisture content at the mean height in the centrifuged sample, corresponding to a lower suction than prevails at that height.

Secondly, one may refer to Russell and Richards' method (1938), of obtaining moisture characteristics in the centrifuge, briefly noted in Section 7.7. A thin soil sample is perched on a porous ceramic or sintered glass support which dips into water in a centrifuge cup. The water is continuous from the base of the support, through the pores of the ceramic or sintered material to the soil sample. If the centrifuge is spun to produce a centripetal acceleration of $Ng$ and the height of the sample above the free water in the cup is $h$, then the suction prevailing in the sample is $Ngh$. Since the value of $N$ is completely under control within the range of the instrument, the suction may be varied at will and the sample subsequently extracted for moisture content determination. A moisture profile is, in fact, set up in the support and sample, although the support material is naturally chosen to be capable of remaining saturated, if possible, so that passage of the soil water is easy as centrifugation progresses. It may be noted that in principle the use of a column of porous material instead of an unobstructed glass tube is possible in the direct manometer method of measuring moisture characteristics (Section 7.3), and such a device would, in theory, extend the range of the instrument beyond 1 atmosphere, thus overcoming the limitation noted in Section 7.4. The necessarily long column of porous material would, however, so restrict water movement that equilibrium would not be attainable in reasonable intervals of time. The device only becomes practicable when the centrifuge is used to increase the gravitational field and so reduce the lengths of ceramic column required.

**8.11 The pressure plate and pressure membrane**

The hydrostatic equilibrium of soil water has been discussed in terms of the difference of pressure between the soil water and the external atmosphere required to maintain a given moisture content. In an inert soil such as sand, the pressure has to be less in the water than in the air in order to maintain the curvature of the air-water interface convex on the water side; and in colloidal material, the pressure has to be less in the water than in the air in order to provide a force keeping the particles together in spite of their mutual repulsions. The absolute values of the pressures inside and outside the soil water are not relevant. Because it is commonly arranged that the air in contact with the soil is at atmospheric pressure and the external mass of water which measures the soil water pressure is at less than atmospheric pressure, one speaks of the soil water suction. It is, however, a simple matter to seal the soil chamber of an apparatus of the kind shown in Figure (7.2) and to raise the air pressure therein, while leaving the soil water to drain through the porous platform and to escape freely at atmospheric pressure. The soil water will be in equilibrium with a moisture content appropriate to the pressure difference between it and the air external to it, whether the external air is at atmospheric pressure and the water under suction, or whether the water is at atmospheric pressure and the external air pressure is artificially raised above atmospheric pressure.

The latter arrangement has been called the pressure plate apparatus, and it has certain advantages. It has been discussed fully by its originator, Richards (1949). The main advantage is that its use may be readily extended to soil water suctions of magnitude much in excess of 1 atmosphere, because the water drains freely at atmospheric pressure and does not require the maintenance of contact between the under side of the porous platform and a hanging column of water under tension. Using suitable construction in steel and with membranes of fine-pored sausage casing instead of porous plates, Croney, Coleman and Black (1958) have in fact pushed the limit of utility up to a thousand atmospheres of suction, and have covered the range usually reserved for vapour pressure methods.

At such very high pressure differences it is possible to have some doubts about the permeability of the membrane. In order to sustain the pressure without permitting the withdrawal of an air-water interface through the pore space of the membrane and thus allowing the passage of air, the pores must be very small. They may in fact be small enough to prevent the free passage of the larger dissolved particles which may be responsible for an appreciable proportion of the osmotic pressure, so that in respect of such particles the membrane constitutes a semi-permeable rather than a permeable membrane. The measured pressure difference in that case will include a contribution due to the osmotic pressure of the excluded particles,

F

and will therefore be directly comparable with the pressure differences obtained by vapour pressure methods.

### 8.12 The measurement of soil moisture content in the field

Tensiometers measure the soil moisture suction directly, but if use is made of the soil moisture characteristic, the moisture content also may be inferred from the indicated suction, and for many practical purposes a knowledge of the content of stored water is as useful as a knowledge of the prevailing suction. Such moisture contents are commonly measured directly by extracting samples, weighing, drying and reweighing them. The consequent disturbance of the sampling site precludes the tracing of changes of moisture content at a given site, and for this purpose the installation of apparatus of tensiometer type has some advantages. In using such apparatus it is common practice to calibrate it directly in terms of moisture content.

The tensiometer principle may be used indirectly by burying in the soil a porous body of which the moisture characteristic is known. Such absorbent bodies have commonly been plaster blocks, but more recently fibreglass and nylon cells have been employed. The soil in which the body is buried has a suction appropriate to its moisture content and transmits this suction to the absorber. This latter then assumes a moisture content which is appropriate to this transmitted suction in accordance with its own moisture characteristic. The moisture content of the buried block may then be measured in a variety of ways, and from it the moisture content of the soil inferred, usually by a direct initial calibration. In one variation, due to Davis and Slater (1942), the buried block is in two parts, one of which is a cone which fits closely into a conical hole bored in the other. This conical portion may readily be raised to the surface, through a permanently installed connecting tube, for weighing. In another method one measures the electrical resistance or capacitance between electrodes buried in the absorbent block, as recommended by Fletcher (1939), Bouyoucos and Mick (1940, 1941, 1948), Anderson and Edlefson (1942b), and Colman and Hendrix (1949). Both the resistance and apparent capacitance depend upon the moisture content of the absorber, and therefore serve as an index ultimately of the soil moisture content. Slater and Bryant (1946) have compared some of these methods with direct sampling, and report that each indirect method has its useful range. Tensiometers are particularly useful at high moisture contents, weighed plugs are accurate over a fairly wide range, while resistance blocks are appropriate for extensive surveys where great accuracy is not a prime consideration. It may be mentioned that the apparent capacitance between electrodes, as commonly measured, is often no more than an indirect reflection of the resistance, which might with benefit be directly measured instead (Childs, 1943a).

The accuracy of measurement of the soil moisture content via the suction depends in part on the shape of the soil moisture characteristic and in part on that of the buried absorbent block, if used. Over those parts of the soil moisture characteristic where considerable changes of suction have but little effect on the moisture content, as at the very wet and very dry ends of the range, the sensitivity may be more than is required, while in the middle ranges, if the characteristic is very steep with much water lost over a small range of suction change, the accuracy may be poor. The method is most convenient for soils of moderately steep moisture characteristic where moisture is lost steadily and appreciably, but not too rapidly, over a fairly extended range of suctions. Where buried absorbent blocks are used, the moisture characteristics of such blocks should match those of the soils in which they are placed, so that both block and soil lose the bulk of their moisture contents over the same range of suctions. If this is not the case, then while the soil is losing the main proportion of its water, the block is changing little and the sensitivity is but poor over the region where it is most needed. It is no compensation that the block loses its moisture over a range where the soil is changing little, for here the sensitivity is great but useless.

It may conveniently be mentioned here, although not strictly relevant to the static equilibrium of soil water, that two other methods for measuring soil moisture content have been developed, which do not involve even indirectly the dependence upon soil water suction. Shaw and Baver (1939) proposed to employ the dependence of the thermal conductivity of soil on the moisture content, and de Vries (1952) established the theory of a practicable method and developed an instrument for the purpose. Briefly, if an electrically heated element is inserted in the soil, its temperature will rise more quickly if heat is retained in it than if heat is conducted rapidly away. Soils at high moisture content are better thermal conductors than when drier, so that the rate of rise of temperature of the heating element for a given power input may be used, by direct calibration, as a measure of soil moisture content.

Another method relies upon the slowing down of fast neutrons by hydrogen nuclei, and has been described by Gardner and Kirkham (1952). If a particle collides elastically with another of much greater mass, it bounces off, in accordance with the laws of conservation of energy and momentum, without much loss of speed; but if the collision is with a particle of comparable mass, the energy becomes shared and the aggressive particle becomes slowed down. The neutron has approximately the same mass as the hydrogen atom, but is light compared with almost all other elements occuring in soils. Hence the effectiveness of a soil in slowing fast neutrons is a measure of the amount of hydrogen present, and since by far the greater proportion of all the hydrogen present is in the form of water, except in highly organic soils such as peat, it is also a measure of the amount of water in the soil. The apparatus used to measure the efficacy of neutron slowing in the soil consists of a source of fast neutrons, such as a mixture of polonium and beryllium or radium and beryllium, mounted in a cylindrical probe next to but well shielded from a detector of slow neutrons. The probe is lowered into position in an auger hole in the soil, and the stream of issuing fast neutrons becomes slowed down to form a cloud of diffusing neutrons in thermal equilibrium with the soil, i.e. with thermal velocities, ultimately becoming absorbed by capture. Thus an equilibrium distribution of slow neutrons is rapidly set up with a concentration which decreases with increasing distance from the probe. In wetter soils, which have the greater slowing ability, the thermal neutrons are more concentrated near the probe, but in drier soils, which are less effective slowing agents, the thermal neutrons extend to greater distances with smaller concentrations in the immediate vicinity of the probe.

The presence of slow neutrons may be detected, and their concentration measured, by a suitable device which is usually an ionization tube containing boron in the form of boron trifluoride gas. The capture of slow neutrons results in an ionizing radiation which produces a current pulse, so that a count of the pulses reflects the concentration of the neutrons. If the slow neutrons lie concentrated near the probe, as in wet soil, the count rate is high, while if they are more widely dispersed as in drier soil, the count rate is low. The method is made quantitative by direct calibration.

Some caution may need to be exercised if the method is used in a soil containing much hydrogen of non-aqueous origin, such as peaty organic soil. It should also be remembered that nuclear radiations constitute a hazard to health and call for extreme caution.

# NOTES

### Note 11. The difference between the pressures on the two sides of an interface between air and water

In Figure (N11.1) is shown a small portion *ABCD* of a curved interface. The more complicated saddle shape has been chosen for greater generality. The remainder of the interface of which it is a portion exerts a tension all round the edges of the element, the direction of the tension lying in the surface (i.e. tangential) and perpendicular to the edge of the element. The arrows indicate the directions of these surface tension forces, the magnitude of which is $T$ per unit length of edge. The element is in equilibrium under these distributed forces.

The edges *AB* and *CD* are approximately equal to the medial line *JK*, while *BC* and *AD* similarly are equal to the medial line *LM*. *JK* is taken to be a part of a circle of radius $R_1$ and with centre at $O_1$, while *LM* has radius $R_2$ and centre at $O_2$. $JK$ subtends the angle $2\theta$ at $O_1$, while $LM$ subtends the angle $2\phi$ at $O_2$. For the sake of clarity of illustration these angles are shown as being of substantial size, but in fact the analysis is generally valid only for very small elements of surface for which $\theta$ and $\phi$ are small, since the radius of curvature of a complicated surface changes from point to point. Since the subtended angles are small one may substitute $\theta$ and $\phi$ for $\sin \theta$ and $\sin \phi$ respectively wherever these trigonometrical functions occur.

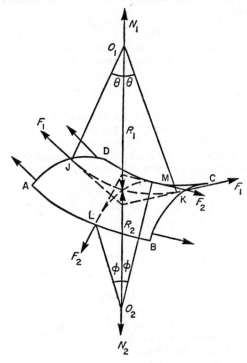

Figure N11.1 A saddle-shaped element of air-water interface illustrating the forces under which it is in equilibrium.

Consider the surface tension forces acting on the opposite edges $AB$ and $CD$ of the element. The length of each of these edges is approximately the same as that of the medial line $JK$, namely $2R_1\theta$. The total force, $F_2$, on each edge is therefore

$$F_2 = 2TR_1\theta \qquad (N11.1)$$

Owing to the curvature of the surface these two forces are not in the same line. The triangle of forces shown in Figure (N11.1) indicates that the resultant force on the element due to this pair of components is $N_2$ through the centre of curvature $O_2$, i.e. normal to the surface, where

$$N_2 = 2F_2 \sin \phi = 2F_2\phi$$

Substitution for $F_2$ from Equation (N11.1) yields

$$N_2 = 4RT_1\theta\phi \qquad (N11.2)$$

Similarly the force $F_1$ on each of the edges $BC$ and $AD$, each of length $2R_2\phi$, is

$$F_1 = 2TR_2\phi \qquad (N11.3)$$

and the resultant force $N_1$ directed to the centre of curvature $O_1$, and therefore normal to the surface, is

$$N_1 = 4TR_2\theta\phi \qquad (N11.4)$$

Since the centres of curvature are on opposite sides, the resultants $N_1$ and $N_2$ oppose each other, and the net force directed towards the centre $O_1$ is, from Equations (N11.2) and (N11.4),

$$N_T = N_1 - N_2 = 4T\theta\phi(R_2 - R_1) \qquad (N11.5)$$

This net force must be neutralized by the excess of pressure on that side of the interface which contains the centre $O_1$. If the pressure on this side is $A$ and that on the other is $P$, then the neutralizing force is the product of $(A-P)$ and the area of the element, which is the product of the lengths of the edges, namely $4R_1R_2\theta\phi$. Thus the force on the element due to the pressure difference is directed from $O_1$ and is of magnitude

$$N_P = 4R_1R_2\theta\phi(A-P) \qquad (N11.6)$$

At equilibrium $N_T$ and $N_P$ are of equal magnitudes, hence from Equations (N11.5) and (N11.6),

$$4T\theta\phi(R_2 - R_1) = 4R_1R_2\theta\phi(A-P)$$

or

$$P = A - T(1/R_1 - 1/R_2) \qquad \begin{cases} (N11.7) \\ \ (8.1) \end{cases}$$

Had the centre $O_2$ been on the same side of the interface as $O_1$, then the net radial force due to surface tension would have been $N_1 + N_2$ and the form taken by Equation (N11.5) would have been

$$N_T = 4T\theta\phi(R_1 + R_2)$$

with the final result,

$$P = A - T(1/R_1 + 1/R_2) \qquad \begin{cases} (N11.8) \\ \ (8.2) \end{cases}$$

CHAPTER 9

# The laws of soil water movement

## 9.1 The cause of water movement; potential difference

Two factors between them determine the rate of flow of water in soil. The first is the force acting on each element of volume of the soil water, and the second is the resistance to flow offered by the soil pore space. These factors will be considered separately.

The force acting on an element of volume of soil water is made up of two items, namely the gravitational force tending to make the element fall to a lower level, and the force due to differences of hydrostatic pressure at different points in the system, the element tending to move from a zone of higher to one of lower pressure. If the temperature varies throughout the system, or if the salt concentration varies, this will affect the vapour pressure of the water and there will be transport of water by diffusion in the vapour phase. In what follows it will be necessary to ignore these vapour contributions for the most part, since the principles, although well known, have not been much developed for the purpose of formulating general laws of soil water movement. Differences of vapour pressure due to differences of hydrostatic pressure (see Section 7.5) produce vapour movements which *can* be included in a general statement of laws of water movement.

The items which contribute to the total force on the soil water will only fortuitously act in the same direction, so that the resultant must be calculated by the geometrical construction of the parallelogram or triangle of forces. In order to avoid this cumbrous inconvenience, it is usual to introduce the concept of potential difference, the items of which are summed by the ordinary laws of algebra, as is shown in Note 12. The potential, which is denoted by $\phi$, is defined to be the amount of work done on a unit quantity of water in moving it slowly from an arbitrarily chosen datum point in the system to the point to which the potential refers, and it is readily measured. One first chooses a convenient point in the system as the arbitrary datum from which all vertical heights are measured, and then one chooses a convenient arbitrary datum of hydrostatic pressure, which is commonly atmospheric pressure. Then if unit volume of water is chosen as the unit quantity, and if $\rho$ represents the density of the water and $g$ the gravitational acceleration of free fall, the potential at a point where the height is $z$ and the hydrostatic pressure is $P$ is, as is shown in Note 14,

$$\phi_{vol} = P + g\rho z \qquad (9.1a)$$

If $P$ is measured by a monometer inserted at the point in question, and the manometric height is $H$ so that $P$ has the value $g\rho H$, Equation (9.1a) takes the convenient form,

$$\phi_{vol} = g\rho(H+z) \qquad (9.1b)$$

Since $z$ is the height of the point at which the potential is being measured and $H$ is the height of the manometer meniscus above this point, the sum $H+z$ is the height of the meniscus above the arbitrary datum level, and this is called the hydraulic head. $H$ itself is called the pressure head. The absolute value of the potential or of the hydraulic head naturally depends upon the choice of datum level, but this is of no consequence since one is always concerned with the difference of potential between two specified points and not the absolute values.

If unit mass of water is chosen instead of unit volume in defining potential, the result is a different measure, $\phi_m$, of potential such that

$$\phi_m = \phi_{vol}/\rho = P/\rho + gz = g(H+z) \qquad (9.2)$$

Again, if unit weight of water is chosen, the result is yet another measure of potential $\phi_{wt}$ such that

$$\phi_{wt} = \phi_{vol}/g\rho = P/g\rho + z = H+z \qquad (9.3)$$

These simple conversion factors permit results derived from one definition to be related to those appropriate to the other definitions.

When the potential is measured at a sufficient number of points in a mass of soil water, it is known everywhere by interpolation, and one may construct a sort of potential contour map. The body of soil water is said to have a known potential distribution. As is shown in Note 13, if the potential distribution is known one may find at any given point the direction in which the potential is increasing most sharply with distance, and this direction is exactly opposite to that in which the force on the soil water at this point acts. The magnitude of the force per unit quantity of water is derived by observing the difference of potential between two points on a line in this direction and dividing this difference by the distance separating the points. A complete specification of this maximum rate of increase of potential and the direction in which it is found is called the potential gradient, and is written as grad $\phi$. If the potential gradient is itself found to increase sharply from point to point, then the two points chosen for measuring it must be close together. Best of all is to plot a curve of potential against distance in the direction of the potential gradient and to measure the slope of this curve at the appropriate point.

A directed quantity such as a force or a gradient of potential is called a vector, and a special notation has to be used in order to specify both the magnitude and direction. Heavy faced type indicates the vector, and its direction, as well as an alternative specification of magnitude, is provided by stating the components of the vector in the directions of Cartesian coordinates, $x$, $y$ and $z$, where $z$ is usually by convention upwards. Thus the magnitudes of the components of grad $\phi$ are $\partial\phi/\partial x$, $\partial\phi/\partial y$ and $\partial\phi/\partial z$, and the full specification of grad $\phi$ is given in the statement,

$$\text{grad } \phi = \mathbf{i}(\partial\phi/\partial x) + \mathbf{j}(\partial\phi/\partial y) + \mathbf{k}(\partial\phi/\partial z)$$

where $\mathbf{i}$, $\mathbf{j}$ and $\mathbf{k}$ are unit vectors in the directions of $x$, $y$ and $z$ respectively, and the addition is understood to be according to the rules of the parallelogram of forces. Similarly the force would be specified by the equation,

$$\mathbf{F} = \mathbf{i}F_x + \mathbf{j}F_y + \mathbf{k}F_z$$

where $F_x$, $F_y$ and $F_z$ are the components of $\mathbf{F}$ in the directions of the indicated axes.

Thus the statements in this section may be summed up in the equations,

$$\mathbf{F} = -\text{grad } \phi$$
$$= [\mathbf{i}(\partial\phi/\partial x) + \mathbf{j}(\partial\phi/\partial y) + \mathbf{k}(\partial\phi/\partial z)]$$

One may thus identify the components of the force as:

$$F_x = -(\partial\phi/\partial x)$$
$$F_y = -(\partial\phi/\partial y)$$
$$F_z = -(\partial\phi/\partial z)$$

An alternative procedure is to express the component of the gradient in the direction of an axis by multiplying the gradient magnitude by the cosine of the angle between the gradient direction and that axis. This cosine is called the direction cosine of the gradient. Thus if the magnitude of the gradient is $G$ and the direction cosines in the $x$, $y$ and $z$ directions are respectively $l$, $m$ and $n$, then one writes

$$\text{grad } \phi = G(l\mathbf{i} + m\mathbf{j} + n\mathbf{k})$$

## 9.2 Water movement in response to potential gradient; Darcy's law

In 1822 Fourier presented his very complete mathematical theory of the transport of heat in conducting materials based on the law that the rate of conduction of heat is proportional to the temperature gradient. In 1827 Ohm enunciated his law to the effect that the rate of transport of electricity (i.e. the strength of an electric current) in a conductor of electricity is proportional to the difference of electric potential between its ends, and from this it is a short step to the law for conductors of different length but of the same material and cross-section, i.e. that the electric current is proportional to the electric potential gradient. In 1822 Navier developed equations describing the flow of viscous fluids in terms of the distribution of hydraulic potential, equations which were later again derived by Stokes (1845) in a more general way. Where the boundaries of the moving fluid are simple, as in the case of flow through a tube of uniform radius, these equations permit one to derive the rate of flow in terms of the dimensions of the conductor and the potential difference between the ends, and indeed Poiseuille's experimentally derived equation of flow of fluid through a tube, published in 1842, can readily be obtained from the earlier theoretical work. Since it will be referred to later, it is quoted here in the form,

$$Q/t = (\Delta\phi/l)(\pi/8\eta)R^4 g\rho \qquad (9.4)$$

where $Q$ is the volume passing in time $t$, $l$ is the length of the tube between the ends of which the potential difference is $\Delta\phi$, (i.e. $\Delta\phi/l$ is the potential gradient), $\eta$ is the viscosity of fluid which may be readily obtained from tables of physical constants, and $R$ is the radius of the tube. Here again the rate of flow is proportional to the potential gradient. The configuration of the pore space in a porous material, such as sand, is far too complicated and unspecifiable in quantitative terms to permit the rate of flow of fluid to be calculated by an application of the Stokes-Navier equations, although the general form of the law of flow can be demonstrated. However, Darcy

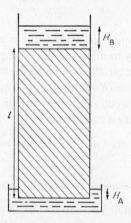

Figure 9.1 Diagram illustrating apparatus used by Darcy to establish his law.

(1856) was able to formulate the law as a result of experiments on the infiltration of water vertically through filter beds of sand. In separate experiments with his apparatus, he showed first that the rate of flow of water through his sample of saturated sand was proportional to what has been defined here to be the potential difference between the ends, and secondly that it was inversely proportional to the length of the column. He took it for granted that it was also proportional to the area of cross-section $S$ of the flow column, for he used only one column of cross-sectional diameter 35 cm, the thickness of the bed of sand varying from 58 cm to 171 cm in different experiments. A schematic diagram of his apparatus is shown in Figure (9.1), where $H_B$ is the depth of water standing on the upper surface of the sand, $H_A$ is the depth of the lower face of the sample column below the surface of the water in the receiving chamber, and $l$ is the thickness of the layer. Darcy expressed the results of his experiments in the form,

$$Q/t = (K_{wt}S/l)(H_B + l - H_A) \tag{9.5}$$

where $K_{wt}$ is a constant characteristic of the porous material. From Section 9.1, Equation (9.3), it will be noted that $(H_B + l)$ is the potential $_B\phi_{wt}$ at the upper surface of the sand, referred to as datum level at the base of the column, while $H_A$ is the potential $_A\phi_{wt}$ at the base of the column, so that the factor $(H_B + l - H_A)$ is the potential difference $(_B\phi_{wt} - _A\phi_{wt})$. Hence Equation (9.5) may be rewritten in the form,

$$Q/t = K_{wt}S(_B\phi_{wt} - _A\phi_{wt})/l \tag{9.6}$$

Alternatively, using the relationship between $\phi_{vol}$ and $\phi_{wt}$ from Equation (9.3),

$$Q/t = K_{vol}S(_B\phi_{vol} - _A\phi_{vol})/l \tag{9.7}$$

where $K_{vol} = K_{wt}/(g\rho)$.

$K_{wt}$ and $K_{vol}$ differ only in the units in which they are expressed, having been derived by the use of different units of potential. They indicate the readiness with which the material permits the passage of water, and in fact express the rate of flow through a column of unit cross-section in the presence of a unit potential gradient. Various names have been proposed for this constant, a very common one having been "permeability", but a degree of controversy has been detectable in this matter in recent years (see Section 9.8). Following the recommendation of the Subcommittee on Permeability and Infiltration of the Soil Science Society of America, the name "hydraulic conductivity" will be adopted here, and where misunderstanding may be possible, both the porous medium and the fluid are to be specified. Of the different units of hydraulic conductivity mentioned, $K_{wt}$ has the dimensions of a velocity and is perhaps most convenient and most commonly in use, but in fundamental work the absolute c.g.s. unit in which $K_{vol}$ is expressed has advantages. In what follows, the symbol $K$ will be used without a subscript, but $K_{wt}$ will be implied unless the contrary is stated.

In the above expressions for Darcy's law the factor $(\phi_B - \phi_A)/l$ represents the potential gradient, which was uniform down the column in the circumstances of Darcy's experiments. A degree of generalization needs to be introduced into the law before it can be applied to conditions other than the very simple ones which Darcy studied. He limited his experiments to water flowing vertically in sand at a constant rate, the sand being at all times saturated. In the form of Equation (9.6) the law is already generalized to flow directions other than the vertical. It may also be written, merely rearranging symbols,

$$(Q/t)/S = K(\phi_B - \phi_A)/l \tag{9.8}$$

The left-hand side is the volume of water entering or leaving unit area of the sand column extremities per unit time, and if the velocity of the water in the chambers external to the column is $v$ (i.e. $v$ is the velocity of the water at the moment just before entering or just after leaving the sand), one has

$$v = (Q/t)/S \tag{9.9}$$

since the rate at which the water moves through the entry and exit chambers must clearly equal the rate at which it moves through the sand, in terms of volume per unit time. One can know nothing precisely about the actual velocity of the water once it enters the tortuous pore space of the sand, but it is certain that since part of the total cross-section is occupied by solids, that part which is still available for water transport must be carrying the water at a speed which, on the average, exceeds that in the unobstructed entry and exit chambers, in order to carry the same *volume per second*. This velocity $v$ is sometimes called the effective velocity, but more often just the velocity of flow. Whatever it is called, one has to remember that it is the hypothetical velocity which the water would have if flowing through the given cross-section quite unobstructed by solid particles.

If the flow of fluid in the porous material is not confined to a parallel-walled column, the velocity of flow must differ at different points, being lower where the cross-section of flow is of greater area. In these circumstances the potential difference $\Delta\phi$ between the ends of column elements of a given length $\Delta l$ will also vary at different parts of the porous body. In such a case one can only state Darcy's law by reference to an experiment carried out on such a small element of volume of the sand that $v$ is sensibly uniform and $\Delta\phi/\Delta l$ is also uniform and has the value which, in the limit of smallness of $\Delta l$, is called grad $\phi$ (see Section 9.1). Using this nomenclature and that of Equation (9.9), and remembering that the velocity of flow is in a direction opposite to that in which the potential is increasing, one has instead of Equation (9.8) the equation,

$$\mathbf{v} = -K \operatorname{grad} \phi \qquad (9.10)$$

The components of the flow velocity in terms of the components of the potential gradient or, from Equation (9.3), in terms of the gradient of the pressure head, are

$$\left.\begin{aligned}
v_x &= -K\,\partial\phi/\partial x = -K\,\partial H/\partial x \\
v_y &= -K\,\partial\phi/\partial y = -K\,\partial H/\partial y \\
v_z &= -K\,\partial\phi/\partial z = -K(1+\partial H/\partial z)
\end{aligned}\right\} \qquad (9.11)$$

In this context of varying directions and magnitudes of grad $\phi$ and $\mathbf{v}$, it is appropriate to use vector notation. This is the most general statement of Darcy's law for isotropic porous materials, i.e. materials which do not have any preferred directions of flow by reason of their structure, and in which, therefore, the flow is in the direction of the potential gradient.

### 9.3 Darcy's law for anisotropic materials

In a simple experiment such as Darcy performed, no information is derived about directions of flow in the porous material, and in any generalizations of the law following from the results, it has been taken for granted that the flow is along the path of the potential gradient. In fact it is not a difficult matter to mark a streamline by injecting dye at a point on the inflow face and, by observing the point of exit on the outflow face or even by observing the whole path of flow adjacent to an inspection window, to confirm this point. Some materials, however, have preferred directions of flow. Suppose, for example, a cylindrical column is permeated by fine parallel capillary tubes which are not parallel to the axis of the column. A potential difference maintained between the ends of the column would constitute a potential gradient parallel to the axis, but flow could take place only in the direction of the capillary tubes, and marker dye would show a flow path making an angle with this axis. Admittedly the true potential gradient in an individual tube is along the axis of that tube and in the direction of flow, but Darcy's law does not concern itself with true flow in the pores of a material, but with apparent or effective flow in the column which is regarded as a uniform substance. Materials of this kind are said to have anisotropic hydraulic conductivity. In general, one cannot know anything very precisely about the internal geometry of such bodies, but one can observe the potential gradient and rate of flow, and infer from these something about the structure.

One way of carrying out Darcy-type experiments with the special material described above would be to cut cylindrical columns from it in many different directions. One would find that for one particular direction one would get a maximum rate of flow for a given potential gradient, i.e. a maximum value of $K$, and in this one direction the path of marker dye would follow the path of the potential gradient. This direction would in fact be one which was parallel to the structural capillaries, although one might have no way of becoming aware of that fact. Those cylindrical samples which were cut in any direction at right angles to the direction of maximum permeability would be found to have zero permeability, and in any other direction the permeability would have an intermediate value and the indicated flow path would differ from the path of the potential gradient.

At a stage further in complexity of structure, one might have a material permeated by sets of capillary tubes or plane laminar fissures in various directions. On cutting samples for experiment in various directions, one would again find maximum permeability $K_\lambda$ in one particular direction, say $\lambda$. Of all the directions perpendicular to this one would find that one, $\mu$, provided a maximum permeability $K_\mu$, in general less than $K_\lambda$, and another, $v$, at right angles to both $\lambda$ and $\mu$, provided a minimum permeability $K_v$. In all three directions, $\lambda$, $\mu$, and $v$, it would be found that the path of marker dye coincided with the path of the potential gradient. In all other directions the permeability would have intermediate values and the direction of the flow path would depart from the direction of the potential gradient. The three directions, $\lambda$, $\mu$ and $v$, are called principal axes. Darcy's law for such a material could then be expressed by a set of three equations,

$$\left.\begin{array}{l} v_\lambda = -K_\lambda(\text{grad } \phi)_\lambda = -K_\lambda(\partial\phi/\partial\lambda) \\ v_\mu = -K_\mu(\text{grad } \phi)_\mu = -K_\mu(\partial\phi/\partial\mu) \\ v_v = -K_v(\text{grad } \phi)_v = -K_v(\partial\phi/\partial v) \end{array}\right\} \quad (9.12)$$

Although it has never been proved that for any unspecified type of structure the principal axes are necessarily mutually perpendicular, this is usually assumed and is almost certainly true.

The experiments could be carried out in a different way. From the slab of material, one could cut three columns with their axes respectively in the mutually perpendicular directions $x$, $y$ and $z$, which would not, except fortuitously, coincide with the directions of the principal axes. In each experiment one would measure the potential gradient, the rate of flow, and the direction of flow; and the latter would be used to resolve the net flow into components in the directions, $x$, $y$ and $z$. With grad $\phi$ in the $x$ direction one could express the results in the form,

$$\left.\begin{array}{l} v_{xx} = -K_{xx}(\text{grad } \phi)_x \\ v_{yx} = -K_{yx}(\text{grad } \phi)_x \\ v_{zx} = -K_{zx}(\text{grad } \phi)_x \end{array}\right\} \quad (9.13)$$

The $v$ and $K$ subscripts indicate first the direction of the flow component, and second the direction of the potential gradient which produces the flow. Similarly, with potential gradients in the $y$ and $z$ directions respectively, the experiments would yield the following expressions for Darcy's law,

$$\left.\begin{aligned} v_{xy} &= -K_{xy}(\text{grad }\phi)_y \\ v_{yy} &= -K_{yy}(\text{grad }\phi)_y \\ v_{zy} &= -K_{zy}(\text{grad }\phi)_y \end{aligned}\right\} \tag{9.14}$$

$$\left.\begin{aligned} v_{xz} &= -K_{xz}(\text{grad }\phi)_z \\ v_{yz} &= -K_{yz}(\text{grad }\phi)_z \\ v_{zz} &= -K_{zz}(\text{grad }\phi)_z \end{aligned}\right\} \tag{9.15}$$

This expression of Darcy's law for anisotropic material in a general framework of Cartesian coordinates, requiring as it does nine equations and nine elements of $K$, may seem cumbrous compared with the three equations of (9.12), referred to the principal axes. However, the introduction of tensor notation results in great simplification. If one defines the conductivity dyadic $\mathscr{K}$ to be

$$\mathscr{K} = \left\{\begin{aligned} &K_{xx}\mathbf{ii} + K_{yx}\mathbf{ji} + K_{zx}\mathbf{ki} + \\ &K_{xy}\mathbf{ij} + K_{yy}\mathbf{jj} + K_{zy}\mathbf{kj} + \\ &K_{xz}\mathbf{ik} + K_{yz}\mathbf{jk} + K_{zz}\mathbf{kk} \end{aligned}\right\} \tag{9.16}$$

then the nine equations of (9.13) to (9.15) may be written in the simple form, as shown in Note 15,

$$\mathbf{v} = -\mathscr{K} \cdot \text{grad }\phi \tag{9.17}$$

Here $\mathbf{v}$ is the vector expression of the velocity of flow, indicating direction as well as magnitude. That is to say,

$$\begin{aligned} \mathbf{v} = \ &\mathbf{i}(v_{xx} + v_{xy} + v_{xz}) + \mathbf{j}(v_{yx} + v_{yy} + v_{yz}) + \\ &+ \mathbf{k}(v_{zx} + v_{zy} + v_{zz}) \end{aligned} \tag{9.18}$$

The dot between the factors $\mathscr{K}$ and grad $\phi$ indicates that any product of vectors which the multiplication involves is carried out according to the rules of scalar products, the scalar product being obtained by multiplying together the magnitudes of the vector factors and the cosine of the angle between their directions. In particular, the product of unit vectors is itself unity if the vectors are in the same direction (e.g. $\mathbf{i.i}$, $\mathbf{j.j}$, $\mathbf{k.k}$), and zero if the vectors are mutually perpendicular (e.g. $\mathbf{i.j}$, $\mathbf{i.k}$, or any scalar product of pairs of vectors $\mathbf{i}$, $\mathbf{j}$ and $\mathbf{k}$ in which the factors are not the same).

The expression (9.16) is said to be the nonion form of the dyadic or tensor $\mathscr{K}$. If the coefficient $K_{xy}$ has the same value as $K_{yx}$, $K_{yz}$ the same as $K_{zy}$, and $K_{zx}$ the same as $K_{xz}$, the dyadic is said to be self-conjugate. Weatherburn (1924) has shown that when this is the case the three principal axes, $\lambda$,

$\mu$ and $\nu$, are mutually perpendicular, while Childs (1957) has shown that the converse also is true. It has already been remarked that it is customary to take self-conjugacy of $\mathcal{K}$ for granted, and a plausible explanation is presented in Note 16.

If the dyadic $\mathcal{K}$ is self conjugate, the directions, $x, y$ and $z$, may be chosen to coincide with the principal axes, $\lambda, \mu$ and $\nu$, when all the coefficients vanish except $K_{\lambda\lambda}$, $K_{\mu\mu}$ and $K_{\nu\nu}$, so that, from Equation (9.16),

$$\mathcal{K} = K_{\lambda\lambda}\mathbf{ii} + K_{\mu\mu}\mathbf{jj} + K_{\nu\nu}\mathbf{kk} \tag{9.19}$$

Equation (9.17), with $\mathcal{K}$ given by Equation (9.19), then becomes,

$$\begin{aligned} \mathbf{v} &= -(K_{\lambda\lambda}\mathbf{ii} + K_{\mu\mu}\mathbf{jj} + K_{\nu\nu}\mathbf{kk}) \cdot [(\text{grad } \phi)_\lambda \mathbf{i} + (\text{grad } \phi)_\mu \mathbf{j} + \\ &\quad + (\text{grad } \phi)_\nu \mathbf{k}] \\ &= -K_{\lambda\lambda}(\text{grad } \phi)_\lambda \mathbf{i} + K_{\mu\mu}(\text{grad } \phi)_\mu \mathbf{j} + K_{\nu\nu}(\text{grad } \phi)_\nu \mathbf{k} \end{aligned} \tag{9.20}$$

This is equivalent to the three component equations (9.12), except that what is there written as $K_\lambda$ is here written as $K_{\lambda\lambda}$, with similar relationships between the other elements of $K$.

## 9.4 Darcy's law for unsaturated materials

An experiment to verify Darcy's law involves the imposition of various potential gradients, other factors remaining constant, and a difficulty arises when the porous material is unsaturated. It was found in Chapter 8 that the state of unsaturation implies that suction prevails, and the moisture content depends upon the magnitude of the suction. If different suctions are now applied at the ends of a column of porous material in different experiments in order to impose different potential gradients, the moisture content will change and the material will therefore be effectively different in different experiments.

It has been very common to assume that Darcy's law holds for a given moisture content, the arguments adduced being of the following kind. The conducting channels in the unsaturated material are those pores which are full of water at the particular suction corresponding to the chosen moisture content. The air-filled pores are ineffective since water can hardly pass through a pore without occupying it. It follows that the air-filled pores could be filled with solids, such as wax, without affecting the rate of flow of fluid through the remaining pores, and the porous material treated in this way must now be regarded as a new *saturated* material with the same flow properties as the original unsaturated material. It can now be put through a series of tests for the verification of Darcy's law using various positive pressure differences in the usual way, the moisture content naturally remaining constant at the value "fixed" by the waxing process and corresponding to the initial unsaturated material. It is a fair assumption that the law would indeed be verified as in any other saturated material.

An experiment for the direct verification of Darcy's law in unsaturated materials has been carried out (Childs and Collis-George, 1950). They devised a method whereby the moisture content and suction down a long column of porous conductor were uniform, the potential gradient being due solely to the gravitational component. Various magnitudes of potential gradient were imposed by suspending the column at various angles of inclination to the vertical. From the results it could be safely inferred that the rate of flow for a given degree of saturation was proportional to the potential gradient, as in the case of saturated materials.

## 9.5 The viscous laminar flow of fluids

Before proceeding to show that Darcy's law is a consequence of more general physical laws of the flow of fluids, it is necessary to define the concepts used. It has already been remarked in passing in Section 1.4 that a fluid cannot withstand the slightest shear stress; it will yield continuously. If it be regarded as consisting of a number of elementary thin layers parallel to the direction of the shear stress, then the yield takes the form of a continuous sliding of the layers over one another. If the shear stress is not too great so that the fluid is not urged along too quickly, the elementary layers will remain well defined and separate from each other, as may be shown by injecting marking dye in the stream. Such a state of flow, with well-defined regular streamlines, is called laminar flow, as distinct from the state of turbulence which supervenes when the driving force becomes sufficiently great.

A layer of fluid sliding over another exerts a frictional drag upon it, and this force is reciprocated; the faster layer tends to drag the slower along with it, the slower tends to hold back the faster. This friction in fluids is called viscosity. The greater the relative velocity, the greater is the mutual viscous drag. The overall effect of an infinitely large number of infinitessimally thin layers, each of which moves at a different speed from its neighbours, is observed as a velocity gradient in the direction, say, $y$, at right angles to the line of movement. At any given point where the velocity gradient is $dv/dy$, the viscous shear stress, $F/A$, in the plane at right angles to the direction $y$ is

$$F/A = \eta \, dv/dy \qquad (9.21)$$

where $\eta$ is a constant called the coefficient of viscosity or, more briefly, the viscosity of the fluid.

**9.6 Darcy's law as a consequence of basic laws of fluid flow**

One first notes that Darcy's law has nothing to say about the true velocity of flow, nor about the potential at each point in the fluid which occupies the pore space. It ignores the internal structure of the conducting body and treats it as a uniform medium with the flow uniformly distributed over the cross-section, solid and pore space alike, and with the appropriate smoothed potential distribution. The rate of flow is usually inferred from a measurement at an inflow or outflow surface and is regarded as distributed over the whole of the area of that surface, while a measurement of the potential within the body is regarded as an indication of the potential of a hypothetical equipotential surface and not as the potential at some unspecifiable point in the fluid. In fact, the study of the true distribution of flow and potential within the pore space would be a very formidable task. Nevertheless, it remains true that the observed macroscopic distribution of flow and potential with which Darcy's law deals is a consequence of the true point-to-point distribution of flow and potential within the pore space, and that if the latter were known, then the former could in principle be deduced. In fact, one can deduce Darcy's law from fundamental physical principles but cannot, except for idealized systems, deduce the value of hydraulic conductivity.

The Stokes-Navier equations, referred to in Section 9.2, are a set of differential equations which describe the movement of viscous fluid in general unspecified space, which in particular may be the pore space of a permeable body. A Darcy type of experiment on such a body concerns itself with the measured potentials of masses of water in contact with the inflow and outflow faces, and these potentials are taken to be the potentials at these surfaces without reference to structure. Similarly the rate at which water leaves the body is regarded as the rate of flow over the outflow face. It is shown in Note 17 that it is a consequence of the Stokes-Navier equations that this rate of flow is proportional to the potential difference between the inflow and outflow faces provided that the flow is not too fast. This proportionality is one of the three elements of Darcy's law.

The remaining features of Darcy's law follow not from the Stokes-Navier equations, but from a description of the porous body. The essential supposition is that the body is sufficiently large, in comparison with the sizes of its pores, for it to be regarded as a uniform body, capable of division into small elements of volume each of which is a fair sample of the whole body and indistinguishable from any other element. Now consider a column of such a porous material set up for a Darcy type of experiment. The cross-section may be regarded as divided into a number of unit cross-sectional elements indistinguishable from one another. They are all subject to the same potential difference so that each will conduct fluid at a rate indistinguishable from any of the others. The total rate of flow will therefore be proportional to the number of unit elements which can be accommodated in the total cross-section, i.e. to the total area of cross-section of the column. This is the second part of a statement of Darcy's law.

Again, imagine the column divided into a number of indistinguishable unit elements of length, so that the flow through each is the same and equal to the rate of flow through the entire column. Since the elements are indistinguishable, they will each have the same potential difference between their inflow and outflow faces in order to have the same rate of flow, and this potential difference will therefore be simply the total potential difference between the ends of the column divided by the number of elements into which it is divided. The potential difference across an element will therefore be inversely proportional to the number of unit elements which can be accommodated in the total length of the column, i.e. it is inversely proportional to the length of the column. It has already been shown as a consequence of the Stokes-Navier equations that the rate of flow is proportional to the potential difference between the faces of an element, so that the rate of flow in an element, and therefore in the whole column, is inversely proportional to the length of the column; and this is the third and last item in the complete statement of Darcy's law.

## 9.7 The circumstances in which Darcy's law holds

The derivation of Darcy's law from basic principles indicates that certain conditions must be satisfied before the law may safely be applied. Since this law is sometimes applied uncritically, these circumstances must be stressed. The first circumstance is that the porous body must be large compared with its microstructure, so that it may indeed be regarded as a uniform material. Thus in a drainage problem one may be concerned with the flow of groundwater from the water table to drains which may be tens of metres apart and a metre or two deep. A cross-section of such dimensions is indeed large compared even with the coarsest of commonly occurring natural structures, and Darcy's law may be applied quite safely from this point of view. At the other extreme one may consider the flow of water to a plant root from a zone of soil specifically associated with it, which may occupy a volume of only a $cm^3$ or two, and except possibly in a fine sand of random structure an application of Darcy's law would be most hazardous.

The second circumstance is that the velocity of flow must be sufficiently small. There exists a dimensionless combination of flow characteristics, known as the Reynolds number, from which one may judge of the safety of applying Darcy's law. This number $R$ is given by

$$R = vr\rho/\eta \tag{9.22}$$

where $v$ is the mean flow velocity, $r$ is a length characteristic of the pore space and may be taken as the mean pore size, while $\rho$ and $\eta$ are respectively the density and viscosity of the fluid. If this number exceeds about 1000, turbulence sets in, as may be demonstrated in the flow in pipes, and since the Stokes-Navier equations on which Darcy's law depends are appropriate only to laminar flow, it is not surprising that Darcy's law breaks down. However, the condition that the acceleration terms in the Stokes-Navier equations be small constitutes an even more stringent restriction, and Fancher, Lewis and Barnes (1933) have demonstrated that Darcy's law may not safely be applied if the Reynolds number exceeds unity. Fortunately, it is not likely that this would be exceeded in any case of natural soil water movement, although this cannot be said of certain circumstances in civil engineering practice. For example, a very coarse sand with a pore size of the order of 0·1 mm would, when saturated, have a hydraulic conductivity of about 1·0 mm per second and permit water to infiltrate vertically under gravitational force at this rate. Substituting in Equation (9.22) these values for $r$ and $v$ together with values for $\rho$ and $\eta$ from tables, one finds that $R$ assumes the value 0·1, which is quite safely less than unity. This is about as unfavourable a case as is likely to occur in natural groundwater movements.

## 9.8 Controversial nomenclature

If one carries out an experiment such as Darcy's with an inert sand, using various liquids in turn with various values of viscosity $\eta$, one obtains various different values of the parameter $K$, which, by agreed convention, one calls the hydraulic conductivity. Thus the hydraulic conductivity is a property which depends on both the material and the fluid, and not on either alone. It would, however, emerge from such results that the rate of flow is inversely proportional to the viscosity $\eta$, other things being equal, so that one could enunciate a somewhat different law, including $\eta$ as a separate factor, in the form,

$$v = -(K'/\eta) \operatorname{grad} \phi \qquad (9.23)$$

This follows also from the Stokes-Navier equations for the steady state, as may be seen by reference to the three Equations of (N17.3), if in these equations the terms expressing differentiation with respect to time are reduced to zero. In Equation (9.23) the factor $(K'/\eta)$ is numerically equal to $K$. The variation of $v$ with the viscosity of the liquid is now taken care of by the inclusion of $\eta$ as a separate factor, so that one would get the same value of $K'$ no matter what liquid were used. $K'$ is thus a property of the particular sample of sand alone, and in particular of its internal pore geometry. It has therefore been proposed by engineers that $K'$ be called the inherent or intrinsic permeability of the sand, or more briefly the permeability. Since it had been common practice to call the parameter $K$ of Darcy's law by this name, some confusion and controversy arose, and it is for this reason that it has seemed desirable to adopt an unambiguous name, hydraulic conductivity, for $K$.

While falling in with current opinion in this respect, one need take no part in the controversy, for the concept of inherent permeability plays no part in soil studies. Soils are in general by no means inert in the physico-chemical sense. It is hazardous in the extreme to suppose that one can change to a different fluid without causing more or less profound changes of the internal pore geometry of the soil. For example, if air is used instead of water, the soil must first be dried, and the high suctions introduced in the process cause shrinkage of the colloidal component and therefore of the soil as a whole. The pore space is thus modified, and quite new shrinkage cracks may be produced. Or again, if a solution of sodium chloride is used instead of pure water, the clay present will become sodium saturated by base exchange. If, subsequently, pure water is again substituted for the Darcy experiment, thus removing free saline solution, the sodium clay will become dispersed, the soil structure will become unstable, and again the pore geometry will become profoundly modified, giving a quite different value of $K$, even though the value of $\eta$ has at no time been greatly altered. There is, in fact, no such concept in soils as an inherent pore geometry independent of the liquid flowing; no such law as that expressed in Equation (9.23) could be invariably demonstrated. At the very most one might introduce the concept of the hydraulic conductivity using a hypothetical liquid of unit viscosity having no other effects than those due to viscosity, and it is by no means self evident that there is anything to gain by so doing.

## NOTES

### Note 12. The summation of potential terms

When a body experiences a force $F$ because it finds itself in a certain environment or system, it is said to be in a field of force, and it may do one of two things. If there is no friction to be overcome, the body will accelerate without limit, but if there is friction of an amount which depends upon the velocity, then the body will settle down to a constant velocity such that the force $F$ just overcomes the friction. The magnitude of the force may be calculated either from a consideration of dynamics, by measuring the mass and acceleration of the body in the first case or the steady velocity in the second, or one may appeal to statics and apply an additional measurable mechanical force which is just sufficient to keep the body stationary. Since in such circumstances there is neither acceleration nor velocity, then whether there is friction or not, the total force on the body is zero, and one knows that the force applied to secure this condition must be equal in magnitude and opposite in direction to the field force which one seeks to determine. Let this externally applied force be $E$, then

$$E = -F \tag{N12.1}$$

The body having been brought to rest, let it be moved very slowly through a distance $s$ in the direction in which $E$ acts. To do this, of course, $E$ must be increased very slightly, but if the movement be sufficiently slow, and the acceleration implied sufficiently small, the necessary change of $E$ will be negligibly small. Then the product of the force $E$ and the distance covered $s$ is said to be the work $W$ done *by* the force $E$ *on* the body in the process, i.e.

$$W = Es \tag{N12.2}$$

Had the body moved in the direction opposite to that of the force $E$, then the work $W$ would have been done *by* the body, or alternatively a *negative* amount of work would have been done *on* the body.

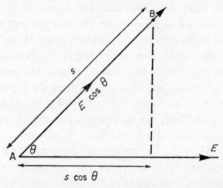

Figure N12.1 The work done by moving a body the distance $s$ when it experiences a force $E$ in a different direction. See text for explanation.

No work is done, by definition, by virtue of a movement perpendicular to the line of action of the force. If the movement makes an angle $\theta$ with the force, as in Figure (N12.1), then either it may be regarded as a movement through a distance $s \cos \theta$ in the direction of $E$ followed by a movement $s \sin \theta$ perpendicular to $E$, giving the work done as equal to $Es \cos \theta$, or the force $E$ may be regarded as a component $E \cos \theta$ in the direction of the movement $s$ plus a component $E \sin \theta$ perpendicular to $s$, again resulting in $Es \cos \theta$ for the work done. Thus, in this general case, the work done is

$$W = Es \cos \theta \tag{N12.3}$$

Neither of the different directions of $E$ and $s$ has greater significance than the other in specifying the work done, and direction has no part in the definition of work, which is a scalar quantity.

Since the force $F$ exerted on the body is by hypothesis due to the environment, it follows that any precisely similar body will also experience a force $F$, and that a conglomerate of $N$ such bodies will experience a force $NF$. If, therefore, one wishes to describe quantitatively the propensity of the environment to urge bodies in it to move, one must specify a particular body and measure the force it experiences. Such a particular body is commonly a unit body, either unit volume, unit mass, unit weight or, in other branches of physics, unit electric or magnetic charge, and the force $F$ which the unit body experiences is called the field strength at the point which the body occupies. The work done on this unit body when it is moved the distance $s$ from $A$ to $B$ in Figure (N12.1) is said to be the amount by which the potential at $B$ exceeds that at $A$, or in other words the potential difference between $B$ and $A$. If one writes $\phi_A$ and $\phi_B$ for the potentials at $A$ and $B$ respectively, then

$$\phi_B - \phi_A = W \qquad (N12.4)$$

or, from Equations (N12.3) and (N12.1),

$$\phi_B - \phi_A = Es \cos \theta$$
$$= -Fs \cos \theta \qquad (N12.5)$$

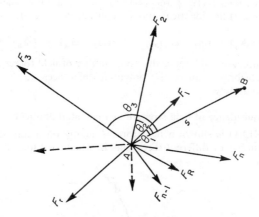

Figure N12.2 Illustrating the algebraic summation of elements of work due to separate components of force in various directions. See text for explanation.

In Figure (N12.2) is shown a unit body acted upon by a number of separate forces, $F_1, F_2, F_3, \ldots F_r, \ldots F_n$, making angles $\theta_1, \theta_2, \theta_3, \ldots \theta_r, \ldots \theta_n$ respectively with the direction in which the body is moved through the distance $s$. If any one of these forces, say the type force $F_r$, is acting by itself, one may, as before, neutralize it by imposing the equal and opposite force $E_r$ and then move the body through the distance $s$ from $A$ to $B$, and as in Equation (N12.5) above, the potential difference $(\phi_B - \phi_A)_r$ due to this force alone is given by

$$(\phi_B - \phi_A)_r = sE_r \cos \theta_r$$
$$= -sF_r \cos \theta_r$$

The algebraic sum of all the contributions to potential difference due to all the forces taken separately is therefore

$$(\phi_B - \phi_A)_1 + (\phi_B - \phi_A)_2 + \ldots + (\phi_B - \phi_A)_n = -s(F_1\cos\theta_1 + \\ + F_2\cos\theta_2 + \ldots + F_n\cos\theta_n) \qquad \text{(N12.6)}$$

When all the forces act together they have the resultant $F_R$ which is the single force equivalent to them all, and the potential difference between $B$ and $A$ due to this resultant is

$$(\phi_B - \phi_A)_R = -sF_R\cos\theta_R \qquad \text{(N12.7)}$$

where the resultant makes the angle $\theta_R$ with the line of movement. Now even vectors, such as forces, must add up according to the simple rules of algebra when they are in the same direction; hence the algebraic sum of the components of the several forces $F_n$ in any given direction, of which $s$ is a special case, must equal the component of the resultant $F_R$ in that same direction, since otherwise the resultant would clearly not be equivalent in all respects to the sum of the individual contributors. Hence,

$$F_1\cos\theta_1 + F_2\cos\theta_2 + \ldots + F_n\cos\theta_n = F_R\cos\theta_R \qquad \text{(N12.8)}$$

Thus from Equations (N12.7) and (N12.8), the right-hand sides of Equations (N12.6) and (N12.7) are identical and so, consequently, are the left-hand sides. Hence

$$(\phi_B - \phi_A)_1 + (\phi_B - \phi_A)_2 + \ldots + (\phi_B - \phi_A)_n = (\phi_B - \phi_A)_R \qquad \text{(N12.9)}$$

Thus the total potential difference due to a number of field forces acting together is simply the algebraic sum of the potential differences due to the individual forces acting separately.

### Note 13. The equivalence of field strength and potential gradient

In Figure (N13.1) is shown a field force $F$ acting on a unit quantity of soil water and a number of different directions in which one may displace the unit

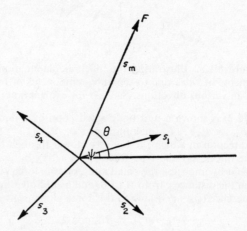

Figure N13.1 Illustrating the equivalence of field force and potential gradient. See text for explanation.

through the distance $s$ in order to calculate the potential difference between the ends of the displacement. Let the force make the angle $\theta$ with the axis $x$ and the displacement make the angle $\psi$. Then by Equation (N12.5)

$$\phi_B - \phi_A = -Fs \cos (\psi - \theta) \qquad (N13.1)$$

When $s$ is in the same direction as $F$ so that $\psi$ and $\theta$ are equal, the potential difference is

$$(\phi_B - \phi_A)_\theta = -Fs \qquad (N13.2)$$

As $\psi$ increases, the cosine factor steadily decreases until it vanishes when $\psi$ exceeds $\theta$ by $90°$, when the potential difference between $B$ and $A$ also vanishes. A continued increase of $\psi$ then causes the cosine term to take increasingly negative values until, when $\psi$ exceeds $\theta$ by $180°$ and the direction of $s$ is opposite to that of $F$,

$$(\phi_B - \phi_A)_{\theta + 180°} = Fs$$

or

$$(\phi_B - \phi_A)_{-\theta} = Fs \qquad (N13.3)$$

A combination of a statement of the direction in which the potential increases most sharply with a statement of the rate of that increase, $(\phi_B - \phi_A)/s$, is a statement of the gradient of the potential, evidently a vector quantity since its direction is an essential part of its specification. The direction of the gradient of potential is evidently the direction $\theta$, the same as that of the field force $F$, since that direction maximizes the difference of potential. The magnitude of the gradient is then $(\phi_B - \phi_A)_\theta/s$. Hence from Equation (N13.2)

$$\text{grad } \phi = (\phi_B - \phi_A)_\theta/s = -F \qquad (N13.4)$$

The negative sign indicates that the direction of the field force is in the direction opposite to that in which the potential increases. The same result naturally attends Equation (N13.3), for if $F$ is written $F_\theta$ to indicate that its direction is $\theta$, then the force in the opposite direction, $-\theta$, is evidently the negative force

$$F_{-\theta} = -F_\theta$$

so that Equation (N13.3) becomes

$$(\phi_B - \phi_A)_{-\theta} = -F_{-\theta} s$$

or

$$\text{grad } \phi = (\phi_B - \phi_A)_{-\theta}/s = -F_{-\theta}$$

When the field strength is uniform, that is to say it is everywhere the same in magnitude and direction, there is no difficulty in measuring it as a gradient of potential, for it does not then matter how short or long one chooses the path $s$. The potential difference is simply proportional to the length $s$. When, however, the field strength varies both in magnitude and direction from point to point, it may be determined at a given point as follows.

At as many points as possible in the body, forming a regular array such as a rectangular grid, the potential is measured as the sum of the pressure and height components in accordance with one or other of Equations (9.1) to (9.3), which are derived in Note 14. By interpolation one can find on any line of the grid network the position which corresponds to a chosen potential, and one can therefore plot a curve joining all points at a given potential. In a three-dimensional body the locus of such points is a surface, but for ease of illustration a two-dimensional potential distribution is shown in Figure (N13.2), in which the locus is a line. Such a locus is called an equipotential. One may draw a series of such equipotentials corresponding to different values of potential, as indicated in the diagram. The two-dimensional case is a set of non-intersecting curves (non-intersecting because an intersection would imply that the point of intersection possessed two different potentials at the same time), and the three-dimensional case, which is the usual one, is a set of non-intersecting surfaces like the layers of an onion. At every point in the body the direction of the field strength is found from the direction in which the potential is increasing most rapidly, i.e.

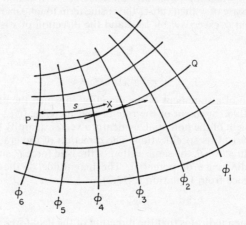

Figure N13.2 The potential distribution in a force field with varying potential gradient. The lines of force cross the equipotentials everywhere at right angles.

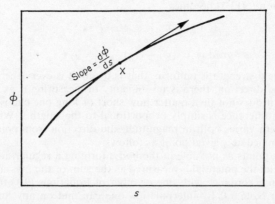

Figure N13.3 The magnitude of the potential gradient in a force field of varying potential gradient.

in the direction of the shortest distance between neighbouring equipotentials.

Figure (N13.2) shows only a few fairly widely spaced equipotentials out of the infinitely large number which could have been drawn with infinitessimally small spacing. The shortest distance between very near neighbours, which must necessarily be almost parallel, lies on the line which crosses them both perpendicularly. The potential gradient, and therefore the line of action of the field force, must lie on a curve which crosses each equipotential in turn at right angles. One may therefore draw a family of curves, each individual of which, such as $PQ$, intersects each equipotential at right angles. The direction of such a curve at any point $X$ on it is the direction of the potential gradient there. If one plots another curve, as in Figure (N13.3), of potential against distance measured along, say, $PQ$, then the slope, or tangent, of the curve at $X$, measured in the units indicated on the axes, is the ratio of the very small increase of potential, $\delta\phi$, to the very small distance traversed, $\delta s$, and is therefore the magnitude of the potential gradient at $X$. In the limit when $\delta\phi$ and $\delta s$ become infinitessimally small the ratio $\delta\phi/\delta s$ becomes the differential coefficient $d\phi/ds$, and is the slope of the curve at the point $X$.

## Note 14. Items of soil water potential

Imagine that water is flowing in a porous body in an orderly manner such that one can trace the mean path. If, for example, one were to introduce a stain at the point $P$ in Figure (N14.1), one would be able to trace a smooth path of flow to the point $Q$. This is because the field strength urges the water in its own direction, which is at right angles to each equipotential in turn. The path is called a streamline. Since the equipotentials cannot cross, neither can the paths of flow,

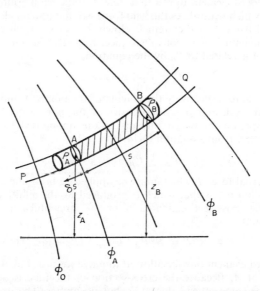

Figure N14.1 A hypothetical scheme to establish that potential is the algebraic sum of pressure and height at a point in the force field.

for at the intersection of two streamlines there would be two intersecting perpendicular equipotentials. Hence each streamline must hold to its ordained course perpendicular to each equipotential. (It has been shown in Section 9.3 that some materials may have preferred directions of water movement and the streamlines may not be in the direction of the field, i.e. they may not be perpendicular to the equipotentials; nevertheless, they are still uniquely determined.) A set of streamlines may be chosen to enclose a tube of flow, as shown, the water flowing inside the tube having no tendency to stray outside. It is proposed to determine, from the definition of potential, what is the potential difference between two cross-sections of the tube, $B$ and $A$, by moving a unit *volume* of water from $A$ to $B$.

Since the stream tube has no tendency to wander outside its boundaries, it is not upset, and therefore neither is the related potential distribution disturbed, if it be enclosed within an impermeable wall shaped to fit it. Neither is the potential distribution modified appreciably by two sufficiently small cavities at $A$ and $B$ respectively in the streamtube, separated by the distance $s$. These cavities simply take up the potentials of the immediately adjacent soil, with pressure $P_A$ at $A$ and $P_B$ at $B$. Enclosure of the cavities within the wall which contains the element $AB$ of the streamtube does not interfere appreciably with the potential distribution outside the walls, provided that the now isolated streamtube is of sufficiently small cross-sectional area, and does not upset the potential distribution within the streamtube if the cavities are maintained at the pre-existing pressures $P_A$ and $P_B$ by connection with external reservoirs at these pressures. Thus at each point the water experiences the same force field after the enclosure of the element as before.

To calculate the potential difference $\phi_B - \phi_A$ as defined in Section 9.1, one brings the enclosed element to rest in a state of static equilibrium by an imposed external force which neutralizes the field force, so that a small element of volume may be moved from $A$ to $B$ extremely slowly so as to introduce no question of changes of momentum or viscous resistance. In a state of static equilibrium the pressure $P$ at $A$ is related to $P_B$ by the equation,

$$P = P_B + g\rho(z_B - z_A) \tag{N14.1}$$

where $z_A$ and $z_B$ are respectively the heights of the points $A$ and $B$ relative to a selected datum level, and $\rho$ is the density of the pore water. The pressure $P$ exceeds the pressure $P_A$ which prevails in the unopposed force field by the amount $\Delta P$ where, from Equation (N14.1),

$$\Delta P = P - P_A = P_B + g\rho z_B - (P_A + g\rho z_A) \tag{N14.2}$$

Hence the external force $F$ which must be applied to bring the enclosed element into static equilibrium is that which is produced by the additional pressure $\Delta P$ acting over the cross-sectional area $S$ of the streamtube at $A$. Thus, from Equation (N14.2),

$$F = S\Delta P = S[P_B + g\rho z_B - (P_A + g\rho z_A)] \tag{N14.3}$$

Let the enclosed element now be allowed to move toward $B$ by the small amount $\Delta s$, measured at $A$. Because the cross-section of the tube is not uniform, the advance will be some other value at $B$, but no matter. The work done by the force $F$ in the process is $\Delta W$ where, from Equation (N14.3),

$$\Delta W = F\Delta s = S\Delta s[P_B + g\rho z_B - (P_A + g\rho z_A)] \tag{N14.4}$$

It is evident from Figure (N14.1) that the movement of the whole tube element by the distance $\Delta s$ is equivalent to removing the element of volume $S\Delta s$ from $A$ only for it to reappear at $B$. Thus the work calculated in accordance with Equation (N14.4) is that which is required to move the element $S\Delta s$ from $A$ to $B$, and therefore the work done per unit volume, which is by definition the potential difference, is

$$(\phi_B - \phi_A)_{vol} = \Delta W/S\Delta s = P_B + g\rho z_B - (P_A + g\rho z_A) \qquad \text{(N14.5)}$$

The subscript vol indicates that the potential is in this case defined on the basis of the unit quantity of water being unit volume.

It is often convenient to specify the potential at any point in a system in absolute terms by reference to an arbitrarily selected datum point at which the pressure $P$ and the height $z$ are zero on the scale so defined. Usually $P$ is in fact atmospheric or conventionally zero pressure. Taking the point $A$ to be such a datum, so that $P_A$ and $z_A$ are zero and the datum potential $\phi_A$ is likewise zero, then from Equation (N14.5)

$$(\phi_B)_{vol} = P_B + g\rho z_B$$

or, in general,

$$\phi_{vol} = P + g\rho z \qquad \left\{ \begin{array}{l} \text{(N14.6)} \\ \text{(9.1a)} \end{array} \right.$$

The pressure $P$ is indicated by a manometer in which the height of water in the open limb is $H$ relative to the point at which the pressure is measured, so that

$$P = g\rho H$$

hence

$$\phi_{vol} = g\rho(H + z) \qquad \left\{ \begin{array}{l} \text{(N14.7)} \\ \text{(9.1b)} \end{array} \right.$$

The mass of the element of volume $S\Delta s$ is $\rho S\Delta s$ so that the work done per unit mass is obtained as in Equation (N14.5), but dividing by a further factor $\rho$, to give

$$\phi_m = W/\rho S\Delta s = \phi_{vol}/\rho = g(H + z) \qquad \left\{ \begin{array}{l} \text{(14.8)} \\ \text{(9.2)} \end{array} \right.$$

Again, the weight of the element of volume is $g\rho S\Delta s$, so that $\phi_{wt}$, the work done per unit weight, is

$$\phi_{wt} = W/g\rho S\Delta s = \phi_{vol}/g\rho = H + z \qquad \left\{ \begin{array}{l} \text{(14.9)} \\ \text{(9.3)} \end{array} \right.$$

### Note 15. The tensor form of Darcy's law in anisotropic materials

Let the potential gradient grad $\phi$ have magnitude $E$ and direction cosines $l$, $m$ and $n$, with respect to the Cartesian coordinates, $x$, $y$ and $z$ respectively. That is to say, $l$ is the expression for $\cos \alpha$ where $\alpha$ is the angle between grad $\phi$ and $x$, and similarly for the other direction cosines. Then

$$(\text{grad } \phi)_x = lE$$
$$(\text{grad } \phi)_y = mE$$
$$(\text{grad } \phi)_z = nE$$

In vector notation, therefore, with $\mathbf{i}$, $\mathbf{j}$ and $\mathbf{k}$ unit vectors in the directions, $x$, $y$ and $z$, respectively,

$$\text{grad } \phi = E(l\mathbf{i} + m\mathbf{j} + n\mathbf{k}) \tag{N15.1}$$

The scalar product of Equation (9.17),

$$\mathbf{v} = -\mathcal{K} \cdot \text{grad } \phi$$

may be expanded, substituting the expression for $\mathcal{K}$ from Equation (9.16) and that for grad $\phi$ from Equation (N15.1). The result is

$$\mathbf{v} = -E \left\{ \begin{array}{l} K_{xx}l\mathbf{ii}.\mathbf{i} + K_{yx}l\mathbf{ji}.\mathbf{i} + K_{zx}l\mathbf{ki}.\mathbf{i} + \\ + K_{xy}m\mathbf{ij}.\mathbf{j} + K_{yy}m\mathbf{jj}.\mathbf{j} + K_{zy}m\mathbf{kj}.\mathbf{j} + \\ + K_{xz}n\mathbf{ik}.\mathbf{k} + K_{yz}n\mathbf{jk}.\mathbf{k} + K_{zz}n\mathbf{kk}.\mathbf{k} + \\ + \Sigma \end{array} \right\} \tag{N15.2}$$

The symbol $\Sigma$ represents the sum of terms, such as $K_{zx}m\mathbf{ii}.\mathbf{j}$, in which the scalar product is the product of unit vectors, in this case $\mathbf{i}$ and $\mathbf{j}$, which are perpendicular to each other. The scalar product of two vectors, say $\mathbf{a}$ and $\mathbf{b}$, which have directions which differ by the angle $\theta$ is defined to be

$$\mathbf{a}.\mathbf{b} = a\,b\cos\theta$$

where $a$ and $b$ are the magnitudes of the vectors. The vector $\mathbf{i}$ has unit magnitude, so that

$$\mathbf{i}.\mathbf{i} = 1$$

since the angle between the two components of the product is zero. Similarly $\mathbf{j}.\mathbf{j}$ and $\mathbf{k}.\mathbf{k}$ are each simply unity. On the other hand,

$$\mathbf{i}.\mathbf{j} = 0$$

since the unit vectors are at right angles to each other, and similarly for the product of any pair of the unit vectors except $\mathbf{i}.\mathbf{i}$, $\mathbf{j}.\mathbf{j}$ and $\mathbf{k}.\mathbf{k}$. Thus $\Sigma$ vanishes, and Equation (N15.2) becomes

$$\mathbf{v} = -E \left\{ \begin{array}{l} l(K_{xx}\mathbf{i} + K_{yx}\mathbf{j} + K_{zx}\mathbf{k}) + \\ + m(K_{xy}\mathbf{i} + K_{yy}\mathbf{j} + K_{zy}\mathbf{k}) + \\ + n(K_{xz}\mathbf{i} + K_{yz}\mathbf{j} + K_{zz}\mathbf{k}) \end{array} \right\} \tag{N15.3}$$

The equations of (9.13) to (9.15) may also be put into vector form thus:

$$\begin{aligned} v &= \mathbf{i}(v_{xx} + v_{xy} + v_{xz}) + \mathbf{j}(v_{yx} + v_{yy} + v_{yz}) + \\ &\quad + \mathbf{k}(v_{zx} + v_{zy} + v_{zz}) \\ &= -E \left\{ \begin{array}{l} \mathbf{i}(K_{xx}l + K_{xy}m + K_{xz}n) + \\ + \mathbf{j}(K_{yx}l + K_{yy}m + K_{yz}n) + \\ + \mathbf{k}(K_{zx}l + K_{zy}m + K_{zz}n) \end{array} \right\} \end{aligned} \tag{N15.4}$$

Equations (N15.3) and (N15.4) are identical in content but differently arranged, so that Equation (9.17) and the equations of (9.13) to (9.15), from which Equations (N15.3) and (N15.4) are respectively derived, are also identical in content. Thus Equation (9.17) expresses Darcy's law for anisotropic materials in brief form just as do the nine equations of (9.13) to (9.15).

## Note 16. The self-conjugacy of anisotropic conductivity

The anisotropy of soils is usually due to structural features such as fissures in more or less characteristic directions. Consider any one set of such fissures which contributes a conductivity component $_r\mathcal{K}$ whose nonion form is

$$_r\mathcal{K} = \left\{ \begin{array}{l} {}_rK_{xx}\mathbf{ii} + {}_rK_{yx}\mathbf{ji} + {}_rK_{zx}\mathbf{ki} \\ + {}_rK_{xy}\mathbf{ij} + {}_rK_{yy}\mathbf{jj} + {}_rK_{zy}\mathbf{kj} \\ + {}_rK_{xz}\mathbf{ik} + {}_rK_{yz}\mathbf{jk} + {}_rK_{zz}\mathbf{kk} \end{array} \right\} \tag{N16.1}$$

The total conductivity due to all such sets of fissures, of which one may suppose there to be $n$, is the sum $\sum_{r=1}^{r=n} {}_r\mathcal{K}$ of all such contributions, i.e.

$$\begin{aligned} \mathcal{K} &= \sum_{r=1}^{r=n} {}_r\mathcal{K} \\ &= \left\{ \begin{array}{l} K_{xx}\mathbf{ii} + K_{yx}\mathbf{ji} + K_{zx}\mathbf{ki} \\ + K_{xy}\mathbf{ij} + K_{yy}\mathbf{jj} + K_{zy}\mathbf{kj} \\ + K_{xz}\mathbf{ik} + K_{yz}\mathbf{jk} + K_{zz}\mathbf{kk} \end{array} \right\} \end{aligned} \tag{N16.2}$$

where

$$\left. \begin{aligned} K_{xx} &= \sum_{r=1}^{r=n} {}_rK_{xx} \\ K_{yx} &= \sum_{r=1}^{r=n} {}_rK_{yx} \end{aligned} \right\} \tag{N16.3}$$

etc.

It follows that if each contributory element of conductivity is self conjugate, then so is the total conductivity, for if

$$_rK_{yx} = {}_rK_{xy}$$

etc.

then

$$\sum_{r=1}^{r=n} {}_rK_{yx} = \sum_{r=1}^{r=n} {}_rK_{xy}$$

etc.

and from the equations in (N16.3) and the above

$$K_{yx} = K_{xy}$$

etc.

Now a plane parallel fissure clearly has three mutually perpendicular principal axes, one perpendicular to the plane of the fissure with zero conductivity, and two in any mutually perpendicular directions in the plane of the fissure with equal conductivities. Cylindrical channels also have three mutually perpendicular principal axes, one in the direction of the axis of the channel with finite conductivity, and two in any two mutually perpendicular directions at right angles to the channel axis with zero conductivity. All systems of such fissures and channels therefore have self-conjugate conductivity, and therefore the total conductivity due to all such contributions is also self conjugate. It follows that the complete system must have three mutually perpendicular axes. Nothing has been said here about determining the directions of these axes from those of the contributory elements, but it is easily shown that in soils, because the process of weathering results in fissures which are predominantly vertical and horizontal, the principal directions, in which the principal conductivities may be specified in the manner of the equations of (9.12), are also vertical and horizontal.

### Note 17. Application of the Stokes-Navier equations to the flow of liquids in porous bodies

To avoid confusion with the components of the effective flow velocity which appears in Darcy's law, namely $v_x$, $v_y$ and $v_z$, let the components of the true velocity at a point in the pore space be $\alpha$, $\beta$ and $\gamma$. Then for relatively incompressible liquids, such as water, the force per unit quantity of water is used in overcoming viscous friction, in acceleration at a given point if the state is not one of steady flow, and in acceleration as the unit quantity moves from a wide channel where the velocity is small to a more constricted one where the velocity is greater, or vice versa. This equilibrium is expressed by a set of three equations known as the Stokes-Navier equations, namely

$$\left.\begin{aligned}
g\rho\ \partial\phi/\partial x + \rho(\partial\alpha/\partial t + \alpha\ \partial\alpha/\partial x + \beta\ \partial\alpha/\partial y + \gamma\ \partial\alpha/\partial z) &= \eta\ \nabla^2\alpha \\
g\rho\ \partial\phi/\partial y + \rho(\partial\beta/\partial t + \alpha\ \partial\beta/\partial x + \beta\ \partial\beta/\partial y + \gamma\ \partial\beta/\partial z) &= \eta\ \nabla^2\beta \\
g\rho\ \partial\phi/\partial z + \rho(\partial\gamma/\partial t + \alpha\ \partial\gamma/\partial x + \beta\ \partial\gamma/\partial y + \gamma\ \beta\gamma/\partial z) &= \eta\ \nabla^2\gamma
\end{aligned}\right\} \quad \text{(N17.1)}$$

where $\nabla^2$ is the Laplacian operator defined by

$$\nabla^2 v = \partial^2 v/\partial x^2 + \partial^2 v/\partial y^2 + \partial^2 v/\partial z^2 \tag{N17.2}$$

A reduction of the point velocity by a factor $1/N$ reduces each of the components, $\alpha$, $\beta$ and $\gamma$, by that same factor and every differential coefficient of each component by that same factor. Hence every term, such as $\alpha\partial\alpha/\partial x$, which is a product of two such reduced components, is itself reduced by a factor $1/N^2$. Therefore, a sufficient reduction of the point velocity must reduce all such product terms to negligible proportions compared with the remainder, which are reduced in the proportion of only $1/N$. Therefore, for sufficiently slow motion, the equations take the form,

$$\left.\begin{aligned}
g\rho\ \partial\phi/\partial x + \rho\ \partial\alpha/\partial t &= \eta\ \nabla^2\alpha \\
g\rho\ \partial\phi/\partial y + \rho\ \partial\beta/\partial t &= \eta\ \nabla^2\beta \\
g\rho\ \partial\phi/\partial z + \rho\ \partial\gamma/\partial t &= \eta\ \nabla^2\gamma
\end{aligned}\right\} \quad \text{(N17.3)}$$

It now follows that if the Stokes-Navier equations are satisfied by one particular distribution of potential $\phi$ and of point velocity, it must also be satisfied when both the potential and the point velocity are altered by the same factor, say $M$, provided that the velocity still remains sufficiently small, since then each term of the equations is multiplied by the same constant $M$. Now the measured outflow or inflow rate at an input or output surface is increased in the same proportion, $M$, as is the point velocity at any given point. The measured potential difference between these inflow and outflow surfaces is increased in the same proportion, $M$, as is the potential at any given point in the pore space, since these surfaces are merely particular points in the general potential distribution. Hence the ratio of measured inflow to measured potential difference is the same at the faster rate as at the slower rate of flow, which is to say that the rate of flow is proportional to the potential difference. This is a statement of Darcy's law as it relates to rate of flow and potential difference.

CHAPTER 10

# The hydraulic conductivity
# of soil to water

## 10.1 The dependence of conductivity on pore size

WHILE it is not yet possible to calculate the absolute value of the hydraulic conductivity of a porous body except in the case of certain idealized models, it is possible to compare the conductivities of two bodies which have geometrically similar pore spaces and differ only in size of corresponding pores. If one designates the bodies $B_1$ and $B_2$, where $B_2$ is simply a magnified version of $B_1$ such that any linear dimension in the former is $N$ times the corresponding linear dimension in the latter, then the respective conductivities, $K_2$ and $K_1$, are related by the expression

$$K_2/K_1 = N^2 \qquad (10.1)$$

The derivation of this ratio from the Stokes-Navier equations is given in Note 18.

## 10.2 The dependence of conductivity on porosity

In contrast with the analysis given in Note 18, it is not possible to derive a relationship between conductivity and porosity on very general grounds, but one must be satisfied with discussing particular idealized models of conducting body. Most models of this kind are based on the flow of fluids in cylindrical tubes or between parallel plane surfaces. If $Q/t$ is the volume of liquid flowing per second through a tube of radius $r$ along which is imposed a potential gradient, grad $\phi$, then as shown in Note 19, the equation of flow is

$$Q/t = (g\rho\pi/8)(r^4/\eta)\text{grad } \phi \qquad (10.2)$$

The corresponding equation for flow per unit width of a plane slit between solid surfaces which are separated by a distance $D$ is, as shown in Note 20,

$$Q/t = (g\rho D^3/12\eta)\text{grad } \phi \qquad (10.3)$$

A body which owed its hydraulic conductivity to its being permeated by

G                                179

parallel tubes of radius $r$, with a concentration of $n$ such tubes per unit area of total cross-section of the body, would be very strongly anisotropic, since only one of its principal axes would present a non-zero conductivity. In the direction of the axes of the tubes one would have, from Equation (10.2),

$$v = n(Q/t) = (g\rho n\pi/8)(r^4/\eta)\text{grad } \phi$$

Because for this model the porosity $f$ is equal to the area of conducting channel per unit area of cross-section, i.e. $n\pi r^2$, the above equation may be rearranged and written:

$$v = (g\rho fr^2/8\eta)\text{grad } \phi \tag{10.4}$$

Comparing this with Equation (9.6) one finds that the conductivity of the body is given by

$$K_T = g\rho fr^2/8\eta \tag{10.5}$$

Treating the case of a body which has $n$ parallel slits per unit thickness, each of elementary thickness $D$, in which case the porosity $f$ is equal to $nD$, one arrives at the result for the conductivity $K_S$

$$K_S = g\rho fD^2/12\eta \tag{10.6}$$

and this would be the conductivity in any direction lying in the plane parallel to the slits.

In Equations (10.5) and (10.6), the proportionality of $K_T$ and $r^2$ in the one case and $K_S$ and $D^2$ in the other is just a special case of the more general result described in Section 10.1. In both cases the conductivity is directly proportional to the porosity, but without further examination, this result must be regarded as specific to the particular models.

Let it now be supposed that, with either of the above models conducting liquids in the direction of a principal axis, one adds an extra set of plane slits, each with its plane perpendicular to the direction of the potential gradient. Each of these slits will lie in an equipotential surface and there will therefore be no tendency for them to facilitate the movement of liquid, except that insofar as, at junctions between intersecting slits, that slit which is the effective conducting element will have a locally increased width and will therefore make a slightly enhanced contribution to the conductivity. On the whole, however, that part of the porosity which is contributed by the equipotential slits is ineffective and may for the purpose of assessing conductivity be regarded as dead space. In any static measurement of total porosity, however, this dead space is indistinguishable from the effectively conducting channels, and the hydraulic conductivity will only be proportional to the total porosity if, fortuitously, the dead space increases in the same proportion as the effective space.

It is possible to give Equation (10.5) a far more general appearance and thereby to hide the fact that it is, by its derivation, applicable only to the capillary tube model. It may be rewritten:

$$K_T = (g\rho f/2\eta)[(n\pi r^2)/(n2\pi r)]^2$$
$$= (g\rho f^3/2\eta)(1/n2\pi r)^2 \tag{10.7}$$

If one assigns the symbol $A$ to the specific surface area developed by the conductor, namely the total surface area of the solid part divided by the volume of *that solid part*, it follows that for this particular model,

$$A = (n2\pi r)/(1-f) \tag{10.8}$$

Combining this with Equation (10.7) gives the relationship,

$$K_T = (g\rho/2\eta)(1/A^2)[f^3/(1-f)^2] \tag{10.9}$$

This is formally Kozeny's equation (1927), as derived also by Fair and Hatch (1933).

Fair and Hatch introduced a further element of generality in the model itself by substituting for $r$ in the square bracket of Equation (10.7) the engineers' concept of the hydraulic radius of a non-cylindrical channel, namely the ratio of the cross-section to the wetted perimeter; but this introduction of an approximation at once renders the final equation only approximately applicable. Kozeny himself (1927) gave an analysis of non-cylindrical conductors based on the Stokes-Navier equations, but his ultimate simplification of the problem is as ruthless as that of Fair and Hatch.

However complicated the pore space of a natural porous material may be, it may be expressed in terms of a specific surface area which is a function only of the fineness of division of the solid material, and a porosity which is independent of that fineness of division. The Kozeny equation, Equation (10.9), can therefore be applied boldly to any kind of porous body. When it is so applied to a geometry for which it was not derived, it is usual to recognize the lapse from strict rectitude by substituting $k\eta$ for the factor $2\eta$ in the denominator of Equation (10.9), where $k$ is an arbitrary or empirically determined pore shape factor which is commonly in the range from 2·0 to 2·5. The generalized Kozeny equation is then

$$K = (g\rho/k\eta)(1/A^2)[f^3/(1-f)^2] \tag{10.10}$$

Even so, since it was derived for a particular model of uniform capillary tubes, one ought not to be surprised to find that it often provides a value of $K$ which is at variance with the observed measurement. It can indeed be shown on theoretical grounds (Childs and Collis-George, 1950) that Equation (10.9) is not even applicable to a bundle of capillary tubes if the radii are distributed over a wide range of sizes. Surprise should be reserved for the fact that the Kozeny formula does indeed give approximately correct values of conductivity for a variety of industrial powders, but here it must be said that in practice it is not possible to vary the porosity over a sufficient range to provide a searching test of the formula in respect of the porosity factor.

## 10.3 The dependence of hydraulic conductivity on structure

The most striking failure of formulas of the Kozeny type occurs in soil with a well-developed structure. The reason becomes clear when one takes as an example a highly idealized case of a material consisting of spherical particles of a size corresponding to the conventional clay fraction, namely with a radius of $10^{-4}$ cm, packed to give a porosity which might commonly be about $0.5$. The specific surface area of such a material is simply the ratio of the surface area to the volume of a sphere of radius $r$, namely $3/r$ or, in this case, $3 \times 10^4$ cm$^2$/cm$^3$. Using these values of $f$ and $A$ in Equation (10.9), one finds that the conductivity $K$ of the material has the value of approximately $(3/\eta) \times 10^{-7}$ cm/sec or about $3 \times 10^{-5}$ cm/sec if the liquid is water with a viscosity of $10^{-2}$ poise. The fact that clay soils with poor structure commonly have low conductivities of this order cannot be regarded as much more than a coincidence bearing in mind the idealization of the model. If one now supposes that the material is fissured by parallel plane cracks of width $0.1$ cm at intervals of 10 cm, the structural porosity is $0.01$, which represents an increase of porosity of no more than $2\%$. The specific surface area is quite unaffected. Hence the hydraulic conductivity as calculated by the use of Equation (10.9) is almost unaffected by the structural fissuring. Equation (10.6), however, is applicable to this case and if one considers the fissuring alone and entirely neglects any contribution that may be made to the conductivity by the textural porosity, one finds that $K$ has a value of about $10^{-2}/\eta$ cm/sec, or about $1.0$ cm/sec when the liquid is water. Thus the structure confers a hydraulic conductivity which is about 30,000 times as large as that which the Kozeny formula would lead one to expect. Experience in the field does in fact provide examples of conductivities of the order of hundreds of times that which would be expected from considerations of texture alone (Childs and collaborators, 1957). It can only be concluded that the rather too common tendency to correlate texture and conductivity is hazardous.

## 10.4 Structure as a cause of anisotropy

An examination of the Kozeny equation shows that it contains no factors which represent directed physical quantities, since neither specific surface area nor porosity are concepts to which direction has relevance. It is of course true that in its derivation, this equation has application only to parallel capillary tubes and in this sense has relevance only to conductivity in the direction of these tubes, but this is invariably ignored in its wide application to random arrays of particles, since the whole point of giving it its very general form is to permit this wide application. It follows that the Kozeny equation in its simplest form can throw no light on anisotropic conductivity. Also it is self-evident that a perfectly random array of particles must be isotropic, since any hint of differentation of properties according to direction is in conflict with the very concept of randomicity. The cause of anisotropic properties must therefore be sought in a departure from randomicity, that is to say, in structure. Here again there can be no generalization in the analysis, since structure is by its nature specific, and each case must be specified geometrically. One can do very little more than refer again to the cases of parallel capillary tubes as examples of conductors in one direction only and to parallel slits which may be regarded as conducting equally in any two mutually perpendicular directions lying in the plane of the slits. In each case, the magnitudes of the highly anisotropic conductivities in the directions of the principal axes may be calculated by Notes 19 and 20 respectively. It has already been demonstrated in Section 9.3 and Note 15 that a body permeated by any specified combination of such slits and tubes will have an anisotropic conductivity which may be specified in terms of the conductivities of the individual sets of slits and tubes. In practice it is, of course, a matter of very great difficulty to specify the geometry of the structural features in the required quantitative terms.

Allied to anisotropy due to structure is anisotropy due to stratification. A plane crack is a space in which the flow of liquid is calculable in terms of the Stokes-Navier equations, bounded by material in which the flow of liquid is relatively negligible. One may instead have alternations of two or more kinds of material in each of which the flow of liquid obeys Darcy's law, but with different values of hydraulic conductivity. Such composite material may have to be treated as a non-uniform medium, each stratum being assigned its proper conductivity. Such would be the case if the thickness of the strata were comparable with the total size of the body, as for example if the strata were some tens of centimetres thick in a drainage problem where the zone of groundwater were some metres in thickness. On the other hand, if the strata were so thin as to be regarded as small scale laminations in comparison with the scale of the whole body, it might be more appropriate to regard the body as one of uniform anisotropic conductivity.

As an example, take the case of strata of thickness $D_1$ and isotropic conductivity $K_1$, interleaved with strata of thickness $D_2$ and isotropic conductivity $K_2$. Then as shown in Note 21, the macroscopic flow in a body consisting of very many such laminations is the same as that in a uniform anisotropic body whose permeability in a direction perpendicular to the plane of lamination is $K_V$ and in any direction parallel to the plane of lamination is $K_H$, where

$$K_V = (D_1 + D_2)/(D_1/K_1 + D_2/K_2) \qquad (10.11)$$
$$K_H = (K_1 D_1 + K_2 D_2)/(D_1 + D_2) \qquad (10.12)$$

In the more general case where there are $n$ layers, the typical layer being the $r$th with conductivity $K_r$ and thickness $D_r$, the horizontal and vertical conductivities respectively are

$$K_H = \sum_1^n K_r D_r \Big/ \sum_1^n D_r \qquad (10.13)$$

$$K_V = \sum_1^n D_r \Big/ \sum_1^n D_r/K_r \qquad (10.14)$$

Since lamination is hardly likely on more than one set of parallel planes, there is no need to complicate matters by adding the conductivity contributions of different sets in accordance with Section 9.3 and more particularly Note 15, although these are applicable.

## 10.5 The dependence of hydraulic conductivity on moisture content, and some more complicated models of porous materials

The validity of Darcy's law when the soil is unsaturated has been established in Section 10.4. A typical soil in its swollen wet state contains pores with an upper size limit of the order of a millimetre, and very little water is lost until the suction exceeds something of the order of ten cm of water. The emptying of a cell at such a suction leaves the solid walls coated with a very thin film of water in which liquid flow can take place but slowly, as compared with the cell when full. It follows that an empty cell contributes only negligibly to the total hydraulic conductivity of the body. A reduction of the moisture content is thus equivalent to a reduction of effective porosity for the purpose of assessing conductivity, and results in a reduction of that conductivity. If one is using the Kozeny equation then the factor $f$ becomes the volumetric moisture content instead of the total porosity.

Secondly, since the moisture content is progressively reduced by a progressive increase of suction, it is the larger pores which are emptied in the earlier stages of unsaturation and the smaller pores which are left full of liquid. Since it is the larger pores which are the more effective conductors, as demonstrated by any of Equations (10.1), (10.5) and (10.6), it follows that the earlier stages of moisture reduction are more effective than the later in reducing the conductivity. This effect may be taken into account in the Kozeny equation in the factor $A^2$, since one may regard a cell which is full of air as being one which is no more effective a conductor than if it were part of the solid phase. One may thus estimate approximately the surface area of the solid water interface from a consideration of the pore size distribution in accordance with Section 8.4, limiting attention to those pores which remain full of water, and assigning the air-filled pores to the solid phase, thus arriving at an effective value of the specific surface area $A$ for the present purpose. This is, of course, a procedure which is very different from that usually adopted when applying Kozeny's equation to saturated materials, for then it is customary to calculate the specific surface area from a consideration of the mechanical composition. In fact a prime reason for the development of formulas of the Kozeny type was that mechanical analysis was a well-recognized technique, while the measurement and analysis of moisture characteristics was not, and this led to formulas in which the emphasis was laid on the geometry of the solid phase although the conductivity is dominated by the geometry of the pore space.

Thirdly, a pore which is full of air is not merely ineffective as a conductor. It becomes an obstacle, so that liquid which originally passed through it when it was water-filled is deflected round it. In effect the true flow paths become more tortuous and therefore longer. The drier the material becomes, the more tortuous are the flow paths.

Lastly, if the material contains a colloidal fraction, then it shrinks as the suction increases, as was described in Section 8.7, so that the pore spaces all become smaller. Again this causes a reduction of conductivity as the moisture content decreases. This last effect is in most cases small compared with the other three, and need only be considered when the material is dominantly colloidal and retains a large proportion of its water at high suctions where the shrinkage is substantial.

The variation of conductivity with moisture content offers both a problem and an opportunity. Since a reduction of moisture content provides a reduction of effective porosity, by this means one may test theoretical formulas over a much greater range of porosity than is otherwise possible. On the other hand, if the severe test shows up deficiencies in the available theories of the relationship between hydraulic conductivity and pore geometry, then it becomes necessary to explore better concepts and models. Such improvements of theory have for the most part been either modifications of the Kozeny equation or analyses of more sophisticated geometrical models.

The most evident lack in the Kozeny equation is a factor to take into account the tortuosity and true length of the actual flow paths, and a means of evaluating this factor if it be included. This lack received attention from Wyllie and Rose (1950), working in the field of petroleum technology. If the capillary tubes upon which the Kozeny model is based are supposed to be so tortuous as to occupy a length $L_e$ in the apparent length $L$ of the porous body, then it may be shown (see Note 22) that the modified Kozeny equation takes the form

$$K = (g\rho/k\eta)(1/A^2)(L/L_e)^2[f^3/(1-f)^2] \qquad (10.15)$$

a result which seems first to have been presented by Carman (1937). It has been proposed that the factor $(L/L_e)$ may be evaluated by measuring the electrical resistivity $\rho_m$ of the material when the pore space contains liquid of known resistivity, $\rho$. If the material is saturated, so that the whole of the pore space is effectively conducting either liquid or electricity, as the case may be, then, as is also shown in Note 22,

$$(L/L_e)^2 = (1/f)(\rho/\rho_m) \qquad (10.16)$$

(A rather different view of effective porosity led Wyllie and Rose to a different form.) If the material is unsaturated, one must replace $f$ by the effective porosity, namely the volumetric moisture content, both in Equation (10.15) and in Equation (10.16), and also calculate $A^2$ on the basis that the air-filled pore space is counted for this purpose as contributing to the solid part of the body, as has already been described. By this procedure, the factor $(L/L_e)$ will emerge as a variable dependent upon the moisture content, for use in Equation (10.15) at the appropriate moisture content. Wyllie and Spangler (1952) have found this modification to be a great improvement on the original Kozeny equation when subjected to the severe test of application to unsaturated porous bodies.

Two criticisms may be levelled at the analysis of Wyllie and Rose. Firstly since appeal is made to a separate experiment for further information about the pore geometry, the modified equation is semi-empirical and not wholly geometrically based. Secondly, it is tacitly assumed that in a material of mixed pore sizes with flow paths of mixed tortuosities, the contributions to tortuosity of the various paths are weighted similarly in both the liquid flow and electrical flow cases, and this is not self-evidently true. Since liquid flow favours paths of larger cross-section, while the flow of electricity, having no dependence on viscosity, treats all paths alike and is simply proportional to the total area of cross-section of conducting channels, the contribution of narrow channels is weighted more favourably for the flow of electricity than for the flow of liquids. However, the error due to using an electrical flow measurement of tortuosity for application to the flow of liquids will be of second order.

Another kind of approach is exemplified by the model adopted by Childs and Collis-George (1950), which is appropriate to a porous body in which the distribution of pores of various sizes in space is entirely random. A somewhat modified but essentially similar treatment was presented later by Marshall (1958). If one takes a column of the porous body of unit cross-section and breaks it in two, each exposed surface at the fracture shows a representative pore size distribution. First fasten attention on a group of pores on one exposed surface whose average size is $\rho$ and whose size range is $\delta r$; that is to say, the size range is from $\rho - \delta r/2$ to $\rho + \delta r/2$. Next consider a group in the other exposed surface whose mean size is $\sigma$ and whose size range is $\delta r$. The area devoted to a particular pore group in the exposure is simply equal to that part of the porosity accounted for by the group, and that in turn is equal to the product of the concentration of pore volume (i.e. pore volume per unit pore size range) about the chosen size, and the width of the range. The concentration of pore volume at the given pore size is a function of the pore size, $r$, known as the distribution function, $F(r)$. Hence the area of exposure of pores of mean size $\rho$ is

$$a_\rho = F(\rho)\delta r$$

while the area of exposure on the other surface of pores of mean size $\sigma$ is

$$a_\sigma = F(\sigma)\delta r$$

Since in the undisturbed column the two exposures come together at random, the area of the junction occupied by pore sequences characterized by a mean pore size $\rho$ on the first side and pores of mean size $\sigma$ on the second side is simply the product of $a_\rho$ and $a_\sigma$ which one may denote by $a_{\rho \to \sigma}$. Thus,

$$a_{\rho \to \sigma} = F(\rho)\delta r F(\sigma)\delta r$$

One now assesses the contribution to the total hydraulic conductivity made by the pore sequence described and computes the total conductivity by summing the contributions made by all the possible sequences covering the whole range of pore sizes in the material. To do this, two assumptions are made. The first is that, in accordance with the Poiseuille equation, Equation (10.2), the resistance to flow increases so rapidly as the pore size decreases that one may neglect the resistance of the coarser pore in the sequence; and the second is that one may ignore all contributions to the conductivity except those due to direct sequences. The two assumptions introduce errors of opposite sign which tend to mutual compensation.

If one takes $\sigma$ to be smaller than $\rho$ in the sequence, the number of pore sequences occupying the area $a_{\rho \to \sigma}$ is proportional to $a_{\rho \to \sigma}/\sigma^2$ and, by Poiseuille's equation, the rate of flow through each, per unit potential gradient, is proportional to $\sigma^4$, so that the contribution $\delta K$ to the total hydraulic conductivity is

G*

$$\delta K = M\sigma^2 F(\rho)\delta r F(\sigma)\delta r$$

and the total conductivity is

$$K = M \sum_{\rho=0}^{\rho=R} \sum_{\sigma=0}^{\sigma=R} \sigma^2 F(\rho)\delta r F(\sigma)\delta r \qquad (10.17)$$

where $M$ is a constant of proportionality to be determined once and for all experimentally and $R$ is the upper pore size limit in the range. The distribution function may be determined from the moisture characteristic in the manner described in Section 7.4, while for unsaturated materials $R$ is taken to be the largest pore size which remains full of liquid at the suction appropriate to the prevailing moisture content. A convenient form of computation table to perform the summation indicated in Equation (10.17) is drawn up in the original paper referred to.

Some experiments with inert materials, such as sand fractions and slate dust, provided curves of variation of conductivity with moisture content, such as those given in Figure (10.1). These curves illustrate the rapid

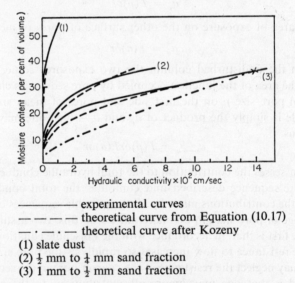

——————— experimental curves
— — — — theoretical curve from Equation (10.17)
—·—·— theoretical curve after Kozeny
(1) slate dust
(2) $\frac{1}{2}$ mm to $\frac{1}{4}$ mm sand fraction
(3) 1 mm to $\frac{1}{2}$ mm sand fraction

Figure 10.1 The dependence of hydraulic conductivity
on the degree of saturation.

diminution of conductivity in the intial stages of unsaturation. For purposes of comparison, the theoretical curves are shown for the Kozeny equation uncorrected for tortuosity and for the random pore sequence model, the matching factor or empirical constant $M$ being derived by matching the experimental and theoretical curves at the one point shown.

## 10.6 Some miscellaneous relationships for hydraulic conductivity

In addition to Kozeny's expression for the hydraulic conductivity in terms of porosity and fineness of division of the solids, and the equivalent formula of Fair and Hatch, others of a similar kind have been proposed from time to time with a more or less empirical basis. Among these is Terzaghi's expression (1925),

$$K = B_T d^2 (f - 0 \cdot 13)^2 / (1 - f)$$

where $B_T$ is a constant. Zunker's proposed form (1933) is

$$K = B_Z (1/A^2) f_0^2 / (1 - f)^2$$

where $B_Z$ is a constant and $A$ and $f$ have the same significance as in Kozeny's formula. $f_0$ is the effectively conducting porosity, namely the porosity after a correction has been applied for tightly bound water and for stagnant water in dead space.

Among early proposals to relate the hydraulic conductivity to the pore size distribution may be noted that of Baver (1938). If the moisture characteristic exhibits a point of inflexion, this may be taken as a mark of division of the pore space into two parts. On the lower suction side the water is held in larger pores, which Baver arbitrarily designates the non-capillary water, and which is the more mobile. The magnitude of the suction at the point of inflexion is a measure of the fineness of division of the pore space, so that the conductivity should be greater for materials characterized by lower suctions at the point of inflexion. Baver takes both factors into account in defining a porosity factor to be the quantity resulting from dividing the non-capillary pore space by the logarithm of the suction at the point of inflexion of the moisture characteristic, and presents a curve which relates this porosity factor empirically to the hydraulic conductivity of the materials that he tested. Later, Nelson and Baver (1940) proposed a much simpler porosity factor for the purpose of such a correlation, namely the pore space which is emptied when the suction is increased from zero to 40 cm of water.

A further elaboration of this kind was presented by Smith, Browning and Pohlman (1944). Instead of dividing the pore space into two parts, one of which is regarded as of negligible importance, they divided it into three parts and attached different weights to each in assessing its contribution to the net porosity factor. Thus if one writes $f_1$ for that part of the pore space which loses water when the suction is increased from zero to 10 cm of water, $f_2$ for that part which empties when the suction is further increased to 40 cm of water, and $f_3$ for that part which empties when the suction is increased still further to 100 cm of water, the net porosity factor in Baver's sense is the sum of $f_1$, a quarter of $f_2$ and one-tenth of $f_3$, all smaller pores being entirely ignored. Bendixen and Slater (1947) returned to the concept of a division at a single suction into a dominant and negligible part of the pore space, but introduced a time factor. Thus their porosity factor is the pore space revealed by draining for a period of one hour only at a suction of 60 cm of water.

### 10.7 Hysteresis of hydraulic conductivity

The hydraulic conductivity of a material depends upon the degree of saturation, and since the moisture content is related to the prevailing suction, the hydraulic conductivity will also incidentally depend upon the suction. It has been noted in Section 8.5 that the relationship between suction and moisture content is subject to hysteresis, so that at a given suction the moisture content can have many different values according to the history of wetting and drying of the sample. It follows that the hydraulic conductivity also can have many different values at the same suction and is characterized by hysteresis.

There is a further minor hysteresis effect in the relationship between the hydraulic conductivity and the moisture content itself. In Section 8.6 it was established that a specification of moisture content is not by itself a unique specification of the identity of the water-filled pores, and that this identity is capable of great variation according to the history of the suction changes leading to the final state. In particular, the pores which are filled in the direct wetting from great suctions to the prevailing suction and moisture content may differ greatly from those remaining filled at the same moisture content reached in drying from zero suction, although a proportion of the filled pores are common to both states. It is therefore to be expected that the curve relating hydraulic conductivity to moisture content should itself exhibit a small hysteresis loop, and this has been confirmed in some unpublished work by Collis-George, privately communicated, and by additional later work by Poulovassilis (1969).

### 10.8 Complementary conduction of water as vapour

In an unsaturated porous material the continuous air space provides a path for the movement of water vapour just as the continuous water-filled space provides for liquid movement. The suction which prevails in the liquid is related to the water vapour pressure in equilibrium with it in accordance with Equation (7.1) of Section 7.5, so that the suction gradient which determines the rate of movement of the liquid also determines, via the vapour pressure gradient, the rate of movement of the vapour. In fact, as was shown by Philip (1955a) and as is shown here in Note 23, the velocity of flow of water in vapour phase, $\mathbf{v}_{vap}$, may be expressed in terms of the gradient of the pressure head, grad $H$, in the liquid water, by a form of Darcy's law as follows:

$$\mathbf{v}_{vap} = -K_{vap} \text{ grad } H \tag{10.18}$$

where

$$K_{vap} = \frac{\alpha(f-c)(\sigma_0 gMD_i)}{RT} \, e^{\, MgH/RT} \tag{10.19}$$

In these expressions $f$ and $c$ are respectively the total and the water-filled porosities, $\sigma_0$ is the mass of water vapour per unit volume of air in contact with water at zero suction, $M$ is the molecular weight of water of density $\rho$, $R$ is the gas constant, and $T$ is the prevailing temperature on the Kelvin scale. Since the vapour movement is by molecular diffusion through the air, it takes place in accordance with Fick's law, which states that the rate of movement is proportional to the gradient of the concentration of water vapour in the air. $D_i$ is the constant of proportionality, which is known as the coefficient of interdiffusion of water vapour through air, and is obtainable from tables. $\alpha$ may for all practical purposes be regarded as a constant of value 0·6. Since only unsaturated materials are in question, $H$ is always negative, being in fact a suction. In Equation (10.18) the gradient of pressure plays the part of a gradient of total potential, since for a vapour of very low density the gravitational component of potential is negligible.

The movements of water vapour and of liquid water are complementary to one another, taking place along parallel paths in the air-filled and water-filled parts of the pore space respectively. Hence the total rate of flow of water, $\mathbf{v}$, is the sum of $\mathbf{v}_{vap}$ and $\mathbf{v}_w$, where the latter is the velocity of flow of liquid water as given by Equation (9.8). Hence,

$$\mathbf{v} = -[K \text{ grad } \phi + K_{vap} \text{ grad } H] \tag{10.20}$$

Defining $\phi$ in accordance with Equation (9.3), one has

$$\text{grad } \phi = \text{grad } H + \mathbf{k}$$

where $\mathbf{k}$ is a unit vector in the direction $z$, so that Equation (10.20) becomes

$$\mathbf{v} = -[\overline{K} \text{ grad } H + \mathbf{k}K] \tag{10.21}$$

where

$$\overline{K} = K + K_{vap} \tag{10.22}$$

The molecular interdiffusion of water vapour is a slow process compared with the mass flow of liquid water, except at very low moisture contents, so that except in fairly dry soils the hydraulic conductivity $K$ may be used instead of $\overline{K}$ with only negligible error.

## NOTES

### Note 18. The ratio of conductivities of geometrically similar bodies

Consider two porous bodies, $B_1$ and $B_2$, such that the latter is a magnified copy of the former. For every point in the former there will be a corresponding point in the latter; and the distance between any two points in the latter will be $N$ times the distance between the corresponding points in the former. Let two of these corresponding points, one in each body, be taken as origin of coordinates. Then any other pair of corresponding points will have coordinates, relative to the origin appropriate to the body, which may be written $x_1$, $y_1$ and $z_1$ for the point in $B_1$, and $x_2$, $y_2$ and $z_2$ for the corresponding point in $B_2$ such that

$$x_2/x_1 = y_2/y_1 = z_2/z_1 = N \tag{N18.1}$$

At the point in $B_1$ let the components of the true velocity of flow in the $x$, $y$ and $z$ directions be $\alpha$, $\beta$ and $\gamma$ respectively, as in Note 17, and let it be further supposed that these same velocity components are found at the corresponding point $(x_2, y_2, z_2)$ in $B_2$. It is of course implied that both points are in the pore spaces of the bodies. The velocity is a unique function of the coordinates, i.e. there can be only one velocity value at any one point. Conversely, although the pattern of flow may repeat itself at intervals of space, it may be taken as true in general that within a small element of space there is only one point at which a particular specified combination of $\alpha$, $\beta$ and $\gamma$ may be found, so that the space coordinates may be regarded as unique functions of $\alpha$, $\beta$ and $\gamma$. Thus one may write,

$$x_1 = F(\alpha, \beta, \gamma) \tag{N18.2}$$

and, by virtue of Equation (N18.1)

$$x_2 = N F(\alpha, \beta, \gamma) \tag{N18.3}$$

where $F$ is an unspecified function of the variables. Hence

$$(\partial x/\partial \alpha)_1 = \partial F/\partial \alpha$$

and

$$(\partial \alpha/\partial x)_1 = 1/(\partial F/\partial \alpha) = \Psi(\alpha, \beta, \gamma) \tag{N18.4}$$

where $\Psi$ is another unspecified function of $\alpha$, $\beta$ and $\gamma$. Similarly treating Equation (N18.3), one has

$$(\partial \alpha/\partial x)_2 = (1/N)\Psi(\alpha, \beta, \gamma) \tag{N18.5}$$

Differentiating Equation (N18.4) again yields

$$(\partial^2 \alpha/\partial x^2)_1 = (\partial \Psi/\partial \alpha)(\partial \alpha/\partial x)_1$$
$$= \Psi\, \partial \Psi/\partial \alpha \tag{N18.6}$$

Similarly treating Equation (N18.5), gives

$$(\partial^2 \alpha/\partial x^2)_2 = (1/N)(\partial \Psi/\partial \alpha)(\partial \alpha/\partial x)_2$$
$$= (1/N^2)\,(\Psi\, \partial \Psi/\partial \alpha) \tag{N18.7}$$

Dividing Equations (N18.6) by (N18.7), one has then that

$$(\partial^2 \alpha/\partial x^2)_1/(\partial^2 \alpha/\partial x^2)_2 = N^2 \tag{N18.8}$$

Similar relationships may be derived for variation of $\alpha$ in the $y$ and $z$ directions so that the net result is

$$(\nabla^2\alpha)_1 = N^2(\nabla^2\alpha)_2 \qquad \text{(N18.9)}$$

and similarly for $\nabla^2\beta$ and $\nabla^2\gamma$. Applying to both bodies the Stokes-Navier equations for the steady state of flow (the equations of (N17.3) omitting the terms $\partial\alpha/\partial t$, $\partial\beta/\partial t$ and $\partial\gamma/\partial t$) as a test that the assumed velocities of flow are possible, one finds that

$$\left.\begin{array}{c}(\partial\phi/\partial x)_1 = \eta(\nabla^2\alpha)_1 \\ \text{etc.} \\ (\partial\phi/\partial x)_2 = \eta(\nabla^2\alpha)_2\end{array}\right\} \qquad \text{(N18.10)}$$

etc.

Comparing the first of the equations of (N18.10) with Equation (N18.9) one sees that the condition that the Stokes-Navier equations are indeed satisfied in $B_2$, given that they are satisfied in $B_1$, i.e. that the flow distribution assumed is a possible one, is that

$$(\partial\phi/\partial x)_1 = N^2(\partial\phi/\partial x)_2 \qquad \text{(N18.11)}$$

and similarly for $(\partial\phi/\partial y)$ and $(\partial\phi/\partial z)$. Hence

$$(\text{grad } \phi)_1 = N^2(\text{grad } \phi)_2 \qquad \text{(N18.12)}$$

From the true point values of potential gradient in the liquid, given by Equation (N18.12), one may derive the relationship between the potential differences between the inflow and outflow of the whole bodies. Let us select a streamline in body $B_1$, with length $L$ between the inflow and outflow faces. The corresponding streamline in $B_2$ will have length $NL$. Corresponding points on the streamlines will be at distances $s_1$ and $s_2$ respectively from the inflow faces of $B_1$ and $B_2$, measured along the streamlines, where

$$s_2 = Ns_1 \qquad \text{(N18.13)}$$

Because $(\text{grad } \phi)_1$ is a function of $s_1$, one may write

$$|(\text{grad } \phi)_1| = \zeta(s_1) \qquad \text{(N18.14)}$$

where $\zeta$ is an unspecified function. From Equations (N18.13) and (N18.14) it follows that

$$|(\text{grad } \phi)_1| = \zeta(s_2/N) \qquad \text{(N18.15)}$$

At the corresponding point in $B_2$ $(\text{grad } \phi)_2$ is related to $(\text{grad } \phi)_1$ by Equation (N18.12) which yields, by combination with Equation (N18.15),

$$|(\text{grad } \phi)_2| = (1/N^2)\zeta(s_2/N) \qquad \text{(N18.16)}$$

Also,

$$(\Delta\phi)_1 = \int_{s_1=0}^{s_1=L} \zeta(s_1)ds_1 \qquad \text{(N18.17)}$$

and

$$(\Delta\phi)_2 = (1/N^2) \int_{s_2=0}^{s_2=NL} \zeta(s_2/N)ds_2 \qquad \text{(N18.18)}$$

where $(\Delta\phi)_1$ and $(\Delta\phi)_2$ are respectively the potential differences between the inflow and outflow faces of $B_1$ and $B_2$. Equation (N18.18) may be rewritten:

$$(\Delta\phi)_2 = (1/N) \int_{s_2/N = 0}^{s_2/N = L} \zeta(s_2/N)\mathrm{d}(s_2/N) \qquad (N18.19)$$

The definite integrals in Equations (N18.17) and (N18.19) are evidently equal so that from these two equations one arrives at the result,

$$(\Delta\phi)_1/(\Delta\phi)_2 = N \qquad (N18.20)$$

Because of the equality of the rates of flow at corresponding points on the outflow and inflow faces of the two bodies, the total rates of flow are in proportion to the areas of the faces, so that in the nomenclature of Equation (9.6)

$$[(Q/t)/S]_1 = [(Q/t)/S]_2 \qquad (N18.21)$$

If the lengths of the two bodies between the inflow and outflow faces (to be distinguished carefully from the true lengths of the streamlines in the pore space) are $l_1$ and $l_2$, then

$$l_2/l_1 = N \qquad (N18.22)$$

One may then write Equation (9.6), which expresses Darcy's law, for the two bodies in the forms,

$$[(Q/t)/S]_1 = K_1(\Delta\phi)_1/l_1 \qquad (N18.23)$$

and

$$[(Q/t)/S]_2 = K_2(\Delta\phi)_2/l_2 \qquad (N18.24)$$

Using Equations (N18.20) and (N18.22), one may rewrite Equation (N18.24) in the form,

$$[(Q/t)/S]_2 = K_2(\Delta\phi)_1/(N^2 l_1) \qquad (N18.25)$$

and finally, dividing Equation (N18.25) by Equation (N18.23) and using Equation (N18.21), one arrives at the relationship

$$K_2/K_1 = N^2 \qquad \left\{\begin{array}{l}(N18.26)\\(10.1)\end{array}\right.$$

Thus if two bodies have geometrically similar pore spaces, but the pore sizes of the one are larger than the corresponding pore sizes of the other by a factor $N$, then the hydraulic conductivity of that one is greater than that of the other by a factor $N^2$.

### Note 19. The rate of flow of a liquid in a cylindrical tube

In Figure (N19.1) is shown a cross-section and a perspective view of a uniform cylindrical tube of radius $R$ and length $l$ containing a fluid which is supposed to be flowing from left to right, the left-hand end of the tube being maintained at a potential $\Delta\phi$ with respect to the right-hand end. The potential gradient is thus $\Delta\phi/l$ from right to left so that, by Note 14, the force experienced by each unit weight of fluid is $\Delta\phi/l$ from left to right. Now consider an internal cylindrical surface of radius $r$ concentric with the containing tube. At this surface the inner plug of liquid is sliding over the outer annular cylinder, so that both will experience mutual viscous drag over the total area of curved surface $2\pi rl$. Since the plug is moving at a steady rate, without acceleration, the net force on it is zero, so that the viscous drag must just be balanced by the force due to the potential gradient. Since the weight of the plug is $\pi r^2 lg\rho$, the total force $F$, which it experiences in the potential field, is

$$F = g\rho\pi r^2 l\,\Delta\phi/l = g\rho\pi r^2\,\Delta\phi \qquad (N19.1)$$

from left to right. The viscous drag on the plug must therefore have the same magnitude, but acts in the opposite direction, i.e. it will be $-F$ if one adopts the

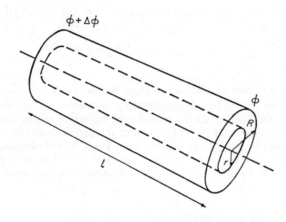

Figure N19.1 An element of flow through a cylindrical tube.

convention of reserving the positive sign for the direction left to right throughout. The viscous drag per unit area of curved cylindrical surface is $-F/(2\pi rl)$. From the definition of viscosity, Equation (9.21), but remembering that the direction at right angles to the sliding surface is $r$, one has

$$-(g\rho\pi r^2\Delta\phi)/(2\pi rl) = \eta\,dv/dr$$

or

$$dv/dr = -(g\rho r\Delta\phi)/(2\eta l) \qquad (N19.2)$$

where $v$ is the velocity of flow from left to right at radius $r$. Bearing in mind that $v$ has the value zero at the wall of the tube, where $r$ has the value $R$, one may integrate Equation (N19.2) directly, and get the result,

$$v = [g\rho\Delta\phi/(4\eta l)](R^2 - r^2) \qquad (N19.3)$$

The next stage is to calculate the total volume of liquid traversing the tube per unit of time from this radial distribution of flow velocity. A thin annular cylinder of mean radius $r$ and wall thickness $\delta r$, flowing at velocity $v$, will contribute an element $\delta(Q/t)$ to the total rate of efflux, where, by substitution for $v$ from Equation (N19.3),

$$\delta(Q/t) = 2\pi rv\ \delta r$$
$$= [(\pi g\rho\Delta\phi)/(2\eta l)]r(R^2-r^2)\delta r \qquad (N19.4)$$

In the limit when $\delta r$ becomes infinitessimally small, Equation (N19.4) may be written in the form,

$$d(Q/t)/dr = [(\pi g\rho\Delta\phi)/(2\eta l)]r(R^2-r^2)$$

which may be integrated between the two limiting values of $r$, namely zero and $R$, to yield finally,

$$Q/t = [(\pi g\rho\Delta\phi)/(8\eta l)]R^4$$

This is a similar form to that quoted as Equation (9.4), but since $\Delta\phi/l$ is the form taken by grad $\phi$ in this case one may write the equation as

$$Q/t = (g\rho\pi/8)(R^4/\eta)\ \text{grad}\ \phi \qquad \left\{\begin{matrix} (N19.5) \\ (10.2) \end{matrix}\right.$$

### Note 20. The rate of flow of a liquid in a plane laminar crack

Let the crack be defined by two parallel plane surfaces separated by a distance $D$. By symmetry the flow velocity is zero at each bounding surface and reaches a maximum at the plane which bisects the crack; at equal distances on each side of this medial plane the flow velocities are equal. Let $y$ be the distance measured from the medial plane in a direction perpendicular to it. One may now repeat the analysis given in Note 19, but this time one considers a portion of the slab of

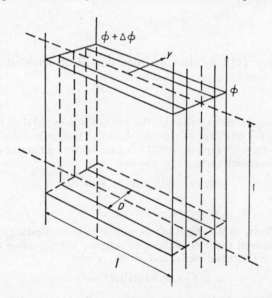

Figure N20.1 An element of flow between parallel plane
plates.

liquid bounded by two planes disposed symmetrically on each side of the medial plane and distant $y$ from it, as shown in Figure (N20.1). The portion selected has unit width and length $l$ in the direction of flow. The force $F$ on the portion of slab in the potential field grad $\phi$ is

$$F = 2g\rho yl \text{ grad } \phi \tag{N20.1}$$

The viscous drag will be $-F$ and the drag per unit area $-F/2l$, so that, applying the definition of viscosity, Equation (9.21) and Equation (N20.1), one has

$$-yg\rho \text{ grad } \phi = \eta \, dv/dy \tag{N20.2}$$

whence, bearing in mind that $v$ has the value zero when $y$ has the value $D/2$, integration results in the equation,

$$v = g\rho[(\text{grad } \phi)/2\eta](D^2/4-y^2) \tag{N20.3}$$

A thin lamina of liquid of thickness $\delta y$ at distance $y$ provides a flow contribution $\delta(Q/t)$ per unit width where

$$\delta(Q/t) = v \, \delta y$$
$$= g\rho[(\text{grad } \phi)/2\eta](D^2/4-y^2) \, \delta y$$

In the limit when $\delta y$ is infinitessimally small, this becomes

$$d(Q/t)/dy = g\rho[(\text{grad } \phi)/2\eta](D^2/4-y^2)$$

whence

$$Q/t = 2 \int_0^{D/2} g\rho[(\text{grad } \phi)/2\eta](D^2/4-y^2)dy$$

$$= g\rho(D^3/12\eta) \text{ grad } \phi \qquad \left\{ \begin{matrix} \text{(N20.4)} \\ \text{(10.3)} \end{matrix} \right.$$

## Note 21. Anisotropy due to lamination

Consider a body consisting of laminations of thickness $D_1$, and isotropic conductivity $K_1$ interleaved with others of thickness $D_2$ and conductivity $K_2$. If a potential gradient is imposed in any direction lying in the plane of lamination, then each lamination of conductivity $K_1$ will accommodate a rate of flow of $-K_1D_1$ grad $\phi$ per unit width, in accordance with Darcy's law, and each lamination of conductivity $K_2$ will similarly accommodate a flow rate of $-K_2D_2$ grad $\psi$ per unit width. A total thickness of $n$ pairs of laminations, i.e. $n(D_1+D_2)$, will therefore provide a total rate of flow, $Q/t$, where

$$Q/t = -n(K_1D_1+K_2D_2) \text{ grad } \phi$$

Regarded as a uniform material of conductivity $K_H$ in this direction, it provides a rate of flow,

$$Q/t = -K_Hn(D_1+D_2) \text{ grad } \phi$$

whence, by comparing these two expressions for the same quantity, one has

$$K_H = (K_1D_1+K_2D_2)/(D_1+D_2) \qquad \left\{ \begin{matrix} \text{(N21.1)} \\ \text{(10.12)} \end{matrix} \right.$$

If now the potential gradient is imposed perpendicular to the plane of lamination, there is a uniform rate of flow $Q/t$ per unit area of cross-section which alternates in material of conductivity $K_1$ and material of conductivity $K_2$. Let $\phi_1$ be the potential at the inflow face of a lamination of the former; $\phi_2$ that at the outflow face and therefore at the inflow face of the neighbouring lamination of conductivity $K_2$; and $\phi_3$ the potential at the outflow face of the latter. Applying Darcy's law to each lamination in succession one has

$$\left. \begin{aligned} Q/t &= K_1(\phi_2 - \phi_1)/D_1 \\ Q/t &= K_2(\phi_3 - \phi_2)/D_2 \end{aligned} \right\} \tag{N21.2}$$

If instead one regards the succession of layers as a single body with conductivity $K_V$ and thickness $(D_1 + D_2)$, one has

$$Q/t = K_V(\phi_3 - \phi_1)/(D_1 + D_2) \tag{N21.3}$$

Eliminating the intermediate potential $\phi_2$ between the pair of equations (N21.2) one has

$$(Q/t)(D_1/K_1 + D_2/K_2) = \phi_3 - \phi_1 \tag{N21.4}$$

Comparison of Equation (N21.3) with Equation (N21.4) provides the relationship,

$$K_V = (D_1 + D_2)/(D_1/K_1 + D_2/K_2) \qquad \left\{ \begin{aligned} &\text{(N21.5)} \\ &\text{(10.11)} \end{aligned} \right.$$

Where there are $n$ layers, the typical layer being designated the $r$th, with thickness $D_r$ and conductivity $K_r$, then with a potential gradient in the direction of the plane of lamination, which is taken to be horizontal, one has

$$(Q/t)_r = K_r D_r \, \mathrm{grad} \, \phi$$

where the rate of flow is per unit width of lamination. Thus

$$Q/t = \sum_1^n K_r D_r \, \mathrm{grad} \, \phi \tag{N21.6}$$

From the point of view of the complete thickness $\sum_1^n D_r$ of the $n$ layers, the horizontal conductivity is $K_H$ where

$$Q/t = K_H \sum_1^n D_r \, \mathrm{grad} \, \phi \tag{N21.7}$$

Comparison of Equation (N21.6) with Equation (N21.7) shows that

$$K_H = \sum_1^n K_r D_r / \sum_1^n D_r \qquad \left\{ \begin{aligned} &\text{(N21.8)} \\ &\text{(10.13)} \end{aligned} \right.$$

When the flow is perpendicular to the plane of lamination so that the rate of flow $Q/t$ per unit cross-section is the same in each layer, one has for the $r$th layer

$$Q/t = K_r \, \Delta\phi/D_r$$

where $\Delta\phi$ is the potential difference between the two faces of the layer. Hence,

$$\Delta\phi = D_r(Q/t)/K_r$$

and the total potential difference between the faces of the complete set of layers is

$$\phi = \sum_1^n \Delta\phi = (Q/t)\sum_1^n D_r/K_r \qquad \text{(N21.9)}$$

From the point of view of the whole array of layers of total thickness $\sum_1^n D_r$ and conductivity $K_V$, the statement of Darcy's law is

$$Q/t = K_V\phi/\sum_1^n D_r \qquad \text{(N21.10)}$$

Again a comparison of the two expressions, Equations (N21.9) and (N21.10), for the same quantity $Q/t$, shows that

$$K_V = \sum_1^n D_r/\sum_1^n D_r/K_r \qquad \begin{cases} \text{(N21.11)} \\ \text{(10.14)} \end{cases}$$

### Note 22. The inclusion of tortuosity in the Kozeny equation

Let it be supposed that a porous body is hydraulically conducting by virtue of being traversed by capillary tubes as in the Kozeny model, but that the tubes are so tortuous that a length $L$ of the body contains capillaries of length $L_e$. Consider a column of unit area of cross-section with a potential difference $\Delta\phi$ maintained between the ends. Then by Darcy's law, the rate of flow $Q/t$ is related to the conductivity $K$ by the equation,

$$Q/t = K \Delta\phi/L \qquad \text{(N22.1)}$$

Now imagine the conductor so moulded as to straighten the capillaries without distorting them in cross-section. The length of the column will now be $L_e$, but since the moulding will not have changed the total volume, the new cross-sectional area will be $S$ where

$$L = SL_e$$

or

$$S = L/L_e \qquad \text{(N22.2)}$$

The moulding process will have altered neither the volume of solids, nor the total volume, nor the pore volume, nor the surface area of the capillaries. Neither will the rate of flow $Q/t$ be altered, since the conducting tubes are the same. Hence the rate of flow $Q/t$ may be related to $K_m$, the hydraulic conductivity of the moulded column, by the equation,

$$Q/t = K_m S \Delta\phi/L_e \qquad \text{(N22.3)}$$

where $K_m$ is given by the Kozeny equation, appropriate to this new parallel straight tube model,

$$K_m = (g\rho/k\eta)(1/A^2)[f^3/(1-f)^2] \qquad \begin{cases} \text{(N22.4)} \\ \text{(10.10)} \end{cases}$$

In this equation $f$ and $A$ have the same values as for the original unmoulded column because of the constancy of the volumes and surfaces referred to above. From Equations (N22.1), (N22.2) and (N22.3), one has

$$K = K_m(L/L_e)^2$$

and substituting the expression for $K_m$ from Equation (N22.3) it follows that

$$K = (g\rho/k\eta)(1/A^2)(L/L_e)^2[f^3/(1-f)^2] \qquad \begin{cases} \text{(N22.5)} \\ \text{(10.15)} \end{cases}$$

If the liquid in the pore space has electrical resistivity, $\rho_l$, then the resistance $R$ of the column, regarded as a set of channels of length $L_e$ and total area of cross-section $s$, is

$$R = \rho_l L_e/s \qquad \text{(N22.6)}$$

However, this same measured resistance may be expressed in terms of the dimensions and overall resistivity $\rho_m$ of the column in the form,

$$R = \rho_m L \qquad \text{(N22.7)}$$

Combining these two equations, one has

$$L/L_e = \rho_l/(s\rho_m) \qquad \text{(N22.8)}$$

Since the total volume of the column of unit cross-section is $L$, while the volume of the contained capillaries is $sL_e$, the fractional porosity $f$ is given by

$$f = sL_e/L \qquad \text{(N22.9)}$$

Substitution in Equation (N22.8) of the expression for $s$ from Equation (N22.9) gives

$$(L/L_e)^2 = (1/f)(\rho_l/\rho_m) \qquad \begin{cases} \text{(N22.10)} \\ \text{(10.16)} \end{cases}$$

## Note 23. The law of movement of water vapour in porous material

In a free unconfined mass of air in which the mass of water vapour per unit volume at a given point is $\sigma$, the mass crossing unit area perpendicular to the direction of the gradient of $\sigma$ in unit time is $v_i$ where, by Fick's law,

$$v_i = -D_i \text{ grad } \sigma \qquad \text{(N23.1)}$$

In this equation $D_i$ is a constant, dependent upon the air pressure, called the coefficient of interdiffusion of water vapour in air. In a porous material of total porosity $f$, of which $c$ is water-filled, a column of unit cross-section has only the fraction $(f-c)$ available for air and therefore for the diffusion of water vapour through air along the column. Furthermore, tortuosity of path through the pore space must increase the effective path length so that the true gradient of vapour density along the column is less than that deduced from the overall length of the column by a factor $\alpha$. The tortuosity must increase, and therefore $\alpha$ must decrease, as the moisture content increases and reduces the paths available for water vapour diffusion. However, since at such high moisture contents the relative contribution of the vapour to the total flow is negligible, one may safely take $\alpha$ to be a constant, of value about $0 \cdot 6$, at all moisture contents at which vapour movement is significant. Hence the rate of movement, $v_{vap}$, of water vapour per unit cross-section of the unsaturated porous material in which there is a macroscopic vapour density gradient, grad $\sigma$ is

$$v_{vap} = -\alpha(f-c)D_i \text{ grad } \sigma \qquad \text{(N23.2)}$$

If $\sigma_0$ is the vapour density in equilibrium with free water at zero hydrostatic pressure, then the vapour density $\sigma$ over water at pressure head $H$ (which will be negative in an unsaturated material in which the water experiences suction) is, in accordance with Equation (7.1),

$$\sigma = \sigma_0 \, e^{MgH/RT} \tag{N23.3}$$

where $M$ is the molecular weight of water, of which the density is $\rho$, so that

$$\text{grad } \sigma = \frac{\sigma_0 Mg}{RT} \, e^{MgH/RT} \tag{N23.4}$$

Substitution of Equation (N23.4) in Equation (N23.2) yields

$$\mathbf{v}_{\text{vap}} = -K_{\text{vap}} \text{ grad } H \qquad \left\{ \begin{array}{l} \text{(N23.5)} \\ \text{(10.18)} \end{array} \right.$$

where

$$K_{\text{vap}} = \frac{\alpha(f-c)(\sigma_0 g M D_i)}{RT} \, e^{MgH/RT} \qquad \left\{ \begin{array}{l} \text{(N23.6)} \\ \text{(10.19)} \end{array} \right.$$

CHAPTER 11

# Some subsidiary expressions of the movement of soil water

### 11.1 A form of Darcy's law when there is a gradient of moisture content

WHENEVER soil is unsaturated, it may naturally arise that the moisture content varies from point to point and with the passage of time. A class of problems presents itself in which one is concerned to trace the changes of moisture distribution with time. An analogous type of problem arises in the theory of heat conduction, and in that case progress is made by transforming the temperature gradient, to which the rate of flow of heat is proportional, into a gradient of heat content. This is rendered possible by the use of the specific heat, which is the ratio between the increase of heat per unit mass and the increase of temperature. The development of the subject from this point is presented by Carslaw and Jaeger (1959), and much of the analysis is helpful in the study of water movement in porous matter.

Just as the heat content depends upon the temperature, so does the water content $c$ of the porous body depend upon the prevailing hydrostatic pressure $H$, which is of course negative in the unsaturated body; and just as the rate of flow of heat depends upon the gradient of temperature, so does the rate of flow of liquid depend in part upon the gradient of the hydrostatic pressure. The chief difference between the two cases is that while the ratio of the increase of heat content to rise of temperature is the sensibly constant specific heat, and may be represented by a symbol, the rate of change of water content per unit volume with increase of pressure is simply the slope of the moisture characteristic, and from the shape of this curve it is evident that the ratio $dc/dH$ varies over a wide range as the moisture content changes. Hence, although it is widely known as the specific water capacity of the material, it is not commonly represented by a generally accepted symbol. A second important difference is that since the moisture characteristic is subject to hysteresis, the specific water capacity is not uniquely related to the moisture content, for it depends also upon the history of wetting and drying by which that moisture content is reached. Thirdly, the movement of water is in response to a gradient of pressure measured relative to an arbitrary datum, while the moisture content depends upon the difference between the pressure in the water and that in the air on the other side of the air-water interface in the pore space. If the air is itself moving, as must be the case in the transient state when the moisture content is varying, then the air pressure must vary from point to point in order to accommodate the air flow, so that the hydrostatic pressure gradient which determines the water flow may differ from the gradient which determines the gradient of moisture content. Lastly, gravity provides a component of potential in addition to the pressure gradient for the purpose of causing the flow of water, and there is no analogous component in the theory of heat flow. These various differences provide the special difficulty of the theory of water movement in porous matter, preventing the simple application of ready-made heat flow theory. They also provide the special interest.

It is shown in Note 24 that provided that the pore air pressure is sensibly uniform and equal to the pressure of the external atmosphere with which the pore air is continuous (a condition which is satisfied either when the moisture content is unchanging or when the pore space is not nearly saturated), then the gradient of the hydrostatic pressure is given by

$$\text{grad } H = (\partial H/\partial c) \text{ grad } c + \sum_{1}^{n} (\partial H/\partial_r c_L) \text{ grad } {_r}c_L \qquad (11.1)$$

where the prevailing moisture profile has been reached via a series of reversals of trend from wetting to drying and vice versa. Here ${_r}c_L$ is the moisture content attained at the $r$th reversal of trend out of the total of $n$ successive reversals. Thus grad $c$ is the slope of the final moisture profile while grad ${_r}c_L$ is the slope of the $r$th reversal profile. The factor $\dfrac{\partial H}{\partial c}$ is the slope of the final scanning curve, while $\dfrac{\partial H}{\partial_r c_L}$ is the proportion by which the pressure at the final moisture content would change with the change of moisture content at the $r$th reversal, all other reversals, and the ultimate moisture content, remaining unaltered.

A situation of practical importance arises when there is just one reversal of trend, as for example when, at the cessation of irrigation, the infiltrated water redistributes itself throughout the profile. Then the upper part of the profile increases in moisture content during the irrigation, but becomes drier during the redistribution as it gives up water to the lower part, into which the water front continues to penetrate. In this case one may drop subscripts and write Equation (11.1) simply as

$$\text{grad } H = (\partial H/\partial c) \text{ grad } c + (\partial H/\partial c_L) \text{grad } c_L \qquad (11.2)$$

where $c_L$ is now the moisture content at the profile point at the moment of the single reversal of trend. Most simply, if there is no reversal of trend at all but continued wetting from dryness or drying from saturation,

$$\text{grad } H = (\partial H/\partial c) \text{ grad } c \qquad (11.3)$$

This is analogous to the heat equation,

$$\text{grad } \tau = (1/S\rho) \text{ grad } \theta$$

where $\tau$ is the temperature, $\theta$ the heat content per unit volume, $S$ the specific heat, and $\rho$ the specific gravity. Equation (11.3) is appropriate also at a point where there has been a reversal of trend, but where the moisture profile at the moment of reversal exhibits a maximum or a minimum, so that grad $c_L$ at this point vanishes.

With the potential $\phi$ given by Equation (9.3), namely,

$$\phi = H + z$$

the gradient of potential is

$$\text{grad } \phi = \text{grad } H + \text{grad } z$$

$$= \text{grad } H + \mathbf{k} \qquad (11.4)$$

Darcy's law in the form,

$$\mathbf{v} = -K \text{ grad } \phi$$

$$= -K(\text{grad } H + \mathbf{k})$$

then takes one of various forms according to the number of reversals of trend, in accordance with which one or other of Equations (11.1) to (11.3) is applicable for the expression of grad $H$.

In the simplest case where there is no complication due to hysteresis, and Equation (11.3) may be applied, one has

$$\mathbf{v} = -K\{(\partial H/\partial c) \text{ grad } c + \mathbf{k}\}$$

or

$$\mathbf{v} = -K(\partial H/\partial c)\text{grad } c - K\mathbf{k} \qquad (11.5)$$

It may be noted that $K$ is a soil property which is dependent on the moisture content and which is known when the moisture content is specified. The same may be said of the slope, $\partial H/\partial c$, of the moisture characteristic, this slope being the inverse of the specific water capacity, $\partial c/\partial H$. Hence the product $K(\partial H/\partial c)$ is also a soil property which is uniquely specified when the moisture content is specified, and may be represented by the single symbol $D$, where

$$D = K(\partial H/\partial c) \qquad (11.6)$$

Thus Darcy's law now takes the form,

$$\mathbf{v} = -(D \text{ grad } c + K\mathbf{k}) \qquad (11.7)$$

The velocity **v** has components represented by

$$v_x = -D \; \partial c/\partial x \qquad (11.8)$$

$$v_y = -D \; \partial c/\partial y \qquad (11.9)$$

$$v_z = -(D \; \partial c/\partial z + K) \qquad (11.10)$$

Equations (11.8) and (11.9) are statements that the rate of flow of water is proportional to the gradient of moisture content in the direction of the flow, a law which is obeyed by the molecular diffusion of solute through its solvent and which is known as Fick's law. In that context, the constant of proportionality $D$ is known as the diffusivity. Hence although there is no suggestion that water moves through the pore space of a porous body by molecular diffusion (for in fact the above equations are merely a restatement of Darcy's law of mass flow in consequence of a potential gradient), the constant $D$ as defined in Equation (11.6) is now commonly referred to as the diffusivity of water through the soil, and one may also, if somewhat loosely, refer to the diffusion of water through the body.

In drier soils, where the movement of vapour provides an appreciable proportion of the total movement of water, it is preferable to use the total hydraulic conductivity $\bar{K}$ in Equation (11.6), as discussed in Section 10.8, resulting in a total diffusivity $\bar{D}$ instead of $D$.

It may be noted that the more complicated Equations (11.1) and (11.2) cannot lead to Equations (11.8) and (11.9), and these more complicated movements cannot be described in terms of "diffusion" in the sense of obedience to diffusion laws. It is not merely that the diffusivity is expressible only in complicated terms; it is simply that the concept of diffusion does not apply. Further, it may be noted that even in the simplest case, the vertical movement described by Equation (11.10) is not strictly in accordance with Fick's law, although one component of such movement may be so described. This has an important bearing on the subsequent analysis of the development of moisture profiles in the soil.

As an example of the construction of a curve showing the dependence of the diffusivity on the moisture content from more basic curves of $H$ and $K$ as functions of $c$, Figure (11.1) is presented. At a specified moisture content $c$, indicated by the common ordinate, a tangent is drawn to the moisture characteristic and $dH/dc$ is derived from its slope. The value of $K$ is read off from the appropriate curve and the product $K(dH/dc)$, that is to say the diffusivity $D$, is plotted in the third diagram.

It is interesting to note that when the conductivity $K$ can be determined from a knowledge of the moisture characteristic as described in Section 10.5, the curve of $D$ versus $c$ may be computed from the moisture characteristic alone.

Since the moisture characteristic is subject to major hysteresis and its slope is a factor in determining $D$, then $D$ itself must also be subject to major hysteresis. The minor hysteresis to which $K$ is subject as a function of moisture content also makes a minor contribution to the hysteresis of $D$. Hence when one is treating a problem in which the diffusivity plays a part, care must be taken that the value of $D$ is chosen correctly in accordance with the history of wetting and drying of the sample.

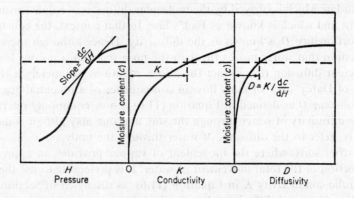

Figure 11.1 The development of diffusivity $D$ from hydraulic conductivity $K$ and specific water capacity $dc/dH$.

## 11.2 Darcy's law with hysteresis

One may proceed a little further with the discussion of Equation (11.2), which is a mathematical statement of Darcy's law when there has been one reversal of trend only, say from wetting to drying, in the history of the water movement, the moisture profile at reversal not having been uniform. Darcy's law then takes the form, if grad $H$ is substituted from Equation (11.2),

$$\mathbf{v} = -K(\text{grad } H + \mathbf{k})$$
$$= -K[(\partial H/\partial c)\text{grad } c + (\partial H/\partial c_r)\text{grad } c_r + \mathbf{k}] \qquad (11.11)$$

The components in the horizontal direction, say $x$, and the vertical direction $z$ are respectively,

$$\left. \begin{array}{l} v_x = -K[(\partial H/\partial c)(\partial c/\partial x) + (\partial H/\partial c_r)(\partial c_r/\partial x)] \\ v_z = -K[1 + (\partial H/\partial c)(\partial c/\partial z) + (\partial H/\partial c_r)(\partial c_r/\partial z)] \end{array} \right\} \qquad (11.12)$$

The physical meaning of $(\partial H/\partial c)_r$ and its relation to $(\partial H/\partial c)$ may be elucidated by reference to hysteresis and independent domain diagrams. In Figure (11.2) are shown two primary drying curves, one starting from the moisture content $c_r$ on the boundary wetting curve and the other starting from $c_r + \delta c_r$. At the common moisture content $c$, the pressure prevailing in the former curve is $H$ and on the latter curve is $H + \delta H$; and it

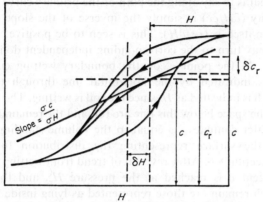

Figure 11.2 Illustrating the difference between

$$\frac{\partial c}{\partial H} \text{ and } \frac{\partial c_r}{\partial H}.$$

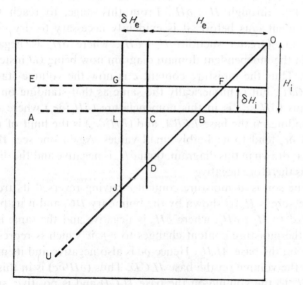

Figure 11.3 Illustrating the relationship between

$$\frac{\partial c}{\partial H} \text{ and } \frac{\partial c_r}{\partial H}. \text{ See text for explanation.}$$

will be seen that $\delta H$ is negative, the suction being greater. The ratio $(\delta H/\delta c_r)$ is thus negative, and in the limit when $\delta c_r$ is chosen to be negligibly small, this ratio tends to the limiting value $(\partial H/\partial c_r)$. At constant $c_r$, that is to say along the path of the primary curve starting from $c_r$, the quantity $(\partial H/\partial c)$ is simply the inverse of the slope of the curve, since the slope itself is $(\partial c/\partial H)$. This is seen to be positive.

Now one may turn to the corresponding independent domain diagram, Figure (11.3). At the point $c_r$ on the boundary wetting curve the pore water state is indicated by the horizontal line through the prevailing pressure, which is indicated as $H_i$ since the soil is wetting. The pore elements indicated by the space below this line are full and the remainder are empty, so that the water content $c_r$ is equal to the volume standing on the base $ABU$ under the surface representing the distribution function $F$, as described in Section 8.6. After reversal of trend from wetting to drying, the moisture content $c$ is reached at the pressure $H_e$, and the pore water elements which remain are those represented as lying inside the boundary $ACDU$, where $DC$ is the vertical line which passes through the axis at pressure $H_e$. The moisture content $c$ is the volume standing on the base $ACDU$ under the distribution function surface.

Similarly the moisture content $c_r + \delta c_r$, which prevails at the wetting suction $H_i + \delta H_i$, is the volume under the base $EFU$, where $EF$ is the horizontal line through $H_i + \delta H_i$. From this stage, to reach the same moisture content $c$ as before, it is evidently necessary to dry out to the lower pressure (greater suction) $H_e + \delta H_e$, where $\delta H_e$ is negative, the boundary in the independent domain diagram now being $GJ$ instead of the earlier $CD$. Thus the moisture content $c$ is now the volume standing on the base $EGJU$, but is numerically the same as that standing on the base $ACDU$. Thus $(\delta H/\delta c_r)$ is in this nomenclature $(\delta H_e/\delta c_r)$ where $\delta c_r$ is the volume standing on the base $EFBA$, and $(\partial H/\partial c_r)$ is the limit of this ratio as $\delta H_i$ and $\delta c_r$ tend to negligibly small values. Again one sees that if $\delta c_r$ is positive, as drawn in this diagram, then $\delta H_e$ is negative and the differential coefficient is therefore negative.

Now if the soil is at moisture content $c$ having reversed its trend at $c_r$, its pore pressure is $H_e$ as shown by the boundary $DC$, and if its pressure is now changed to $H_e + \delta H_e$, where $\delta H_e$ is negative and the same increment as before, the moisture content changes to $c + \delta c$ which is represented by the volume on the base $ALJU$. Hence $\delta c$ is also negative and its magnitude is given by the volume on the base $JLCD$. Thus $(\delta H/\delta c)$ is in this case the ratio of $\delta H_e$ to the volume on the base $JLCD$ and is positive, since both elements are negative.

If one writes the volume standing on the base $EFBA$ as $V_{EFBA}$, and similarly for other elements, then

$$\delta H_e/\delta c_r = \delta H_e/V_{EFBA}$$
$$= \delta H_e/(V_{EGLA} + V_{GFBL}) \qquad (11.13)$$

But

$$V_{EGJU} = V_{EGLA} + V_{ALJU} = c$$

and

$$V_{ACDU} = V_{ALJU} + V_{LCDJ} = c$$

Hence,

$$V_{EGLA} = V_{LCDJ}$$

Thus Equation (11.13) becomes

$$\delta H_e/\delta c_r = \delta H_e/(V_{LCDJ} + V_{GFBL}) \qquad (11.14)$$

Similarly,

$$\delta H_e/\delta c = -\delta H_e/V_{LCDJ} \qquad (11.15)$$

It is evident from Equations (11.14) and (11.15) that $\delta H_e/\delta c_r$ and $\delta H_e/\delta c$ are of opposite sign and that the magnitude of the latter exceeds that of the former. In the limit of smallness of the changes of $H$ and $c$, one has that $\partial H/\partial c$ and $\partial H/\partial c_r$ are of opposite sign and

$$|\partial H/\partial c| > |\partial H/\partial c_r| \qquad (11.16)$$

where $|\partial H/\partial c|$ indicates the magnitude of the differential coefficient.

One intuitively recognizes that the redistribution of water in a moisture profile after infiltration operates to reduce the gradient of the water content, and this will appear in Section 12.13, so that the prevailing moisture gradient $\partial c/\partial x$ at a particular point and moment tends to be less than the moisture gradient, $\partial c_r/\partial x$, at the moment of reversal. Thus,

$$\partial c/\partial x < \partial c_r/\partial x \qquad (11.17)$$

Thus it appears from Equations (11.16) and (11.17) that the product $(\partial H/\partial c)(\partial c/\partial x)$ tends to become of the same order as $(\partial H/\partial c_r)(\partial c_r/\partial x)$, and the two products are of opposite sign. Hence the terms in square brackets in the equations of (11.12) tend to neutralize each other as redistribution proceeds, so that in principle it would seem that there is no reason why the flow velocity $v_x$ should not vanish while an appreciable gradient of moisture content still prevails.

In the case of vertical infiltration and redistribution, $v_z$ cannot be reduced to zero until the product $(\partial H/\partial c_r)(\partial c_r/\partial z)$ exceeds in magnitude that of $(\partial H/\partial c)(\partial c/\partial z)$ by unity, which is less probable.

## 11.3 The equation of continuity

When the pattern of flow of the fluid in the porous medium is complicated, varying in magnitude and direction both with time and with position, it is usually found that Darcy's law cannot be applied directly with useful results. By incorporating with it a statement which expresses the balance sheet of water entry into and exit from a small element of volume, one derives from it a differential equation which contains only a scalar property such as the moisture content, the hydraulic potential or the hydrostatic pressure. This equation is known as the equation of continuity. Solutions of it, when obtainable, are in the form of distributions of the scalar variable both in space and in time. Thus the most general form of the equation of continuity for pore water is, as is shown in Note 25,

$$\partial c/\partial t = \partial[K_x(\partial\phi/\partial x)]/\partial x + \partial[K_y(\partial\phi/\partial y)]/\partial y + \\ + \partial[K_z(\partial\phi/\partial z)]/\partial z \tag{11.18}$$

which has the alternative forms, according to the variable chosen instead of $\phi$,

$$\partial c/\partial t = \partial[K_x(\partial H/\partial x)]/\partial x + \partial[K_y(\partial H/\partial y)]/\partial y + \\ + \partial[K_z(\partial H/\partial z) + K_z]/\partial z \tag{11.19}$$

and, in the absence of hysteresis effects,

$$\partial c/\partial t = \partial[D_x(\partial c/\partial x)]/\partial x + \partial[D_y(\partial c/\partial y)]/\partial y + \\ + \partial[D_z(\partial c/\partial z) + K_z]/\partial z \tag{11.20}$$

These most general forms become simpler in particular cases, as for example when the material is isotropic, or saturated, or when the flow has attained a steady state, or is restricted to a single direction. Thus it may be said of both the steady state and of flow in saturated soil that $\partial c/\partial t$ has the value zero, and in the latter case the hydraulic conductivity does not change from point to point in the flow pattern. In that case, Equation (11.18) takes the form,

$$K_x\, \partial^2\phi/\partial x^2 + K_y\, \partial^2\phi/\partial y^2 + K_z\, \partial^2\phi/\partial z^2 = 0 \tag{11.21}$$

If the material is isotropic so that subscripts may be omitted when writing $K$, one is left with the very simple form,

$$\partial^2\phi/\partial x^2 + \partial^2\phi/\partial y^2 + \partial^2\phi/\partial z^2 = 0 \tag{11.22}$$

This is the well-known Laplace's equation, which is one of the much-studied basic equations of physics.

Even when the material is anisotropic in respect of conductivity, one may transform Equation (11.21) into Laplace's equation by a suitable distortion of the coordinates. Thus new coordinates, $\lambda$, $\mu$, and $\nu$, are chosen, such that

$$\left.\begin{array}{l} \lambda = x \\ \mu = y(K_x/K_y)^{\frac{1}{2}} \\ \nu = z(K_x/K_z)^{\frac{1}{2}} \end{array}\right\} \tag{11.23}$$

Then

$$\left.\begin{array}{l} \partial^2\phi/\partial x^2 = \partial^2\phi/\partial\lambda^2 \\ \partial^2\phi/\partial y^2 = (\partial^2\phi/\partial\mu^2)(K_x/K_y) \\ \partial^2\phi/\partial z^2 = (\partial^2\phi/\partial\nu^2)(K_x/K_z) \end{array}\right\} \qquad (11.24)$$

Substitution of the equations of (11.24) in Equation (11.21), followed by division through by $K_x$, leads to Laplace's equation in the transformed coordinates, namely,

$$\partial^2\phi/\partial\lambda^2 + \partial^2\phi/\partial\mu^2 + \partial^2\phi/\partial\nu^2 = 0 \qquad (11.25)$$

Thus a problem of flow in a saturated anisotropic material may be transformed into one of flow in an isotropic body by a distortion of the shape of the body in accordance with the equations of (11.23), and if a solution is obtainable in this distorted body, then a converse distortion of the solution to restore the original boundary shape gives the solution in the original undistorted region. The isotropic conductivity of the distorted region is shown in Note 26 to be $K_{\lambda\mu\nu}$ where

$$K_{\lambda\mu\nu} = (K_y K_z)^{\frac{1}{2}} \qquad (11.26)$$

Hence no special attention to anisotropic saturated materials is called for.

The symmetry of a problem may sometimes simplify even Laplace's equation. For example, a problem that will be dealt with at some length is that of the control of groundwater by a system of parallel drains of which the separation is small compared with the length. In such a case the flow is, except near the ends of the drain lines, restricted to directions perpendicular to the drain lines, so that one is concerned only with water movements in a plane containing, say, the $y$ and $z$ axes, $x$ being the direction of the drain lines. Equation (11.22) then takes the two-dimensional form,

$$\partial^2\phi/\partial y^2 + \partial^2\phi/\partial z^2 = 0 \qquad (11.27)$$

Chapter 12 will deal particularly with those circumstances where the movement of water is confined to a single direction in the soil, usually the vertical, as when water is precipitated uniformly over a great extent of level land surface and percolates downward, or moves upward to be evaporated uniformly at the surface. For movement in a horizontal direction, say $x$, Equations (11.19) and (11.20) become respectively,

$$\partial c/\partial t = \partial[K_x(\partial H/\partial x)]/\partial x \qquad (11.28)$$

and

$$\partial c/\partial t = \partial[D_x(\partial c/\partial x)]/\partial x \qquad (11.29)$$

For vertical movement, in the $z$ direction, one has instead,

$$\partial c/\partial t = \partial[K_z(\partial H/\partial z) + K_z]/\partial z \qquad (11.30)$$

and

$$\partial c/\partial t = \partial[D_z(\partial c/\partial z) + K_z]/\partial z \qquad (11.31)$$

H

Even when flow is restricted to the vertical direction, it may often occur that the term $K_z$ is very small compared with $D_z(\partial c/\partial z)$, as for example when water movement is taking place in rather dry soil in which both $K$ and $D$ are small but locally steep moisture gradients $\partial c/\partial z$ are possible. When it is safe to ignore the gravitational term for this reason, Equation (11.31) assumes the simpler form of Equation (11.29).

Equations (11.28) to (11.31) show how the moisture content at a fixed point of observation changes with the passage of time, the rate of change being a function of the shape of the moisture profile or curve of distribution of $c$ as a function of $x$ or $z$ as the case may be. Solutions of these equations provide curves of $c$ versus $x$ or $z$ at any given moment of time. The changes of the moisture profiles as time passes are in some cases more conveniently followed by fastening attention upon a particular specified moisture content and by tracing the movement of the point at which this moisture content is found. A particular example is the tracing of the progress of a moisture profile by observation of the descent of the water front. In such a case, one is interested in $(\partial x/\partial t)_c$ or $(\partial z/\partial t)_c$ rather than $(\partial c/\partial t)_x$ or $(\partial c/\partial t)_z$. The conversion is made from the equation which expresses that $c$ is a function of both $z$, say, and $t$ when both $z$ and $t$ are variables. Thus when $z$ varies by an increment $\delta z$ and $t$ increases by an increment $\delta t$, the total increase, $\delta c$, of moisture content is

$$\delta c = (\partial c/\partial t)\ \delta t + (\partial c/\partial z)\delta z \tag{11.32}$$

Now if attention is fastened upon a spot at moisture content which remains constant at $c$, then as this spot moves through the distance $\delta z$ in time $\delta t$, that is to say with velocity $\delta z/\delta t$, the change of moisture content is zero, so that Equation (11.32) becomes in this case

$$(\partial c/\partial t)\delta t + (\partial c/\partial z)\delta z = 0$$

Hence when the increments of $z$ and $t$ are very small, so that the average velocity $\delta z/\delta t$ during the passage is the momentary velocity $\partial z/\partial t$, the relationship becomes

$$\partial z/\partial t = -(\partial z/\partial c)(\partial c/\partial t) \tag{11.33}$$

Substitution for $\partial c/\partial t$ from Equation (11.33) in, say, Equation (11.30), yields,

$$-(\partial c/\partial z)(\partial z/\partial t) = \partial[K_z(\partial H/\partial z) + K_z]/\partial z$$

or

$$-\partial z/\partial t = \partial[K_z(\partial H/\partial z) + K_z]/\partial c \tag{11.34}$$

This is the required form of the equation of continuity to give $\partial z/\partial t$. The equivalent form of Equation (11.31) is

$$-\partial z/\partial t = \partial[D_z(\partial c/\partial z) + K_z]/\partial c \tag{11.35}$$

In similar fashion Equations (11.28) and (11.29) develop into

$$- \partial x / \partial t = \partial [K_x (\partial H / \partial x)] / \partial c \qquad (11.36)$$

$$- \partial x / \partial t = \partial [D_x (\partial c / \partial x)] / \partial c \qquad (11.37)$$

## NOTES

### Note 24. The transformation of the pressure gradient to the moisture gradient in unsaturated porous material

In the transient state of an unsaturated porous material, the difference of air pressure at various parts of the pore space prevents one from relating the water content to the prevailing hydrostatic pressure alone. One must be concerned with two separate potential functions, $\phi_w$ for the water moving in the water-filled part of the pore space and $\phi_a$ for the moving air. In this case it is advantageous to adopt the definition of potential based on unit volume of fluid, Equation (9.1a). Thus,

$$\phi_w = P_w + g \rho_w z$$

$$\phi_a = P_a + g \rho_a z$$

Because the density of air $\rho_a$ is negligibly small in the context, the gravitational contribution to $\phi_a$ may be neglected so that

$$\phi_a = P_a$$

Applying Darcy's law to each fluid separately, one has

$$\mathbf{v}_w = -K_w \text{grad } \phi_w = -K_w (\text{grad } P_w + g \rho_w \mathbf{k}) \qquad (N24.1)$$

$$\mathbf{v}_a = -K_a \text{grad } \phi_a = -K_a \text{grad } P_a \qquad (N24.2)$$

The interfacial pressure $P_c$ which determines the water content is given by

$$P_c = P_w - P_a \qquad (N24.3)$$

From Equations (N24.1) and (N24.3), one has

$$\mathbf{v}_w = -K_w (\text{grad } P_c + \text{grad } P_a + g \rho_w \mathbf{k}) \qquad (N24.4)$$

Combination of Equation (N24.4) with Equation (N24.2) gives

$$\mathbf{v}_w = K_w (\text{grad } P_c + g \rho_w \mathbf{k}) + \mathbf{v}_a K_w / K_a \qquad (N24.5)$$

If the pore air is freely in contact with the external atmosphere, then a steady state of flow of water implies unchanging moisture contents and therefore constant air contents everywhere, so that the velocity of air flow is zero and the last term on the right-hand side of Equation (N24.5) disappears. Similarly if the air-filled part of the pore space is comparable with that containing water, then the much lower viscosity of air as compared with that of water ensures that $K_a$ is greatly in excess of $K_w$ and again the last term in Equation (N24.5) may be neglected even for the non-steady state. When, however, the pore space is nearly saturated with water and the air flow is consequently greatly restricted, then in spite of the low viscosity of air, the ratio $K_w / K_a$ may not be neglected and the analysis to follow may not be applied.

When the movement of air is either negligible or else almost unrestricted, because of the high value of $K_a$ in appreciably unsaturated pore space, there is little difference of air pressure throughout the material. Hence $P_a$ is everywhere equal to the external atmospheric pressure with which the pore air is continuous, and this is arbitrarily taken to be zero by convention. Hence Equation (N24.3) operates in this case to show the identity of $P_c$ and $P_w$, in agreement with the identity of Equation (N24.1) with Equation (N24.5) when the ratio of the conductivities $K_w/K_a$ is negligibly small. Hence the pressure $P_c$ which is related to the moisture content $c$ is the same as the pressure $P_w$, the gradient of which appears in Darcy's law, and in particular $dP_c/dc$ is identical with $dP_w/dc$. Hence one may proceed without distinguishing between subscripts, which will therefore be dropped.

The moisture gradient is defined in terms similar to those used for defining potential gradient in Section 9.1, so that its components in the $x$, $y$ and $z$ directions respectively are $\partial c/\partial x$, $\partial c/\partial y$ and $\partial c/\partial z$. Then

$$\operatorname{grad} c = \mathbf{i}\, \partial c/\partial x + \mathbf{j}\, \partial c/\partial y + \mathbf{k}\, \partial c/\partial z \tag{N24.6}$$

Now only exceptionally will the various points on the moisture profile have had the same history of wetting and drying. Hence the prevailing negative hydrostatic pressure will not only depend upon the prevailing moisture content $c$, but also on the particular prevailing scanning curve, and this latter may be specified by the sequence of moisture contents at the reversal of trend from wetting to drying or vice versa. If the moisture content at the $r$th reversal, of which there have been $n$ in all, is indicated by $_rc_L$, one may write, reverting to pressure head $H$ instead of pressure $P$,

$$H = f(c, {}_1c_L, {}_2c_L \cdots, {}_rc_L \cdots, {}_nc_L)$$

where $f$ is the abbreviation of "function", so that

$$\partial H/\partial x = (\partial H/\partial c)(\partial c/\partial x) + \sum_{r=1}^{r=n}(\partial H/\partial_r c_L)(\partial_r c_L/\partial x) \tag{N24.7}$$

Similarly differentiation with respect to $y$ and $z$ gives

$$\left.\begin{aligned}
\partial H/\partial y &= (\partial H/\partial c)(\partial c/\partial y) + \sum_{1}^{n}(\partial H/\partial_r c_L)(\partial_r c_L/\partial y)\\
\partial H/\partial z &= (\partial H/\partial c)(\partial c/\partial z) + \sum_{1}^{n}(\partial H/\partial_r c_L)(\partial_r c_L/\partial z)
\end{aligned}\right\} \tag{N24.7a}$$

Thus by combining the components specified in Equations (N24.7) and (N24.7a), using the symbolism of the vector notation, one arrives at the gradient of $H$ in the form,

$$\operatorname{grad} H = \mathbf{i}\, \partial H/\partial x + \mathbf{j}\, \partial H/\partial y + \mathbf{k}\, \partial H/\partial z$$

$$= (\partial H/\partial c)(\mathbf{i}\, \partial c/\partial x + \mathbf{j}\, \partial c/\partial y + \mathbf{k}\, \partial c/\partial z) +$$

$$+ \sum_{1}^{n}(\partial H/\partial_r c_L)(\mathbf{i}\, \partial_r c_L/\partial x + \mathbf{j}\, \partial_r c_L/\partial y + \mathbf{k}\, \partial_r c_L/\partial z)$$

By introducing Equation (N24.6), one then arrives at

$$\text{grad } H = (\partial H/\partial c) \text{ grad } c + \sum_{1}^{n}(\partial H/\partial_r c_L) \text{ grad }_r c_L \qquad \left\{ \begin{array}{l} \text{(N24.8)} \\ \text{(11.1)} \end{array} \right.$$

**Note 25. The derivation of the equation of continuity**

Consider an anisotropic soil whose conductivity has components $K_x$, $K_y$ and $K_z$ in the directions of the principal axes, which are also chosen as the axes of coordinates. In soils it is reasonable to assume that the development of structure determines that $z$ is the vertical direction and $x$ and $y$ are two mutually perpendicular horizontal directions. Now examine the flow in a small element of volume in the shape of a rectangular parallelepiped with its centre at the point $(x,y,z)$, the lengths of its edges being $\delta x$, $\delta y$ and $\delta z$, as shown in Figure (N25.1). Let the components of the velocity of flow in the directions of the axes be $v_x$, $v_y$ and $v_z$). Then the velocity of flow issuing out of the element over the face of area $\delta y \delta z$ at $x + \frac{1}{2}\delta x$ exceeds that of the inflow over the face at $x - \frac{1}{2}\delta x$ by the amount

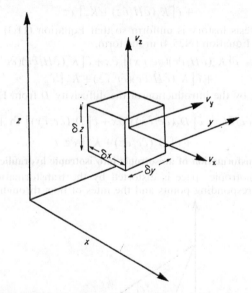

Figure N25.1 The flow of water into and out of a cubic element of material.

$(\partial v_x/\partial x)\delta x$. Hence the rate of loss of water suffered by the element on this account is $(\partial v_x/\partial x)\delta x\delta y\delta z$. The rate of gain of water is given by attaching the minus sign to this expression. Similar expressions may be derived for the rate of gain of water by the element in respect of the flow components $v_y$ and $v_z$, the total rate of gain being given by the sum of the three terms so obtained. An alternative expression for this same rate of gain is to be had in terms of the rate of increase of the moisture concentration, $c$, where $c$ is the volume of water per unit apparent volume of the soil. Thus the rate of gain is $(\partial c/\partial t)\delta x\delta y\delta z$ for the element of volume $\delta x\delta y\delta z$. Hence one has the equation,

$$(\partial c/\partial t)\delta x\delta y\delta z = -(\partial v_x/\partial x + \partial v_y/\partial y + \partial v_z/\partial z)\delta x\delta y\delta z \qquad \text{(N25.1)}$$

Now apply Darcy's law for anisotropic materials in the form of the equations of (9.11), with appropriate subscripts, namely

$$v_x = -K_x(\partial\phi/\partial x)$$

with similar equations for $v_y$ and $v_z$. Differentiation of this with respect to $x$, in the knowledge that in general $K_x$ is a function of $x$, gives

$$\partial v_x/\partial x = -\partial[K_x(\partial\phi/\partial x)]/\partial x \qquad (\text{N25.2})$$

Substituting Equation (N25.2) in Equation (N25.1), together with similarly derived expressions for $\partial v_y/\partial y$ and $\partial v_z/\partial z$, one has

$$\partial c/\partial t = \partial[K_x(\partial\phi/\partial x)]/\partial x + \partial[K_y(\partial\phi/\partial y)]/\partial y + \qquad \begin{cases} (\text{N25.3}) \\ \\ (11.18) \end{cases}$$
$$+ \partial[K_z(\partial\phi/\partial z)]/\partial z$$

From Equation (9.3) for $\phi$, it follows that Equation (N25.3) may be written:

$$\partial c/\partial t = \partial[K_x(\partial H/\partial x)]/\partial x + \partial[K_y(\partial H/\partial y)]/\partial y + \qquad \begin{cases} (\text{N25.4}) \\ \\ (11.19) \end{cases}$$
$$+ \partial[K_z(\partial H/\partial z) + K_z]/\partial z$$

When the hysteresis history is uniform so that Equation (11.3) is appropriate, one may rewrite Equation (N25.4) in the form,

$$\partial c/\partial t = \partial[K_x(\partial H/\partial c)(\partial c/\partial x)]/\partial x + \partial[K_y(\partial H/\partial c)(\partial c/\partial y)]/\partial y +$$
$$+ \partial[K_z(\partial H/\partial c)(\partial c/\partial z) + K_z]/\partial z$$

and this in turn, by the introduction of the diffusivity $D$ from Equation (11.6), becomes

$$\partial c/\partial t = \partial[D_x(\partial c/\partial x)]/\partial x + \partial[D_y(\partial c/\partial y)]\ \partial y + \qquad \begin{cases} (\text{N25.5}) \\ \\ (11.20) \end{cases}$$
$$+ \partial[D_z(\partial c/\partial z) + K_z]/\partial z$$

**Note 26. The transformation of anisotropic into isotropic hydraulic conductivity**

When the anisotropic space is distorted by the transformation (11.23), the potentials at corresponding points and the rates of flow through corresponding

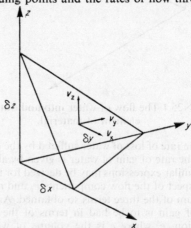

Figure N26.1 The flow across an element of a plane. The flow is invariant when the coordinates are distorted when transforming from an anisotropic to an isotropic space.

elements of area remain unchanged by this merely mathematical device. Consider the flow, with velocity components $v_x$, $v_y$ and $v_z$, through the element of plane area which intersects the $x$, $y$ and $z$ axes at $\delta x$, $\delta y$ and $\delta z$ respectively, as shown in Figure (N26.1). These intercepts define a triangular element, and the flow of fluid through it remains unchanged when the element is distorted. This rate of flow, $Q$, has components $Q_x$, $Q_y$ and $Q_z$.

The component $Q_x$ is perpendicular to the projection of the element of area on the $y,z$ plane, and this projection has an area of $\frac{1}{2}\delta y\delta z$, so that the rate of flow due to the component $Q_x$ through the projected area, which is the same as that through specified element itself, may be calculated by Darcy's law. Thus

and similarly

$$
\left.
\begin{aligned}
Q_x &= -\tfrac{1}{2}\delta y\delta z\, v_x = -\tfrac{1}{2}\delta y\delta z\, K_x\, \partial\phi/\partial x \\
Q_y &= -\tfrac{1}{2}\delta z\delta x\, K_y\, \partial\phi/\partial y \\
Q_z &= -\tfrac{1}{2}\delta x\delta y\, K_z\, \partial\phi/\partial z
\end{aligned}
\right\}
\tag{N26.1}
$$

The total rate at which liquid crosses the specified element of area is thus

$$
\begin{aligned}
Q &= Q_x + Q_y + Q_z \\
&= -\tfrac{1}{2}(K_x\, \delta y\delta z\, \partial\phi/\partial x + K_y\, \delta z\delta x\, \partial\phi/\partial y + \\
&\quad + K_z\, \delta x\delta y\, \partial\phi/\partial z)
\end{aligned}
\tag{N26.2}
$$

When the space is transformed by the equations of (11.23) and the now isotropic conductivity is indicated by $K_{\lambda\mu\nu}$, the corresponding relationships become

$$
\left.
\begin{aligned}
Q_\lambda &= -\tfrac{1}{2}K_{\lambda\mu\nu}\, \delta\mu\delta\nu\, \partial\phi/\partial\lambda \\
Q_\mu &= -\tfrac{1}{2}K_{\lambda\mu\nu}\, \delta\nu\delta\lambda\, \partial\phi/\partial\mu \\
Q_\nu &= -\tfrac{1}{2}K_{\lambda\mu\nu}\, \delta\lambda\delta\mu\, \partial\phi/\partial\nu
\end{aligned}
\right\}
\tag{N26.3}
$$

$$
Q = Q_\lambda + Q_\mu + Q_\nu
$$

With the aid of the equations of (11.23), these equations may be written in terms of the original coordinates, and they then take the form,

$$
\left.
\begin{aligned}
Q_\lambda &= -\tfrac{1}{2}K_{\lambda\mu\nu}\{K_x/(K_yK_z)^{\frac{1}{2}}\}\delta y\delta z\, \partial\phi/\partial x \\
Q_\mu &= -\tfrac{1}{2}K_{\lambda\mu\nu}(K_y/K_z)^{\frac{1}{2}}\delta z\delta x\, \partial\phi/\partial y \\
Q_\nu &= -\tfrac{1}{2}K_{\lambda\mu\nu}(K_z/K_y)^{\frac{1}{2}}\delta x\delta y\, \partial\phi/\partial z
\end{aligned}
\right\}
\tag{N26.4}
$$

$$
\begin{aligned}
Q &= Q_\lambda + Q_\mu + Q_\nu \\
&= -\tfrac{1}{2}\{K_{\lambda\mu\nu}/(K_yK_z)^{\frac{1}{2}}\}(K_x\, \delta y\delta z\, \partial\phi/\partial x + K_y\, \delta z\delta x\, \partial\phi/\partial y + \\
&\quad + K_z\, \delta x\delta y\, \partial\phi/\partial z)
\end{aligned}
\tag{N26.5}
$$

Equations (N26.2) and (N26.5) provide alternative expressions for the same quantity $Q$, and a comparison between them shows that

$$
K_{\lambda\mu\nu}/(K_yK_z)^{\frac{1}{2}} = 1
$$

or

$$
K_{\lambda\mu\nu} = (K_yK_z)^{\frac{1}{2}}
\qquad
\left\{
\begin{aligned}
&\text{(N26.6)} \\
&\text{(11.26)}
\end{aligned}
\right.
$$

# CHAPTER 12

# The movement of water in the soil profile

## 12.1 The nature of the problem

IF a pit be dug to expose a vertical face extending from the land surface to a depth beyond the influence of surface weathering, one may usually distinguish more or less sharply the variations of certain properties. Some of these, such as colour and texture, may be distinguished either visually or manually, while others may require the application of analytical techniques. A zone which is distinguishable in its properties from its neighbours above and below is called a soil horizon, and the sequence of horizons from the surface to the limiting depth of differentiation is called the soil profile. Any single property, such as the moisture content, may be singled out and the curve showing the variation of this property with depth may then be called the profile of this property, as for example the moisture profile.

A variation of the moisture content of the soil with depth from the surface carries with it, as has been described, a variation of the associated suction, hydraulic potential and hydraulic conductivity, even in a soil profile which is homogeneous in respect of the solid phase. Where the solid part itself varies in its properties with depth, as for example, by reason of a variation of texture or structure or both, there will be additional variations of suction and conductivity on this account for a given moisture content.

In these circumstances there are water movements in accordance with Darcy's law which will in general tend to alter the moisture profile with the passage of time and therefore to alter the conditions which are responsible for the movement. These movements must usually link up with imposed conditions at known levels in the soil profile. For example, irrigation from a flooded surface imposes at that level the condition of saturation and the accompanying conditions of maximum hydraulic conductivity, and a hydrostatic pressure equal to the depth of ponded water. In another case, rainfall or spray irrigation may impose a known rate of percolation at the surface. In addition there may be groundwater present with a water table at a known depth where the hydrostatic pressure is therefore known to be zero. Alternatively, the soil at great depth may be known to be at a specifiable low moisture content. All these specifiable boundary circumstances are known as boundary conditions.

The general problem posed in this chapter is the formulation of the appropriate equations of flow in these complicated circumstances and the derivation of solutions which conform to the boundary conditions. These solutions will be in the form of equations or computed curves which present the moisture content, or an associated property, as a function of the depth and time measured from a moment when the moisture profile is known. In addition, the rate of infiltration will emerge, if it was not an imposed boundary condition, either as the rate of change of the amount of water stored in the profile or as the rate of flow which satisfies Darcy's law at the surface where the conductivity and the potential gradient are known. In the following sections some idealized cases will be considered in order to elucidate the principles of procedure.

## 12.2 The steady-state moisture profile with surface precipitation

Let it be supposed that rain falls uniformly and at a constant rate on the land surface in such a way that, after the lapse of a period of time sufficient for initial perturbations to die down, the moisture profile settles down to a constant shape. Since it is postulated that there is neither accumulation nor depletion of water at any depth in the profile, the rate of entry of water into any element must equal the rate of exit, so that the velocity of flow must be the same at all depths and equal to the rate of precipitation at the surface. The only circumstance in which this can happen for a moisture profile of

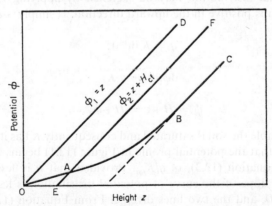

Figure 12.1 The potential distribution $OABC$ in a profile with uniform percolation to a water table at $O$.

finite depth is when that depth is constant and the profile terminates at the bottom at a water table which is maintained at that constant depth by drainage. The land surface is supposed to be horizontal and to extend in all directions to so great a distance that, by symmetry, there can be no sideways flow and no variation of properties in any horizontal direction. The flow is thus restricted to the vertical direction, which is designated $z$, and it will be convenient to choose the water table as the datum level from which $z$ is specified. With this choice, the potential at the water table is zero, since both the height $z$ and the hydrostatic pressure are separately zero at that level.

The main features of the resulting moisture profile may be demonstrated qualitatively by tracing the shape of the potential distribution as in Figure (12.1). The downward flow is everywhere equal to $q$, the rate of rainfall at the surface. The water table has zero height and zero potential, so the potential profile begins at the origin $O$. The straight line $OD$ represents the relationship,

$$\left. \begin{array}{r} \phi_1 = z \\ d\phi_1/dz = 1 \end{array} \right\} \tag{12.1}$$

and below it the parallel straight line $EF$ represents the relationship,

$$\left. \begin{array}{r} \phi_2 = z + H_{cf} \\ d\phi_2/dz = 1 \end{array} \right\} \tag{12.2}$$

where $H_{cf}$ is the pressure head at which the soil first begins to lose appreciable amounts of water. That is to say, $H_{cf}$ is the pressure head at the upper boundary of the capillary fringe as defined in Section 8.9. Since $H_{cf}$ is negative, being a suction, $EF$ lies below $OD$.

The potential distribution may be sketched by applying Darcy's law. Since $\mathbf{v}$, which is positive in the upward direction, is simply $-q$, the law is written:

$$q = K \, d\phi/dz$$

or

$$d\phi/dz = q/K \tag{12.3}$$

where

$$\phi = H + z = H + \phi_1 \tag{12.4}$$

At the water table the soil is saturated and consequently $K$ has its maximum value $K_{sat}$, so that the potential profile in Figure (12.1) begins at $O$ with a slope, from Equation (12.3), of $q/K_{sat}$. Provided that $q$ is less than $K_{sat}$, therefore, the slope of the curve of potential distribution is less than that of the line $OD$, and the two lines diverge. From Equation (12.4),

$$\phi_1 - \phi = -H$$

so that at any given value of $z$, the vertical distance of the potential profile in Figure (12.1) below the line $OD$ is a measure of the soil water suction, which evidently increases steadily with height $z$.

Provided that the profile lies between $OD$ and $EF$, this suction is less than that, $H_{cf}$, at the capillary fringe boundary, which is to say that the profile is inside the capillary fringe and the soil is sensibly saturated. Hence from $O$ to the intersection with $EF$ at $A$ the conductivity is constant at its maximum value $K_{sat}$ and, from Equation (12.3), the slope of the profile $d\phi/dz$ must also be constant, so that $OA$ is a straight line. Beyond $A$ the potential profile falls below the line $EF$ and the suction exceeds that at the capillary fringe boundary, so that the soil becomes more and more unsaturated with increasing height. It follows that the hydraulic conductivity decreases with increasing height, and in order to accommodate the constant rate of flow $q$ the potential gradient must become increasingly steep. This is indicated in Figure (12.1) by the increasing steepness of the profile along the section $AB$. As long as the profile is less steep than $OD$ and $EF$, the suction continues to increase and the hydraulic conductivity to decrease with increasing height, so that the steepness of the profile must continue to increase until at some point $B$ it becomes parallel to $OD$. Beyond $B$ any further increase of steepness must cause the profile to approach $OD$ and result in a decrease of suction and an increase of hydraulic conductivity. This is not consistent with an increase of potential gradient with unchanged $q$, and is therefore not possible. Similarly a decrease of steepness beyond $B$ implies an increase of suction and therefore a decrease of conductivity, and again this is not consistent with a decrease of potential gradient with unchanged $q$. Hence the only possible course of the potential profile beyond $B$ is to continue as a straight line $BC$ parallel with $OD$, so that the suction, and therefore the moisture content and hydraulic conductivity, are all uniform beyond this point. Along the section $BC$ the potential gradient is uniform and the same as is indicated by $OD$, namely,

$$(d\phi/dz)_{BC} = d\phi_1/dz = 1 \qquad (12.5)$$

It follows from Darcy's law that the moisture content must be such as to confer a conductivity $K_{BC}$ such that

$$q = K_{BC} \qquad (12.6)$$

Along the section $OA$ with constant $q$ and $K_{sat}$, the greater the value of $q$, the steeper must be the potential gradient $d\phi/dz$, in accordance with Equation (12.3), and the steeper must be the line $OA$. Along the section $BC$, Equation (12.6) applies, so that the greater the value of $q$ the greater must be $K_{BC}$ and the moisture content, and therefore the smaller must be the suction. Hence the greater the rainfall $q$, the nearer must $BC$ lie to $OD$. The steepening of $OA$ also lengthens it, increasing the value of $z$ at the intersection with $EF$, and the physical interpretation of this is that increasing $q$ results in a thickening of the capillary fringe. Ultimately the steady increase of $q$ will result in a potential profile which lies along $OD$ for the whole of its length with zero suction and complete soil saturation at all depths. This rate of flow must be the maximum, $q_{max}$, which can be accepted without the help of a depth of standing water at the surface, and this rate of acceptance is given by

$$q_{max} = K_{sat}(d\phi/dz)_{OD} = K_{sat}d\phi_1/dz = K_{sat} \qquad (12.7)$$

If the soil profile is not of sufficient depth to contain the whole range of the potential profile $OABC$ as drawn, then just so much of it as can be contained will be observed.

When the relationship between the moisture content and the hydraulic conductivity is known quantitatively, whether in the form of tabulated corresponding values or of an empirical mathematical expression, the precise course of the moisture profile may be computed. A combination of Equations (12.3) and (12.4) gives

$$q = K(1 + dH/dz)$$

or

$$dz/dH = 1/(q/K - 1) \qquad (12.8)$$

On the right-hand side, the only variable is $K$, which is known as a function of the moisture content $c$. Provided, therefore, that the history of wetting and drying is known, so that one knows which scanning curve to use to relate $c$ to $H$, the conductivity $K$ is known as a function of $H$. For example, it may be known that rainfall or irrigation is wetting up initially quite dry soil, so that it is appropriate to use the boundary wetting curve of the hysteresis loop to relate $H$ to $c$ and thereby to $K$. The whole of the right-hand side of Equation (12.8) may then be computed at a given value of $H$, and by choosing successive values of $H$, it may be expressed as a function of $H$ in tabulated form.

Integration of Equation (12.8) between the water table, where both $z$ and $H$ vanish, and the height $z$ where the pressure head is $H$, gives

$$z = \int_0^H dH/(q/K - 1) \qquad (12.9)$$

If $K$ may be expressed as a function of $H$ by an empirical equation, the integration may conceivably be carried out analytically, but in general numerical or graphical methods must be used. Thus the tabulated function $1/(q/K - 1)$ may be plotted as a curve against the variable $H$, when the integral of Equation (12.9) is the area under this curve between the specified limits of $H$, the area being measured, of course, in units indicated by the units of the plotted variables. In this way one computes the value of $z$ at which a specified pressure head prevails, or, in other words, one compiles the pressure profile by calculating a sufficient number of values of $z$ over a required range of values of $H$.

It will be noted that when the moisture content becomes reduced to a value such that $K$ is reduced to $q$, the stage in the integral of Equation (12.8) becomes infinitely great so that the increment $dz$ also becomes infinitely great. That is to say, the suction and moisture gradient at the surface become zero and the conductivity equals the infiltration rate, in confirma-

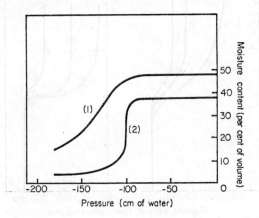

Figure 12.2 The moisture characteristics of (1) slate dust and (2) grade 15 Ballotini.

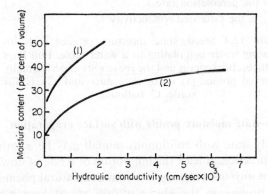

Figure 12.3 The relationship between hydraulic conductivity and moisture content of (1) slate dust and (2) grade 15 Ballotini.

tion of Equation (12.6). Referring back to the scanning curve, one may then transform pressures into moisture contents and plot the moisture profile. In this way Youngs (1957) took material with the moisture characteristics shown in Figure (12.2) and the curves of conductivity versus moisture content shown in Figure (12.3), in order to compute the moisture profiles at various rates of downward flow, shown in Figure (12.4). This last diagram also indicates the impressive agreement obtained between the calculated and the observed profiles.

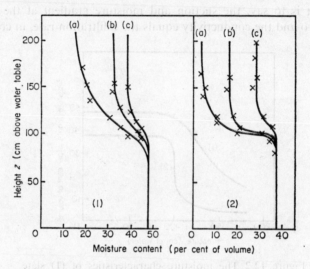

curve (a), $q = 0$;   curve (b), $q = K_{sat}/10$;
curve (b′), $q = K_{sat}/25$; curve (c), $q = K_{sat}/5$
$q$, is the percolation rate
$K_{sat.}$ is the saturated conductivity

Figure 12.4 Steady-state moisture profiles with in-filtrating water percolating to a water table. Full lines are theoretical curves and the crosses are experimentally observed points. (1) is for slate dust and (2) is for grade 15 Ballotini.

## 12.3 The steady-state moisture profile with surface evaporation

While a steady-state with continuous rainfall may be a rather unlikely contingency in practice, at least the imposition of a rate of downward flow independently of any soil property is a common natural phenomenon and the calculated profiles in the above section are at least possible. The imposition of a rate of upward flow by evaporation at the surface at a rate determined by meteorological circumstances only, independent of soil conditions, may, however, be a fictitious boundary condition, since a dry soil surface will impose its own limitations on the rate of upward movement and therefore of evaporation. If, therefore, one attempts to carry out the procedures of Section 12.2 with a uniform specified rate of upward flow $q$, one may experience a frustration which is evidence of the futility of the attempt, and of the impossibility of the existence of a moisture profile which satisfies the requirements.

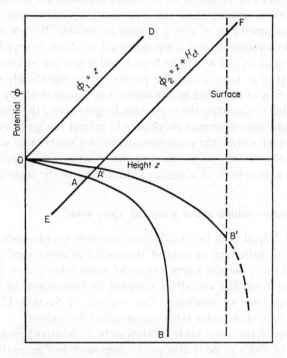

Figure 12.5 The potential distribution $OA'B'$ in a profile with uniform flow of water from a water table to evaporation at the surface. $OAB$ is the hypothetical potential distribution with excessive attempted evaporation.

Let Figure (12.5) be the equivalent of Figure (12.1), but with upward flow at the uniform rate $q$. Since the velocity of flow $\mathbf{v}$ is now $+q$, Darcy's law takes the form,

$$q = -K \, d\phi/dz \qquad (12.10)$$

With both $q$ and $K$ positive quantities, it follows that $d\phi/dz$ is necessarily negative, as is indicated by the downward slope of the potential profile in Figure (12.5). As before, this profile begins with a straight portion $OA$

from the origin to the intersection with *EF* at *A*, and beyond this point progressive desaturation with increasing height *z* requires a progressive steepening of the potential profile, for the same reason as before. This time, however, the steepening bends the potential profile away from the line *OD* and not towards it, so that the suction increases at an ever increasing rate instead of tending to a limiting value. Thus the suction may tend to infinity, as at *B*, at a level but little above the water table and well below the land surface. In effect one is unable to construct the moisture profile from the water table to the surface, where the supposed evaporation rate is imposed, because the assumed rate of flow *q* is quite impossible. With a smaller rate of upward flow requiring less steep potential gradients everywhere, some other profile such as *OA'B'* will be found, and if this can be constructed to the surface, it is at least a possible profile. By systematically computing potential profiles of this kind by the method of Equation (12.9), Wesseling (1957) was able to calculate the maximum heights above the water table to which a steady-state potential profile could extend for given evaporation rates, or in other words, the maximum depth of a water table which could sustain a steady surface evaporation at the specified rate. Gardner (1958) has reported calculations of a similar kind leading to the same conclusion.

## 12.4 The moisture profile above a moving water table

Section 12.2 dealt with the steady-state moisture profile where the water table was held stationary in spite of the arrival of downward percolating water. It will more usually happen that the water table is only imperfectly controlled, and will rise or fall in response to fluctuations of the rate of percolation in spite of drainage. The analysis of Section 12.2 may be extended to take such water table movements into account.

The velocity of the water table is taken to be *V*, positive upward, and the rate of flow of water to be *v*, also positive upward. For generality, there is assumed to be surface precipitation at the rate *q*. Provided that the water table has been moving at its specified speed for a sufficient length of time, one may suppose that the moisture profile has settled down to a constant form which moves with the water table at the same speed *V* everywhere. If such an assumption leads to a solution, it is valid. Lastly, it is supposed that the water table is sufficiently deep for the surface moisture gradient to be negligibly small, so that there is a finite depth at uniform moisture content $c_u$. It is convenient to measure the height *z* at a given point relative to the position occupied by the water table at the given moment. It has been shown by Childs and Poulovassilis (1962) and is also demonstrated in Note 27, that in these circumstances the height *z* at which the pore water pressure is *H*, which is of course negative, is given by

$$z = \int_0^H dH/[(K_u/K)\{1+(dH/dz)_u\}-1-(V/K)(c-c_u)] \quad (12.11)$$

Here $K_u$ is the hydraulic conductivity at the surface moisture content $c_u$ and $K$ is the conductivity at specified moisture content and corresponding pore pressure.

The value of $(dH/dz)_u$ at the surface may not be assumed to be negligibly small simply because $(dc/dz)_u$ is assumed to be negligibly small, since from the shape of commonly observed moisture characteristics at low moisture contents, quite small changes of moisture content may correspond to very large changes of suction. However, provided that a known value may be assigned to $(dH/dz)_u$, the integration may be carried out numerically or graphically in the manner of Equation (12.9) since all the quantities on the right-hand side are either constants or known functions of $H$. If the integration is carried out to a sufficient number of limits $H$, the complete suction profile of $H$ versus $z$ may be plotted, and from this the moisture profile of $c$ versus $z$ may be plotted by the use of the moisture characteristic.

The particular circumstances of a given situation will determine the value to be assigned to $(dH/dz)_u$. For example, when the surface precipitation is constant at the rate $q$, then as was shown in Section 12.2, the suction gradient at a sufficient height above the water table is zero and the moisture content has such a value, $c_u$, as confers on the soil a conductivity $K_u$ which is equal to the rate of infiltration $q$. Thus if the water table is sufficiently deep to allow the development of the necessary length of moisture profile, Equation (12.6) operates to give

$$q = K_u \quad (12.12)$$

Then Equation (12.11) becomes

$$z = \int_0^H dH/[q/K-1-(V/K)(c-c_u)] \quad (12.13)$$

with $c_u$ determined from the value of $q$.

It is evident that $z_H$ goes to infinity as $c$ approaches the value $c_u$ and $K$ simultaneously tends to the value $q$. Furthermore, provided that $V$ is positive, that is to say, the water table is rising, the denominator of Equation (12.13) is negative at all stages up to the moisture content at which it vanishes and causes the positive infinity of the integral, positive because of the negative values of $dH$. Hence the height steadily increases as $H$ steadily decreases until it ultimately goes rapidly to infinity. That is to say, $H$, $c_u$ and $K_u$ approach uniform values asymptotically at the surface, consistent with the assumed boundary conditions.

When the rate of rise of the water table is known to be due to the complete absence of drainage, $V$ can be expressed in terms of the rate of precipitation

$q$ as follows. The rate of arrival of water at a given moment at the top of the moisture profile is $q$, while the rate of removal below the water table at a level $l$, say, beneath the soil surface is zero, since it has been specified that there is no drainage. If the water stored in a column of unit cross-section in this length $l$ is $S$, then

$$S = \int_0^C z \, dc = lc_u + \int_{c_u}^C z \, dc \qquad (12.14)$$

where in this case $z$ is measured from the depth $l$ and $C$ is the saturation moisture content which prevails at the depth $l$. Accordingly, the rate of increase of storage of water in the profile is

$$dS/dt = \int_{c_u}^C (dz/dt) dc$$

$$= \int_{c_u}^C V \, dc = V(C - c_u) \qquad (12.15)$$

In terms of the difference between the rate of arrival $q$ at the surface and the zero rate of removal at the base, the rate of storage of water in the profile is

$$dS/dt = q \qquad (12.16)$$

Hence, from Equations (12.15) and (12.16), one has

$$V = q/(C - c_u) \qquad (12.17)$$

Substitution for $V$ in Equation (12.13) by the use of Equation (12.17) gives

$$z = \int_0^H dH/[(q/K)(C-c)/(C-c_u)-1] \qquad (12.18)$$

Figure (12.6), after Childs and Poulovassilis (1962), presents some moisture profiles for various rates of percolation and speeds of movement of the water table. A rising water table tends to compress the profile into a shorter space, and conversely a falling water table extends it.

If the water table is falling, the negative values of $V$ limit the application of Equation (12.13). The denominator of this equation may be reduced to zero at any values of $c$ and $K$ in excess of $c_u$ and $K_u$ that one may choose, simply by taking $V$ negative and sufficiently large. Hence the integral, and therefore $z$, tends to infinity at this value of $c$, which is inconsistent with the known boundary condition that the moisture profile approaches the lower moisture content $c_u$ asymptotically with increasing height. Thus the analysis is in this case self-inconsistent and therefore discredited.

The condition that Equation (12.13) may be applied is evidently that

$$(V/K)(c-c_u)>q/K-1 \quad \Big\} \atop \text{for } c>c_u \quad \quad (12.19)$$

which, upon rearrangement and the use of Equation (12.12), reduces to

$$V > -(K-K_u)/(c-c_u) \quad \Big\} \atop \text{for } c>c_u \quad \quad (12.20)$$

Reference to a typical curve of the relationship between $K$ and $c$, such as Figure (10.1), shows that $d^2K/dc^2$ is positive, so that $(dK/dc)_c$ is greater than $(dK/dc)_u$ where $c$ exceeds $c_u$, and $(K-K_u)/(c-c_u)$ lies intermediately between these two values. Hence,

$$(K-K_u)/(c-c_u)>dK/dc_u \quad \Big\} \atop \text{for } c>c_u \quad \quad (12.21)$$

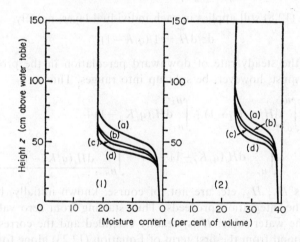

curve (a), $V = 69$ cm per hour; curve (b), $V = 34 \cdot 5$ cm per hour; curve (c), $V = 17$ cm per hour; curve (d), $V = 0$ cm per hour
$q$, the percolation rate
$V$, the speed of fall of the water table

Figure 12.6 Moisture profiles with steady percolation to a falling water table, where (1) $q = 6$ cm per hour; and (2) $q = 20$ cm per hour.

It follows that if $V$ exceeds $-(dK/dc)_u$ it must also exceed the ratio $-(K-K_u)/(c-c_u)$ at all stages of the integration short of the ultimate one, so that the two conditions stated in (12.20) may be expressed as the single necessary and sufficient requirement,

$$V > -(dK/dc)_u \qquad (12.22)$$

This condition is satisfied by all positive values of $V$, that is to say, for all rising water tables, and for all water tables which fall at a speed which does not exceed the magnitude of the right-hand side of the condition (12.22).

## 12.5 The steady-state moisture profile in a composite soil profile

It has been shown by Childs (1967) and Bybordi (1968) that the procedure of Section 12.2 may be applied equally when the soil profile comprises a succession of layers of conductivities respectively $K_1, K_2, \ldots K_r, \ldots K_n$, counting upwards from the water table. Each $K$ is of course a different function of the moisture content. Let the junctions between the successive layers be at $z_1, z_2, \ldots z_r, \ldots z_{n-1}$.

Equation (12.8) still applies in each individual layer, namely,

$$dz/dH = 1/(q/K-1)$$

where $q$ is the steady rate of downward percolation in the profile. The integration must, however, be split up into ranges. Thus

$$z = \int_0^{H_1} dH/(q/K_1-1) + \int_{H_1}^{H_2} dH/(q/K_2-1) + \ldots$$

$$+ \int_{H_{r-1}}^{H_r} dH/(q/K_r-1) + \ldots + \int_{H_{n-1}}^{H} dH/(q/K_n-1) \qquad (12.23)$$

The limits $H_1$, $H_2$, etc. are not, of course, known initially, but must emerge as the integration proceeds. Thus starting from zero values of $z$ and $H$ at the water table, $H$ is steadily extended and the corresponding value of $z$ found from the first term of Equation (12.23) alone for as long as $z$ thus found proves to be less than $z_1$, the known height of the first junction. The value of $H_1$ is thus found when the junction is reached, and since the pressure is continuous across the junction, $H_1$ is the lower limit of the second term of the integration. In this second range the conductivity is appropriately changed to that of the second layer. The integration then proceeds by extending $H$ further until the value of $z$ reaches $z_2$, at which stage the second boundary pressure $H_2$ emerges, and provides the lower limit for the third term of the integration, for which the conductivity of the third layer is used. The integration proceeds in this manner until the chosen value of $H$ is reached, at which stage the height at which this pressure is to be found emerges. In Equation (12.23) this pressure appears as lying in the uppermost layer, and of course it usually happens that one is interested in the whole moisture profile to the surface so that ultimately all terms of the integration will be used. But of course the integration may be stopped at any desired height in whatever term this height happens to lie.

As in Section 12.2, the pressure profile thus determined may readily be transformed into the moisture profile by use of the appropriate moisture characteristic of the section in which the moisture profile lies. While the pressure is continuous across a junction, since any discontinuity would produce an infinitely great potential gradient at such a point, the moisture content is only fortuitously continuous since one and the same pressure corresponds, in general, to quite different moisture contents in the two different materials on the two sides of the junction. In Figure (12.7) are shown some moisture profiles in a composite column of two different materials after Bybordi (1968), both observed and calculated in the manner described here.

When the conductivity of the profile is continuously variable with height, a reiterative procedure of approximation may be used to compute the moisture profile. The differential equation, Equation (12.8), is first replaced by the equivalent finite difference equation,

$$\Delta z = \Delta H / [2q/(K_H + K_{H + \Delta H}) - 1] \tag{12.24}$$

In this equation $H$ is the pressure head at $z$ and $H + \Delta H$ is the pressure head at $z + \Delta z$. $K_H$ and $K_{H + \Delta H}$ are the conductivities at the boundaries of this interval, so that half the sum of these conductivities is the average value over the interval, provided that the latter is not too large.

The chosen pressure $H$ for which the height $z$ is required is divided into a number of intervals $\Delta H$, and the integration begins at the lowest element for which $H$ is known to be zero and the conductivity is the known saturation conductivity at the base of the column. In order to calculate the height $\Delta z$ at which the pressure is $\Delta H$, it is necessary to estimate a value of $K_{\Delta H}$ for insertion in Equation (12.24). Since in fact $\Delta z$ is not known, it is not known in what soil horizon the pressure $\Delta H$ lies, so that the appropriate value of $K_{\Delta H}$ cannot be read off. A value of $\Delta z$ is therefore guessed, the soil at this horizon is known, and its conductivity at pressure $\Delta H$ read off. Then from Equation (12.24) the value of $\Delta z$ may be calculated and compared with the guessed value. A discrepancy indicates an error in the guessed value, and a second trial is made with a fresh guess, guided by the results of the first trial. A few repetitions result in a compatibility between the guessed and emergent values of $\Delta z$, which is therefore the height in the moisture profile at which the pressure is $\Delta H$.

Attention is then turned to the next element, with pressure head $\Delta H$ at the base and $2\Delta H$ at the top. Again the height $\Delta z$ at the base is known from the calculation of the first element, so that the soil properties here are known, and in particular the conductivity $K_{\Delta H}$. In a manner similar to that for the lowest element, a second height element $\Delta' z$ is assumed, the conductivity here is read off and inserted as $K_{2\Delta H}$ in Equation (12.24), the pressure being known to be $2\Delta H$, and a value of $\Delta' z$ calculated. Again the

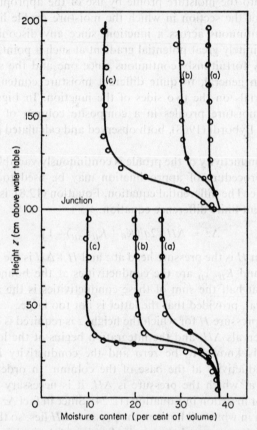

curve (a), $q = 0.9$ cm per min; curve (b), $q = 0.35$ cm per min; curve (c), $q = 0.035$ cm per min

Figure 12.7 Moisture profiles with uniform percolation in a composite column to a water table. Finer-grained material lies above coarser-grained material. Full lines are theoretical curves and circles are observed points.

discrepancy between the guessed and emergent values is removed by trial and error, and the true value of $\Delta' z$ found. In this way, by treating each element in turn, the whole pressure profile is calculated, and turned into the moisture profile by the use of the moisture characteristics at the appropriate levels.

## 12.6 The horizontal diffusion of water into a semi-infinite column at initially uniform moisture content

The case to be discussed here is not of great intrinsic interest, but it provides an introduction to the more important, but more difficult, case of vertical moisture profile development. In addition, as was pointed out in Section 11.2, the equations governing horizontal movement of water are also approximately true for vertical movement where the hydraulic conductivity is inherently low, as in heavy clay soils and more generally where the soil moisture content is everywhere low.

The particular problem to be treated is that where one has a horizontal column of porous material with its axis in, say, the $x$ direction, and with an exposed cross-sectional surface at a point on the axis which may for convenience be taken as the $x$ origin. It is supposed that the column extends to infinity in the positive direction of $x$. The initial state is taken to be one of uniformly distributed moisture content $c_0$. At a given moment, which may conveniently be taken as the arbitrary zero of time $t$, a different moisture content $C$ is imposed and thereafter maintained at the exposed surface. This may be done, for example, by suddenly flooding the free surface of the initially dry column, thereby maintaining a state of saturation at this surface. The maintenance of any other constant state than saturation is attended by technical difficulties which need not be discussed here, but one may suppose these difficulties to be surmountable. The problem then is to trace at any given subsequent moment the variation of moisture content in the column with distance from the exposed surface.

The appropriate equation of continuity, the solution of which provides the required information, is Equation (11.29), namely,

$$\partial c/\partial t = \partial[D(\partial c/\partial x)]/\partial x \qquad (12.25)$$

The boundary conditions which describe the particular problem are that the concentration $c$ has the value $c_0$ for all values of $x$ other than zero when $t$ has the value zero; and it has the value $C$ when $x$ has the value zero for all values of $t$. Further it is known intuitively that it will take an infinite time for the effect of the sudden surface flooding to be felt at an infinite distance, and also that after an infinite lapse of time all points at a finite distance from the surface will be at the same moisture content $C$. Hence it is known that $c$ has the value $c_0$ at all finite times when $x$ has the value infinity; and it has the value $C$ at all finite values of $x$ when $t$ has the value infinity.

These boundary conditions may be tabulated as follows:

$$c = c_0 \text{ for } t = 0 \ (1/t = \infty) \text{ and } x > 0$$

$$c = c_0 \text{ for } 1/x = 0 \ (x = \infty) \text{ and } t < \infty \ (1/t > 0)$$

$$c = C \text{ for } x = 0 \ (1/x = \infty) \text{ and } t > 0$$

$$c = C \text{ for } 1/t = 0 \ (t = \infty) \text{ and } x < \infty \ (1/x > 0)$$

The mathematical equivalence between zero distance and infinite time and vice versa is strikingly brought out, and led Boltzmann to combine the two independent variables $x$ and $t$ in a single variable $\chi$, according to the relationship,

$$\chi = x/t^{\frac{1}{2}} \tag{12.26}$$

and thence to transform the partial differential equation, Equation (12.25), into an ordinary differential equation. From Equation (12.26), it follows that

$$\left. \begin{array}{l} \partial \chi/\partial x = 1/t^{\frac{1}{2}} \\ \partial \chi/\partial t = -\frac{1}{2}x/t^{1\frac{1}{2}} \end{array} \right\} \tag{12.27}$$

Equation (12.25) is now rewritten in the form,

$$(\partial \chi/\partial t)(dc/d\chi) = (\partial \chi/\partial x)^2 d[\,D(dc/d\chi)\,]/d\chi$$

By substitution of the appropriate expressions for the partial differentials from the equations of (12.27), one arrives at the equation,

$$-(\tfrac{1}{2}x/t^{\frac{1}{2}})dc/d\chi = d[\,D(dc/d\chi)\,]/d\chi$$

Finally, using Equation (12.26), one has

$$\chi dc/d\chi + 2d[\,D(dc/d\chi)\,]/d\chi = 0 \tag{12.28}$$

This equation has only the one independent variable $\chi$ and the dependent variable $c$, which means that the solution, if and when found, is a curve or a mathematical expression of $c$ as a function of the variable $\chi$. It may be noted from Equation (12.26) that at a given moment the curve of $c$ versus $\chi$ may be regarded as one of $c$ versus $x$ in which the $x$ scale depends upon the time chosen; the greater the lapse of time, the greater is the distance $x$ at a given value of $\chi$ and $c$, which is to say the greater is the penetration of a given value of moisture content $c$. Similarly, the curve may be regarded as a plot of $c$ versus $1/t^{\frac{1}{2}}$ for a chosen value of $x$, in which the scale depends upon the value of $x$ chosen. It may also be noted that during the course of an experiment to observe the development of the moisture profile, the observer would record one and the same moisture content at different distances of penetration at different times, but if the movement is in accordance with

Equation (12.28), the ratio of the penetration and the square root of the time interval will be constant. It follows that if such an invariance is in fact observed, this is evidence that the water movement is in accordance with the diffusion equation, irrespective of the manner in which the diffusivity $D$ depends upon the moisture content, for the manner of such dependence plays no part in the derivation of Equation (12.28) from Equation (12.25).

The solution of Equation (12.28) will obviously depend upon the way in which $D$ varies with moisture content and until this is specified, no solution can be attempted. To some extent, however, the nature of the solution can be demonstrated in general terms. First expand Equation (12.28) and write,

$$\chi \frac{dc}{d\chi} + 2\frac{dD}{dc}\left(\frac{dc}{d\chi}\right)^2 + 2D\frac{d^2c}{d\chi^2} = 0 \tag{12.29}$$

where $dD/dc$ is the parameter which represents the variation of diffusivity with moisture content. In the special case where the diffusivity is constant and therefore independent of the moisture content, the second term of Equation (12.29) disappears, and it follows from what is left that, if the first term is negative, the third is positive and vice versa. Since $\chi$ and $D$ are both inherently positive, then if $dc/d\chi$ is positive, $d^2c/d\chi^2$ must be negative. Hence, the curve of $c$ versus $\chi$ must begin with its steepest gradient and become less and less steep with increasing $\chi$. In fact this case has a known analytical solution. It is, as is shown in standard texts (e.g. Carslaw and Jaeger, 1959),

$$c - c_0 = (C - c_0)\operatorname{erfc}(\chi/2D^{\frac{1}{2}}) \tag{12.30}$$

where

$$\operatorname{erfc}(z) = (2/\pi^{\frac{1}{2}})\int_z^\infty \exp(-\eta^2)d\eta$$

and is a readily available tabulated function. The solution of Equation (12.30) is shown plotted in Figure (12.8) in the form of the ratio $(c - c_0)/(C - c_0)$ versus $(\chi/2D^{\frac{1}{2}})$, and exhibits the feature described above. If it is remembered that for this solution $D$ is constant, then Figure (12.8) may be interpreted as a plot of moisture content versus distance from the exposed surface for a given lapse of time and for the chosen value of $D$, the scale of the abscissa being determined by these factors. The scale indicated in that figure indicates $x$ directly for all combinations of $D$ and $t$ for which the product $2(Dt)^{\frac{1}{2}}$ has the value unity.

One may postulate other forms of moisture profile and try to discover what kinds of variation of diffusivity with moisture content are responsible for them. A simple solution to try is that in which the moisture content decreases linearly from $C$ at the surface to $c_0$ at $\chi_0$ and then remains

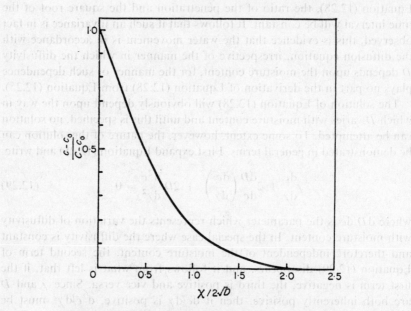

Figure 12.8. Distribution of moisture content $c$ as a function of $\chi$. Water is diffusing from a surface at moisture content $C$ into a semi-infinite profile initially at uniform moisture content $c_0$. $D$ is the uniform and constant diffusivity.

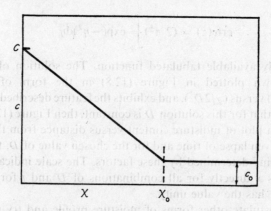

Figure 12.9 Hypothetical distribution of moisture content $c$ as a function of $\chi$. This distribution determines the dependence of the diffusivity on the moisture content.

constant at $c_0$ for all greater values of $\chi$, as indicated in Figure (12.9). This solution is written in the form,

$$c = C - (C - c_0)(\chi/\chi_0) \qquad (12.31)$$
$$\text{for } 0 < \chi < \chi_0$$

$$c = c_0 \qquad (12.32)$$
$$\text{for } \chi > \chi_0$$

From Equation (12.31), one has

$$\left. \begin{aligned} \mathrm{d}c/\mathrm{d}\chi &= -(C - c_0)/\chi_0 \\ \mathrm{d}^2c/\mathrm{d}\chi^2 &= 0 \end{aligned} \right\} \qquad (12.33)$$

while from Equation (12.32), it follows that, over the range of $\chi$ greater than $\chi_0$,

$$\mathrm{d}c/\mathrm{d}\chi = \mathrm{d}^2c/\mathrm{d}\chi^2 = 0 \qquad (12.34)$$

Substitution of Equation (12.34) in Equation (12.28) shows at once that the diffusion equation is satisfied over the range of $\chi$ in excess of $\chi_0$. Substitution of Equation (12.33) in Equation (12.28) results in the equation,

$$\mathrm{d}D/\mathrm{d}\chi = -(\chi/2) \qquad (12.35)$$

which integrates directly to give

$$D = A - \chi^2/4 \qquad (12.36)$$

where $A$ is a constant of integration which remains to be found. At the point of transition from the linear part to the constant value of $c$, namely when $\chi$ has the value $\chi_0$, the value of $\mathrm{d}^2c/\mathrm{d}\chi^2$ tends to infinity as the zone of transition becomes infinitesimally short, and Equation (12.28) in its expanded form, Equation (12.29), can only be satisfied if $D$ becomes zero at this point. Hence Equation (12.36) becomes at this point,

$$O = A - \chi_0^2/4$$

or
$$A = \chi_0^2/4$$

so that the general expression for $D$ is, from Equation (12.36) with the above value of $A$,

$$D = (\chi_0^2 - \chi^2)/4 \qquad (12.37)$$

Finally to express $D$ as a function of the moisture content which determines its variation directly, one uses Equation (12.31) to eliminate $\chi$ from Equation (12.37), and arrives at the result,

$$D = (\chi_0^2/4)[1 - (C - c)^2/(C - c_0)^2] \qquad (12.38)$$

Thus if $D$ increases with increasing $c$ in this manner, the resulting moisture profile will be given by Equation (12.31) as illustrated by Figure (12.9). The indicated scale of $\chi$ will also be that of $x$ at unit time.

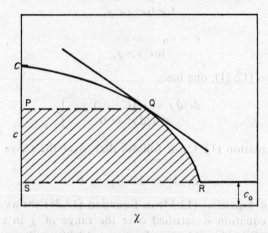

Figure 12.10 A hypothetical distribution of moisture content as a function of $\chi$ illustrating Matano's method of determining the diffusivity as a function of moisture content.

A graphical method of plotting the diffusivity as a function of the moisture content for any known shape of moisture profile has been presented by Matano (1932-33) and described by Crank (1956). Take a general profile, such as is shown in Figure (12.10), which satisfies the boundary conditions that have already been specified, namely that the limits of $c$ are $C$ and $c_0$, corresponding respectively to the zero and infinity limits of $\chi$. Equation (12.28), which governs the case, may be integrated between the limits $c_0$ and $c$, where the latter is any chosen value of the moisture content corresponding to a known chosen value of $\chi$. The result is

$$\int_{c_0}^{c} \chi \, dc = -2 \int_{c_0}^{c} d(D \, dc/d\chi) \qquad (12.39)$$

Since it is the nature of the boundary condition for $dc/d\chi$ to tend to zero as $\chi$ tends to infinity, where $c$ has the value $c_0$, Equation (12.39) may be written:

$$\int_{c_0}^{c} \chi \, dc = -2D_c (dc/d\chi)_c \qquad (12.40)$$

where the subscripts indicate that the quantities to which they are attached are measured at the moisture content $c$, that is to say at the point $Q$ in

Figure (12.10). The left-hand side of Equation (12.40) is simply the measurable area *PQRS*, shown shaded in Figure (12.10), so that the diffusity $D_c$ is obtained by dividing this area by twice the measurable slope of the profile at $Q$. By choosing a sufficient number of different points, such as $Q$, one may trace the whole of the course of the dependence of $D$ upon $c$. It may readily be shown, as an example of the method, that Equation (12.38) above is also to be derived as a special case of Equation (12.40).

The converse procedure of calculating a moisture profile when one knows the manner of the dependence of $D$ upon $c$ cannot be carried out directly. For this reason Philip (1960) has proposed the collection of pairs of moisture profiles and corresponding expressions of $D$ as a function of $c$ as calculated analytically by Matano's method, and has listed some such pairs as a start. Given a sufficiently extensive library of this kind, one might hope to match from it any imposed diffusivity function with which one might be confronted in practice and to pick out the corresponding paired moisture profile.

The calculation of a moisture profile for a specified dependence of $D$ upon $c$ is, in general, an exercise in integration by reiterative approximation, using the finite difference equivalents of the appropriate differential equations. Crank and Henry (1949) devised such a procedure in connection with problems unrelated to soils, and Klute (1952) applied the method to a case of water moving in a porous material of which the diffusivity was computed by the method of Childs and Collis-George (1950), but no experiments were reported in confirmation of the results. The commonly observed sharply defined water front was shown to be predicted as a consequence of the marked reduction of diffusivity with moisture content, as may readily be deduced from Matano's analysis.

A more rapidly converging and therefore much less laborious routine was evolved by Philip (1955b), consisting essentially of an inversion of Matano's method of treating Equation (12.40). By reference again to Figure (12.10) and Equation (12.40), it is evident that if the area *PQRS* and the diffusivity $D_c$ at moisture content $c$ are known, one may calculate the value of $dc/d\chi$ and therefore determine approximately the interval $\Delta\chi$ corresponding to the moisture range $\Delta c$. Thus if $\chi$ is known at $c$, it may be determined at $c + \frac{1}{2}\Delta c$ and at $c - \frac{1}{2}\Delta c$, so that these additional points on the profile may be plotted. Since of course the whole object of the exercise is to plot the profile from a knowledge of no more than the dependence of $D$ upon $c$ and the moisture content $C$ at the inflow face where $\chi$ is known to vanish, areas under the curve are *not* known initially, and the difficulty is to effect a beginning of the routine.

The procedure described here differs slightly from that recommended by Philip, but is in all essentials the same. The moisture range from $c_0$ to $C$ is divided into a number, as large as is practicable, of equal intervals $\Delta c$, and the Equation (12.40) is written in the finite difference form,

$$-(\chi_{c-\Delta c} - \chi_c)/\Delta c = 2\, D_{c-\frac{1}{2}\Delta c}/A_{c-\frac{1}{2}\Delta c}$$

where the subscripts indicate relevant moisture contents and

$$A_{c-\frac{1}{2}\Delta c} = \int_{c_0}^{c-\frac{1}{2}\Delta c} \chi \, dc$$

The integration is begun at the inflow face with vanishing $\chi_C$ and a moisture content $C$. The value of $A_{C-\frac{1}{2}\Delta c}$ is frankly guessed. A bad guess prolongs the reiterative procedure, and a guide for making a good initial guess is provided in the paper referred to. [Since $D_{C-\frac{1}{2}\Delta c}$ is known, as also is $\Delta c$ and the zero value of $\chi_C$, the finite difference equation permits the calculation of $\chi_{C-\Delta c}$, and a second point on the profile is now known.

The area under the profile between the horizontals through $C-\frac{1}{2}\Delta c$ and $C-\Delta c$, that is to say $A_{C-\frac{1}{2}\Delta c} - A_{C-\Delta c}$, can now be calculated and hence, applied as a correction to the initial guess, provides a measure of $A_{C-\Delta c}$. The finite difference equation is entered again with $A_{C-\Delta c}$, $D_{C-\Delta c}$ and $\chi_{C-\frac{1}{2}\Delta c}$ which emerged as a result of the preceding step. Thus $\chi_{C-1\frac{1}{2}\Delta c}$ is found and another point plotted on the profile.

This procedure is continued until the range of moisture content is covered, and if the initial guess was good, the area under the curve will be eliminated in the last step as the moisture content $c_0$ is reached. If this simultaneous event is not experienced, the initial guess was inadequate and a new one must be made, guided by the discrepancy shown in the first trial. The nature of this guidance may also be found in Philip's paper, which should be consulted for working details. The rapid convergence of the approximations ensures that an acceptable profile is obtained after only a few successive reiterations of the procedure.

## 12.7 The vertical infiltration of water into infinitely deep uniform soil

### (a) The uniform profile

As a preliminary, take the case of a uniform moisture profile with moisture content $c_0$ at all depths. It follows that the hydraulic conductivity will be uniform at the value $K_0$. Provided that the past history is also uniform, Darcy's law may be expressed in the form of Equation (11.10) which, with a vanishing gradient of moisture content, is

$$v_z = -K_0 \tag{12.41}$$

where $v_z$ is the vertical flux of water, positive upward. From Equations (11.8) and (11.9) there is no movement in any other direction. It follows from Equation (12.41) that $v_z$ is uniform throughout the profile, and since this means that the rate of entry of water into any element just equals the rate of exit, the rate of storage water is zero and the moisture content remains constant. This result is expressed formally by Equation (11.20) in which, in the present case, the right-hand side vanishes because all the quantities to be differentiated are either zero or constant.

Hence the uniform moisture profile is a steady-state, but the condition that it remains so is, of course, that $v_z$ is maintained at the value given by Equation (12.41) even at the boundary surface by an appropriate rate of infiltration at that surface. It has still to be shown that the profile is stable, that is to say, that a small perturbation does not trigger off an even greater one but tends to be reduced by compensating effects automatically called into being.

Such a local perturbation would introduce a maximum or a minimum of moisture content at a certain point. It is not enough to show that the perturbation would subsequently be reduced at the point $z$ at which it was introduced, for such a result would ensue from the movement of the perturbation away from $z$ without reduction. It needs to be shown that in such a case the perturbation would be removed as it moved.

Suppose that the moisture profile is moving at such a speed that the velocity of a particular point on it, on which attention is fastened, and which may be designated by its moisture content $c$, is $(\partial z/\partial t)_c$. Let the rate of change of moisture content at this point, as it moves, be $(\partial c/\partial t)_c$, and let the gradient of the moisture content be $(\partial c/\partial z)_c$. Then as the profile moves past a point which is stationary in the soil profile, i.e. of constant $z$, the moisture content must change by a rate which is the sum of two components, the first being due to the change of moisture content $(\partial c/\partial t)_c$ of the profile as it moves, and the second being due to the movement of the profile past the point, so that a greater moisture content is replaced by the lower content at a point lower down the gradient. Thus the rate of change of moisture content $(\partial c/\partial t)_z$ at the fixed point $z$ is given by

$$(\partial c/\partial t)_z = (\partial c/\partial t)_c - (\partial c/\partial z)_c(\partial z/\partial t)_c \qquad (12.42)$$

In the present case the point in the profile upon which interest is centred is the maximum of the induced perturbation, and at this point in the moisture profile the gradient $(\partial c/\partial z)_c$ vanishes. Hence Equation (12.42) becomes in this special case,

$$(\partial c/\partial t)_z = (\partial c/\partial t)_c \qquad (12.43)$$

Thus at the maximum of the perturbation there is no distinction to be made between the rate of change of moisture content at the fixed point in the soil and at the point fixed in the profile. The distinctive subscript will therefore be omitted in what follows.

The restriction of attention to the maximum of the perturbation also allows one to apply Equation (11.3), whether or not hysteresis operates, and therefore Equation (11.20) is also applicable. The one-dimensional form of this in the $z$ direction is Equation (11.31), namely,

$$\partial c/\partial t = \partial[D(\partial c/\partial z)+K]/\partial z \qquad (12.44)$$

and this may be expanded thus:

$$\partial c/\partial t = (\mathrm{d}D/\mathrm{d}c)(\partial c/\partial z)^2 + D(\partial^2 c/\partial z^2)+(\mathrm{d}K/\mathrm{d}c)(\partial c/\partial z) \qquad (12.45)$$

Again the fact that $(\partial c/\partial z)$ vanishes at the maximum of the perturbation simplifies Equation (12.45) to the form,

$$\frac{\partial c}{\partial t} = D\frac{\partial^2 c}{\partial z^2} \qquad (12.46)$$

If the perturbation presents a maximum of moisture content, it follows that

$$\partial^2 c/\partial z^2 < 0$$

so that, from Equation (12.46), it follows that

$$\partial c/\partial t < 0$$

and the maximum tends to be reduced. Conversely, it may be shown in the same way that a minimum tends to be increased while the straight portions of the profile on each side of the perturbation remain unchanged. Hence the perturbation, whether a minimum or a maximum of moisture content, tends to be eliminated and the profile is stable. The slightest perturbation at once calls into play a compensating reaction and no perturbation in fact develops, so that one is spared the difficulties of argument that are associated with regions on each side of a large perturbation where the disturbance of the moisture content is substantial, but where one cannot advance arguments which are valid only at a maximum or minimum.

### (b) A qualitative description of the development of the moisture profile

Suppose that the uniform moisture profile of the preceding section is suddenly disturbed at a moment which may conveniently be taken as the arbitrary time zero from which subsequent developments are reckoned. The disturbance consists of a sudden change of moisture content imposed at the surface to a value $C$, at which it is thereafter maintained. The technical difficulty of doing this except when the new moisture content is saturation due to surface flooding has already been mentioned in Section 12.6, but the theoretical analysis need not be so restricted.

The initial moisture profile is thus shown at stage 1 in Figure (12.11), and is characterized by an abrupt change of moisture content at the surface

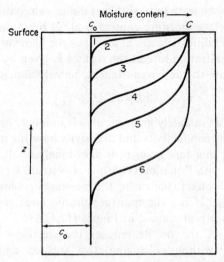

Figure 12.11 Hypothetical stages of the penetration of the moisture profile vertically from a surface maintained at moisture content $C$ into a profile initially at moisture content $c_0$. The figures indicate the order of development of the stages.

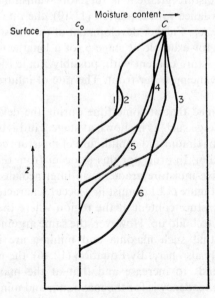

Figure 12.12 Illustrating some forbidden shapes of moisture profile stages. See text for explanation.

I

from $c_0$ immediately beneath to $C$, with an infinitely steep moisture gradient $dc/dz$. Since the surface moisture content is at all times the high value $C$, the conductivity and diffusivity are at all times the high values $K_C$ and $D_C$ respectively. The infiltration rate at the surface is given by Darcy's law in the form appropriate to the circumstances, namely Equation (11.10), or

$$v_z = -[D_C(dc/dz) + K_C] \tag{12.47}$$

At stage 1 with $dc/dz$ infinitely great, $v_z$ also is infinitely great. Just below the surface where the conductivity and diffusivity have the minimum values $K_0$ and $D_0$ and the moisture gradient is vanishingly small, the velocity of flow in accordance with Equation (11.10) is also small, so that the surface infiltration rate is devoted to increasing the moisture content in the immediately subsurface soil. Hence the moisture profile must rapidly assume a shape such as is shown at stage 2 in Figure (12.11).

The penetration of the profile reduces the steepness of the surface moisture gradient without changing the surface conductivity and diffusivity, so that the infiltration rate as shown by Equation (12.47) is reduced with the passage of time. Hence subsequently developed profiles, such as are shown at stages 3 and 4, require progressively longer times for their attainment. Ultimately the profile must reach a stage such as is shown at stage 5, where the moisture content is uniformly $C$ for a finite depth below the surface. The moisture gradient is therefore vanishingly small in this region and, in accordance with Equation (11.10), the rate of infiltration is reduced to the value $K_C$. Further development of the moisture profile takes the form, as shown for example at stage 6, of a lengthening of the upper zone of uniform moisture content with, possibly, some further changes in the shape of the advancing water front. The rate of infiltration remains at the value $K_C$.

It might be imagined that at some time during the development of the moisture profile, a stage such as is shown at stage 2 in Figure (12.12) might occur, exhibiting a maximum and a minimum of moisture content separated by a point of inflection. In order for this to occur from the initial stage of the uniformly positive moisture gradient, an intermediate stage such as is shown at stage 1 in Figure (12.12) must first occur, characterized by a short zone of uniform moisture content at the region where the maximum and minimum are about to build up. However, the same arguments that proved in Section 12.7(a) that such maxima and minima are inhibited in the uniform profile apply also here. By Equation (12.46), the moisture content at the minimum tends to increase and that at the maximum tends to decrease, so that the tendency is for the maximum and minimum feature to be eliminated. In fact, since the inhibition operates at once as soon as the stage 1 is reached, the maximum and minimum do not in fact develop at all.

Similarly, there can be no development which results in any part of the profile having a moisture content in excess of $C$, since, as shown in the hypothetical stage 3 of Figure (12.12), such a feature necessarily implies the existence of a maximum moisture content at some point in the profile, and this has been shown to be impossible. Finally, the ultimate stage, with an upper part of unchanging shape and increasing length joined to a lower part which may possibly change in shape as it descends, cannot be of such a shape as is shown at stage 4 in Figure (12.12), where the zone of uniform moisture content and increasing length is at a moisture content less than $C$. For such a zone must be joined to the surface by a zone above it; this zone must not only be of positive moisture gradient, but somewhere in it the rate of change of gradient must be positive, i.e. the gradient becomes steeper with height. Thus the equation of continuity as expanded in Equation (12.45) has every term positive, so that $\partial c/\partial t$ must be positive and the moisture content is increasing. Hence this is not the ultimate steady shape.

It may be noted here that if there is any question of an ultimate profile of constant shape descending as a whole without further changes, it can only be a profile attained subsequently to stage 5 of Figure (12.11) where the zone of uniform moisture content has developed at the surface. Any earlier profile would, if descending as a whole, join a surface zone of uniform moisture content discontinuously or with a kink, as shown by curves at stages 5 and 6 of Figure (12.12). The following sections will deal further and quantitatively with the various stages of development.

If the profile is developed by the infiltration of rain or irrigation spray at a constant rate instead of by maintenance of a surface flood, Equation (12.47) operates with a constant value of $v_z$ instead of constant diffusivity and conductivity. Hence at the initial moment when, because the surface is suddenly wetted, the moisture gradient $dc/dz$ is very steep, the moisture content must adjust itself at the surface to provide fairly low values of $D$ and $K$ to satisfy Equation (12.47) with the imposed value of $v_z$. The initial profile is thus of the form shown at stage 1 of Figure (12.13), with a steep moisture gradient at the surface ending at a not very high moisture content.

As infiltration proceeds to stage 2, the moisture gradient becomes less steep as deeper soil becomes wetted, so that in order to accommodate the constant flow rate $v_z$, the values of $D$ and $K$ must increase, so that the moisture content at the surface must become greater. This process continues, the surface becoming wetter and the surface moisture gradient decreasing steadily so as to maintain the constancy of $v_z$, until at last the gradient becomes vanishingly small and the surface moisture content settles to a constant value equal, as given by Equation (12.47), in these circumstances, to such a value as endows the soil there with a conductivity $K$ which is equal to the rate of precipitation $v_z$. The ultimate stage 5 is thus similar to that shown in Figure (12.11) for maintenance of constant surface moisture content, but it is reached by a different course, shown by stages 3 and 4.

The stages shown in Figure (12.11) are capable of mathematical analysis, but those of Figure (12.13) have not yet yielded to such treatment.

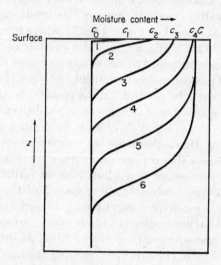

Figure 12.13 As Figure (12.11), but infiltration maintained at a constant rate instead of surface being maintained at constant moisture content.

### (c) Late stages following surface infiltration

In this section is considered the moisture profile at stages subsequent to stage 5 of Figure (12.11). It has been suggested tentatively that an ultimate profile shape may be attained, which thereafter descends steadily without further change of shape. An analysis of developing moisture profiles commonly exploits the equation of continuity. However, the equation of continuity, in any of its forms presented in Section 11.3, is applicable only if hysteresis plays no part, and the absence of such complications must therefore first be demonstrated.

It was shown in Section 12.7(b) that the infiltration profile initiated by the sudden surface flooding of a hitherto uniformly dry profile cannot develop maxima and minima of moisture content at any stage of its progress, but must show either uniform or increasing moisture content with increasing height from the water front. Hence as such a profile moves downward past a given point in the soil, the moisture content at that point must either increase or remain constant. Thus there is at no time any reversal of trend from wetting to drying and hysteresis plays no part. It follows that any of the forms of the equation of continuity are applicable. The appropriate form, derived only for the circumstances of the uniformly descending moisture profile, is Equation (11.35), namely,

$$-\partial z/\partial t = \partial[D(\partial c/\partial z)+K]/\partial c \qquad (12.48)$$

The postulate that the profile has reached a stage at which it descends at constant speed as a whole, without change of shape, may now be tested simply by equating $-\partial z/\partial t$ to a constant value, say $V$, independent of $c$. If this leads to a solution of the equation in the form required, namely of an expression for $c$ unambiguously as a function of $z$, or vice versa, the validity of the postulate is demonstrated. Thus Equation (12.48) becomes

$$d[D(dc/dz)+K]/dc = V \qquad (12.49)$$

This transformation turns the partial differential equation of continuity into an ordinary differential equation with two variables, $c$ and $z$, as does the Boltzmann transformation, Equation (12.26), described in Section 12.6. It seems first to have been proposed by Irmay (1956). The solution, giving the moisture profile, has been presented by Philip (1957a) and by Youngs (1957), who also carried out experiments to confirm its validity. As shown in Note 28, it is

$$z-z_f = (C-c_0)\int_{c_0}^{c} D \; dc/[K_C(c-c_0)+K_0(C-c)+K(c_0-C)] \qquad (12.50)$$

where $z_f$ is the height of the advancing water front at the particular moment,

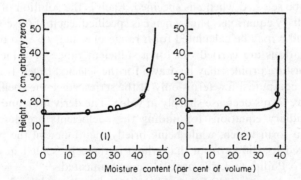

Figure 12.14 The ultimate shape of the descending moisture profile in (1) slate dust and (2) grade 15 Ballotini. Full lines are theoretical curves and circles are experimental points.

measured from an arbitrary datum level. $z - z_f$ is thus the height of the point at moisture content $c$, measured from the moving water front.

The right-hand side of Equation (12.50) contains only the known constant values, $C$, $c_0$, $K_C$ and $K_0$, and the properties $K$ and $D$ which are known functions of the moisture content $c$. Hence the expression to be integrated may be tabulated at intervals of $c$, and the integration may be carried out either numerically or graphically in the manner described in Section 12.2 to give the moisture profile. It will be noted that when $c$ is given the value $C$, the expression within the integral sign becomes infinitely great, so that the last increment of $c$ occupies an infinitely long increment of $z$. This means, of course, that the uppermost part of the profile is of uniform moisture content, confirming what was said on this point in Section 12.7(b), namely that an ultimate profile of constant speed of descent is not attained until the surface moisture gradient is reduced to zero. A comparison of an observed profile with a calculated profile, after Youngs, is shown in Figure (12.14).

### (d) Early stages following surface change

The analysis of the early stages of the descending moisture profile as functions of both depth and time has been described by Philip (1957b). The equation of continuity, Equation (12.48), is solved to give $z$ as the sum of a series of terms of ascending powers of $t^{\frac{1}{2}}$ in the form,

$$-z = \lambda t^{\frac{1}{2}} + \mu t + v t^{1\frac{1}{2}} + \xi t^2 + \dots \qquad (12.51)$$

Each of the coefficients, $\lambda$, $\mu$, $v$ and so on, is a function of the moisture content alone, presented in the form of a curve from which numerical values may be selected when $c$ is specified. Each is the solution of one of a set of subsidiary equations. Thus when $c$ is specified, each of the coefficients is known and $z$ may be calculated for a range of values of $t$, so that when these calculations are carried out for a sufficient range of values of $c$, a series of moisture profiles may be drawn for the selected times. It is usually sufficient to compute a few terms only of the series, since the coefficients in the successive terms decrease sharply in value. The derivation and solution of the subsidiary equations for finding the coefficients are described in Note 29 in a form which, while being briefly mentioned in the published paper referred to, differs from that which is there presented in detail. It also is due to Philip, and was privately communicated.

The form of Equation (12.51) may be seen to be a logical development of the equation, Equation (12.26), for horizontal diffusion. Firstly, it is expected that the solution will express $z$ as a function of $c$ at any particular moment, rather than vice versa, for that has been the form in which all other profiles so far examined have emerged. The solution Equation (12.26) may therefore be written:

$$x = \lambda t^{\frac{1}{2}}$$

where $\lambda$ is a function, written $\chi$ in Equation (12.26), of $c$ alone, and which may be determined and presented as a curve of $\lambda$ versus $c$ as described in Section 12.6.

Next one may note that the equations of continuity, Equations (11.9) and (11.10), for horizontal and vertical movement of water respectively, become indistinguishable when $t$ is very small, for at the initial stage the moisture gradient $dc/dz$ is infinitely steep, so that the term $K$ in Equation (11.10) may be neglected in comparison with $D(\partial c/\partial z)$. This equation is then the same as Equation (11.9). Hence in the very early stages, the development of the vertical profile is the same as that of the horizontal, and the solution will therefore be

$$-z = \lambda t^{\frac{1}{2}} \qquad (12.52)$$

The negative sign occurs because the profile advances downward in the negative direction of $z$, as compared with the horizontal advance in the positive direction of $x$.

At later times when the term $D(\partial c/\partial z)$ does not greatly outweigh $K$, the latter term in Equation (11.10) operates to make the downward advance of the moisture profile more rapid than the horizontal, so that the solution may be expected to consist of Equation (12.52) with the addition of

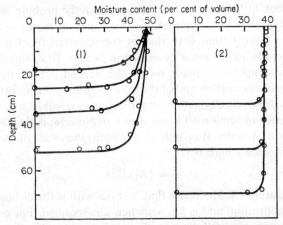

Figure 12.15 Stages in the development of the descending moisture profile in (1) slate dust and (2) grade 15 Ballotini. Full lines are theoretical curves and circles are experimental points.

correcting terms which are smaller for smaller $t$. Such terms are ascending powers of $t$, and since the first term contains $t^{\frac{1}{2}}$, it seems logical to try ascending powers of $t^{\frac{1}{2}}$. The result is Equation (12.51).

Equation (12.51) has self-evident faults when $t$ becomes very large, and is therefore appropriate to the early stages only. For example, differentiation with respect to $t$ to obtain the speed of descent gives

$$-\partial z/\partial t = \tfrac{1}{2}\lambda t^{-\frac{1}{2}} + \mu + 1\tfrac{1}{2}vt^{\frac{1}{2}} + 2\xi t + \dots \tag{12.53}$$

As $t$ becomes very large, the first term on the right-hand side becomes negligibly small compared with the others, which, with the exception of the constant $\mu$, become increasingly large with the lapse of time. Hence the speed of descent so predicted does not become constant at all moisture contents, as is known to be the case, but continues to increase as a function both of time and of moisture content, the profile thus changing shape continuously. However, within the limits of validity in respect of lapsed time, Youngs (1957) has demonstrated good agreement with experimentally observed moisture profiles in random granular materials. His curves are reproduced in Figure (12.15).

## 12.8 Horizontal infiltration into inhomogeneous material

Natural soil is rarely even approximately uniform in its properties, but exhibits profile variations of texture, structure, water retentivity and hydraulic conductivity. The development of rigorous analytical methods to trace the development of advancing moisture profiles in such material has made but little progress, but one tentatively probing study of the diffusion of water into a horizontal column may be mentioned.

If the variation is limited to that of the scale length of geometrically similar materials, in the sense described in Section 10.1, and if, further, a particular assumption is made as to the dependence of the hydraulic conductivity on the suction and of the pore suction on the moisture content, then as was shown by Philip (1967), a solution is possible.

As was stated in Section 10.1 and shown in Note 18, if $\overline{K}$ is the hydraulic conductivity of a material of unit scale length, then the conductivity $K$ of a material of scale length $N$ is

$$K = \{N(x)\}^2 \overline{K} \tag{12.54}$$

The nomenclature $N(x)$ indicates that $N$ varies with $x$, the distance from the surface of infiltration, and is known when $x$ is specified. For convenience, $N$ will be written in what follows without this functional indication.

Similarly, at equal moisture contents corresponding pores will be filled, and the radii of curvature of the interfaces will be in the proportion of $N$, the scale length. Since from Section 8.2, and in particular from Equations

(8.1a), (8.2a) and (8.3a), the prevailing suction is inversely proportional to the radius of curvature, it is therefore in this case inversely proportional to the scale length. Hence,

$$H = \bar{H}/N \tag{12.55}$$

where $\bar{H}$ is the pressure (negative since suction prevails) in the material of unit scale length. The appropriate equation of continuity, Equation (11.28), is thus, by substitution from Equations (12.54) and 12.55),

$$\frac{\partial c}{\partial t} = \frac{\partial}{\partial x}\left[N^2 \bar{K}\frac{\partial(\bar{H}/N)}{\partial x}\right] \tag{12.56}$$

where $N$ is quite unspecified as a function of $x$.

The particular hypothetical relationships between $\bar{K}$, $\bar{H}$ and $c$ which allow further progress are

$$\bar{K} = A/\bar{H}^2 \tag{12.57}$$

$$\bar{H} = -Be^{-\beta c} \tag{12.58}$$

where $A$, $B$ and $\beta$ are constants. From Equation (12.58), one has

$$\partial H/\partial t = B\beta\, e^{-\beta c}\partial c/\partial t$$
$$= -\beta\bar{H}\, \partial c/\partial t \tag{12.59}$$

Substitution of Equations (12.57) and (12.59) into Equation (12.56) yields

$$-\left(\frac{\partial \bar{H}}{\partial t}\right)/A\beta\bar{H} = \partial\left[\frac{N^2}{\bar{H}^2}\frac{\partial(\bar{H}/N)}{\partial x}\right]/\partial x \tag{12.60}$$

Since $N$ is a function of $x$ only and not of time, differentiation of Equation (12.55) with respect to time gives

$$\partial\bar{H}/\partial t = N\, \partial H/\partial t$$

Substituting both this and Equation (12.55) itself back into Equation (12.60), one arrives at the equation,

$$-(\partial H/\partial t)/A\beta H = \partial[(1/H^2)\,\partial H/\partial x]/\partial x \tag{12.61}$$

The boundary conditions to be taken are that, before infiltration, the column is in static equilibrium with the pressure head uniform at $H_0$, and that infiltration is initiated by changing the surface pressure head to $H_1$ and thereafter maintaining it at that value. Thus,

$$\left.\begin{array}{l} H = H_0 \text{ for all values of } x \text{ at } t = 0 \\ H = H_1 \text{ at all times for } x = 0 \end{array}\right\} \tag{12.62}$$

One now makes the transformations,

$$\tau = -A\beta t/H_0 \tag{12.63}$$

$$\Pi = \ln(H/H_0)/\ln(H_1/H_0) \tag{12.64}$$

1*

Differentiation of the new time and pressure variables, $\tau$ and $\Pi$, from Equations (12.63) and (12.64) gives

$$d\tau/dt = -A\beta/H_0 \tag{12.65}$$

$$(1/H)\partial H/\partial t = (\partial\Pi/\partial\tau)(d\tau/dt)\ln(H_1/H_0)$$

$$= -(A\beta/H_0)(\partial\Pi/\partial\tau)\ln(H_1/H_0) \tag{12.66}$$

$$(1/H)\partial H/\partial x = (\partial\Pi/\partial x)\ln(H_1/H_0) \tag{12.67}$$

Substitution of these values in Equation (12.61) leads to

$$\partial\Pi/\partial\tau = \partial[(H_0/H)(\partial\Pi/\partial x)]/\partial x$$

From Equation (12.64),

$$H_0/H = (H_0/H_1)^\Pi$$

so that finally,

$$\partial\Pi/\partial\tau = \partial[(H_0/H_1)^\Pi \partial\Pi/\partial x]/\partial x \tag{12.68}$$

This equation is of the same form as Equation (12.11), with the factor $(H_0/H_1)^\Pi$ playing the part of the moisture dependent diffusivity $D$ of that equation, for when $\Pi$ is specified the factor is known. Hence Equation (12.68) may be solved in the same way as Equation (12.25), by adopting the Boltzmann transformation,

$$\chi = x/\tau^{\frac{1}{2}}$$

and finding $\chi$ as a function of $\Pi$ by the reiterative procedure described in Section 12.6. In view of the very restricted nature of the conditions in which the solution is applicable, further discussion will not be undertaken here.

## 12.9 Moisture profiles during non-steady seepage to a water table

In Section 12.6 Boltzmann's transformation was used with success because no physical limit was imposed upon the ultimate depth of penetration of the water front. The equivalence between the effects of infinitessimally short times and infinitely great distances of penetration for a given moisture content permitted the boundary conditions as well as the variables to be expressed in terms of a single compound variable, $xt^{-\frac{1}{2}}$. Where a soil profile is effectively bounded not only above, by the land surface, but also below, as for example, by a water table, this equivalence no longer exists and Boltzmann's transformation is no longer useful. Such a case arises when an excessively wet moisture profile, such as might be maintained by infiltration from a flooded surface, is subsequently allowed to subside to an equilibrium profile by drainage to the groundwater following the removal of the surface flood. The problem is to trace the course of the varying moisture profile during this subsidence.

A method of solution has been demonstrated by Day and Luthin (1956), and is a direct numerical integration of the equation of continuity, Equation (11.30), namely,

$$\partial c/\partial t = \partial[K(\partial H/\partial z)+K]/\partial z \tag{12.69}$$

They adopt the usual device of dividing the profile column arbitrarily into a number, $n$, of elements, each of length $\Delta$, and substituting for Equation (12.69) the corresponding finite difference equation for a single element.

Following them, one labels the boundaries between the elements from $O$ to $n$, $O$ being the water table and $n$ the land surface. The label $r$ is thus attached to the upper surface of the $r$th element counting from the water table. The hydraulic conductivity and potential at a specified boundary are labelled similarly, so that $K_r$ is the conductivity at boundary $r$, and $\phi_r$ is the potential at this same boundary.

Provided that the elements are sufficiently short, one may suppose that the variations of both $K$ and $\phi$ are linear, so that the conductivity $K_{r-\frac{1}{2}}$ at the midpoint of the $r$th element is the arithmetic mean of the conductivities at the ends, and the potential gradient is the difference between the terminal potentials divided by $\Delta$, the length of the element. Then,

$$K_{r-\frac{1}{2}} = (K_{r-1}+K_r)/2$$
$$(\partial H/\partial z)_{r-\frac{1}{2}} = (H_r-H_{r-1})/\Delta$$

and similarly,

$$K_{r+\frac{1}{2}} = (K_r+K_{r+1})/2$$
$$(\partial H/\partial z)_{r+\frac{1}{2}} = (H_{r+1}-H_r)/\Delta$$

Also,

$$\{\partial[K(\partial H/\partial z)]/\partial z\}_r = [K_{r+\frac{1}{2}}(\partial H/\partial z)_{r+\frac{1}{2}}-K_{r-\frac{1}{2}}(\partial H/\partial z)_{r-\frac{1}{2}}]/\Delta$$

and

$$(\partial K/\partial z)_r = (K_{r+\frac{1}{2}}-K_{r-\frac{1}{2}})/\Delta$$
$$= (K_{r+1}-K_{r-1})/2\Delta$$

Substitution of these expressions for the differential coefficients in Equation (12.69) gives, after a little rearrangement,

$$H_r = \frac{[(H_{r+1}+\Delta)(K_{r+1}+K_r)+(H_{r-1}-\Delta)(K_{r-1}+K_r)-2\Delta^2(\partial c/\partial t)_r]}{(K_{r+1}+2K_r+K_{r-1})} \tag{12.70}$$

As always, the values of $K$ and $H$ must be known at every specified value of the moisture content $c$, and the procedure is then as follows. One supposes an initial state, which may, for example, be one of saturation at zero pressure head everywhere, as during infiltration from a flooded surface. This is the chosen arbitrary zero of time. The next stage of the profile to be calculated is then defined to be that at which the head $H_n$ at the surface has a specified value and one makes a trial guess at the distribution $H_r$ and the accompanying distributions of $K_r$ and $c_r$. At the water table $H_0$ is known to be zero at all times, and $K_0$ and $c_0$ to have the saturation values. The distribution of the moisture content thus assumed enables one to calculate the total water content in the profile and therefore the water $\delta Q$ lost during the interval since the preceding stage of the profile. The distribution of head enables one to discover the potential gradient at the water table and therefore the rate of passage of water across the water table in accordance with Darcy's law, given the known saturated conductivity. Thus,

$$\delta Q/\delta t = K_0 \, (\text{grad } \phi)_0$$

The time interval $\delta t$ between the successive stages is thus determined, and therefore $(\partial c/\partial t)$ at any given part of the profile is calculated from this value of $\delta t$ and the difference between the stage water contents at the beginning and end of the interval.

In the next step of the calculation, one takes the top two elements and, using Equation (12.70), computes a value of $H_{n-1}$ from the guessed values $H_n$, $H_{n-2}$ and the corresponding values of $K_n$, $K_{n-2}$, $c_n$ and $c_{n-2}$. The value of $H_{n-1}$ so obtained is then used, instead of the initially guessed value, for the purpose of calculating $H_{n-2}$, using the second pair of elements and equation Equation (12.70), in the same way. This process is repeated until the water table is reached, and then one goes back to the top and repeats the whole process using the corrected profile. Reiteration of this procedure

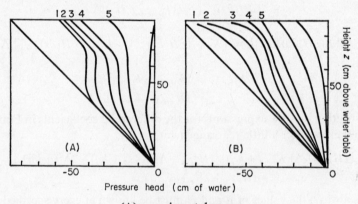

(A) experimental curves
(B) theoretical curves

Figure 12.16 Stages in the development of the suction profile in a vertical column which is draining to a water table from an initial state of saturation.

results ultimately in a profile which is not altered by further repetitions. This profile is then used as the initial profile for the purpose of computing the next successive stage, and in this way the whole course of development of the moisture profile is traced to the ultimate stage of equilibrium with the water table.

Agreement with observed moisture profile development was found to be no more than fair, as may be seen in Figure (12.16), and in fact great accuracy is only possible by the use of quite fine subdivision of the profile into elements and quite short intervals of time. The great labour associated with such measures can now, of course, be greatly eased by programming the procedure for the digital computer.

## 12.10 The redistribution of infiltrated water

Let it be supposed that infiltration has been proceeding and a moisture profile has been developing, as described in Sections 12.7(b), (c) and (d),

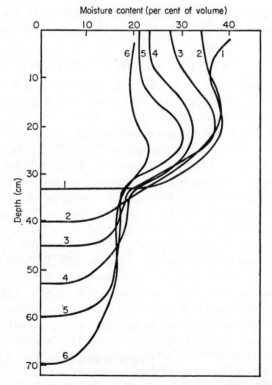

Figure 12.17 Moisture profiles showing stages of re-distribution following an initial infiltration, after Childs.

until an appreciable depth of soil has been wetted from the surface flood. The resulting profile may be of some such form as is shown at stage 1 in Figure (12.17). At this moment the irrigation is brought to a halt by the removal of the surface flood. The moisture profile is not instantaneously changed and the potential gradients must for a brief time be unaltered, so that the water movement and the penetration of the profile must continue. Now, however, since there is no longer infiltration, the continuation of the wetting of deeper dry soil must be at the expense of the surface saturated soil, which must from this time on decrease in moisture content. This change of trend from moistening to drying in the surface layers introduces the complications of hysteresis, and the quantitative theory of redistribution has for this reason not yet reached a presentable stage. Certain qualitative and approximate approaches may, however, be described.

The equation of water movement which governs the development of the moisture profile, namely Equation (11.10), is

$$v_z = -K - D(dc/dz) \tag{12.71}$$

over that part of the profile where wetting is maintained, while over the upper part where wetting has changed to drying, the appropriate equation is Equation (11.12), namely,

$$v_z = -K - D(dc/dz) - K(\partial H/\partial c_r)(dc_r/dz) \tag{12.72}$$

where $c_r$ is the moisture content which prevailed at the time that wetting changed to drying. As has been noted, the former of these equations provides for appreciable water movement even when dryness of soil dictates low values of $K$ and $D$ because the moisture gradient at the wetting front, where wet soil is immediately adjacent to dry soil, can be very steep. However, in the draining part of the profile, neither $(dc/dz)$ nor $(dc_r/dz)$ can be very steep, since there is no possibility of dry soil being adjacent to wet soil. Hence as the moisture content here decreases with a consequent sharp decrease of $K$, $D$, and $K(dH/dc_r)$, the velocity of flow $v_z$ must also decrease steadily. Thus the moisture content of the upper zone decreases continuously, but at an ever decreasing rate, while the further penetration of the wetting profile similarly persists, but at an ever decreasing rate.

The manner in which the shape of the moisture profile develops during redistribution has been the subject of considerable conflict of experimental evidence. As a result of laboratory observations on the distribution of water at various times, following the placing together of soil columns at different moisture contents, Alway and Clark (1911) obtained curves of the type sketched in Figure (12.18), while experiments of a somewhat similar kind carried out by Veihmeyer (1927) produced redistribution curves such as are sketched in Figure (12.19). Experiments of this kind are hardly capable of theoretical interpretation, since a soil block at a uniform initial moisture content, obtained by mixing increasing amounts of water into dry soil until the desired content is obtained, results in an essentially unknown but certainly non-uniform initial state of soil water. The added water overwets those parts of the soil which receive it first and it is then redistributed to the drier parts by the mixing process, so that even in the stage of preparation hysteresis is called into play. The resulting state is unspecifiable. In contrast to this Shaw (1927) studied the redistribution profile following infiltration and presented curves such as are sketched in Figure (12.20), but in similar experiments with sand columns Childs, in unpublished work, obtained curves such as are presented as the redistribution phases in Figure (12.17). While curves such as those of Shaw have

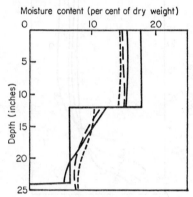

Figure 12.18 As Figure (12.17), but after Alway and Clark.

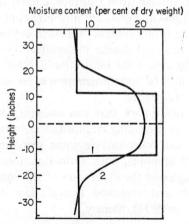

Figure 12.19 As Figure (12.17), but after Viehmeyer.

been observed also in natural profiles, insofar as such non-uniform profiles are capable of presenting typical redistribution profiles, those shown in Figure (12.17) have been confirmed by Youngs (1958) and, in some cases, by Biswas, Nielsen and Biggar (1966). The latter present also some profiles similar to Figure (12.20). Vachaud's profiles (1966) are somewhat similar to those of Figure (12.17). The characteristic features of this type of redistribution are (a) that the water front at the conclusion of the infiltration stage persists recognizably through the subsequent redistribution stages without appreciable change of location; (b) that further penetration of the

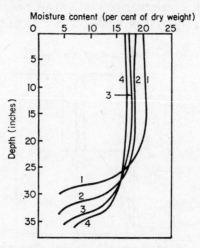

Figure 12.20 As Figure (12.17), but after Shaw.

deeper soil proceeds as though from a surface, namely the infiltration water front, which is maintained at a constant moisture content less than saturation; and (c) that this moisture content of further wetting depends upon the depth of penetration achieved during the infiltration stage, increasing as this depth increases. In fact, if the infiltration depth is less than a characteristic amount, there may be no redistribution at all, the infiltration profile remaining indefinitely suspended.

It is evident that any theory that can account for all these various profiles according to the prevailing circumstances, which have themselves hardly yet been clarified, must survive some searching tests. Up to the present time only two attempts, both approximations, can be described.

Youngs (1958) abandoned the expression of the equation of continuity in the diffusion form and exploited Darcy's law in the form of the third of the equations of (9.11), namely,

$$v_z = -K[1+(\mathrm{d}H/\mathrm{d}z)]\tag{12.73}$$

It is evident that there must be some point on the moisture profile at the cessation of infiltration which separates the lower part, which continues to get wetter, from the upper part, which from this moment begins to get drier. This is called in this context the transition point and may be indicated by the symbol $c_m$. At profile points deeper than this, there is no hysteresis complication and the modified Fick's law equation, Equation (12.71), applies. At points between the surface and the transition point, Equation (12.73) remains true, but not Equation (12.71).

Next, arguments were set out to show that, in the stages of redistribution immediately following the cessation of infiltration, the only part of the profile to lose water is the near-surface layer. These arguments are quali-tative in nature and it is perhaps more satisfactory to accept the experi-mental evidence that this is so. Two consequences follow from this supposi-tion. The first is that the pressure at the surface falls almost instantaneously from zero to the known air entry pressure $H_f$ as the redistribution phase begins; and the second is that the rate of flow of water is constant from very near the surface down to the transition point, since between these two points there is neither storage nor loss of soil water, which must therefore be passed on from horizon to horizon without change.

Hence at the moment of transition from infiltration to redistribution, one may write

$$v_m = -[K + D(\mathrm{d}c/\mathrm{d}z)]_m$$

where the subscript $m$ indicates that all the quantities so marked are those measured at the transition point $m$. Furthermore this value $v_m$ is the con-stant velocity of flow down the upper part of the moisture profile. Hence Equation (12.73) may be written:

$$[K + D(\mathrm{d}c/\mathrm{d}z)]_m = K[1 + (\mathrm{d}H/\mathrm{d}z)]$$

The left-hand side is constant and $K$ is known as a function of $z$, since the infiltration profile is known and $K$ is a known function of $c$. Hence this equation may be integrated simply to give the result,

$$[K + D(\mathrm{d}c/\mathrm{d}z)]_m \int_{z_m}^{0} \mathrm{d}z/K = \int_{z_m}^{0} \mathrm{d}z + \int_{H_m}^{H_f} \mathrm{d}H$$

or

$$[K + D(\mathrm{d}c/\mathrm{d}z)]_m \int_{z_m}^{0} \mathrm{d}z/K = -z_m + H_f - H_m \qquad (12.74)$$

The soil surface is taken as the origin of $z$.

The procedure is to take a trial value of $z_m$, and from this the value of $c_m$ is read off from the infiltration profile as also is the slope $(\mathrm{d}c/\mathrm{d}z)_m$. From $c_m$ the value of $H_m$ is read off from the boundary wetting moisture characteristic, and the values of $K_m$ and $D_m$ are read off from the curves of these properties as functions of moisture content. The value of $H_f$ is obtained from the boundary drying moisture characteristic. By plotting $(1/K)$ as a function of $z$, as described above, the integral in the left-hand side of Equation (12.74) may be computed graphically or numerically as described in Section 12.2. Hence the left-hand and right-hand sides of Equation (12.74) may be calculated independently, and if $z_m$ has been guessed correctly, the two will agree. By making such comparisons for various systematically chosen values of $z_m$, the correct value will emerge as that which satisfies Equation (12.74).

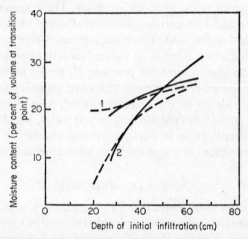

Figure 12.21 The dependence of the moisture content at the transition point of the moisture profile during re-distribution, as a function of the depth of the initial infiltration. Curves 1 are for slate dust and curves 2 are for grade 15 Ballotini. Solid lines are experimental curves, broken lines are calculated.

In this way it was shown that the transition point can be calculated with fair agreement with observed values, although the form of the redistribution profile cannot itself be deduced. Moreover, as shown in Figure (12.21), the way in which the moisture content at the transition point depends upon the depth of initial infiltration is in fair agreement with observation.

More recently, Staple (1966) has programmed a digital computer to integrate the equation of continuity, in the form of Equation (11.30), in the finite difference form, for the case of redistribution in which hysteresis must be taken into account. This form avoids the invalid procedure of transforming the conductivity into a diffusivity. The moisture content at which successive depths change from wetting to drying emerges during the computation and the appropriate scanning curve is interpolated in the programme from a series of such curves stored in the computer. The resulting redistribution moisture profiles are strongly reminiscent of those of Shaw (1927), whereas an essentially similar procedure carried out by Rubin (1967) produced moisture profiles confirming those of Childs (unpublished work) and of Youngs (1958). In neither case was there reported any attempt to carry out experiments to compare observed with computed profiles. The conflict between theoretically derived profiles, which repeats the discrepancies between observed profiles, indicates that much remains to be done to elucidate this matter.

Whatever may be the detailed progress of the stages of the redistribution moisture profile, it is evident that the reduction of the moisture content of the upper initially wetted zone must eventually result in a more or less uniform profile at a moisture content which is low enough to make the rate of further change of the profile almost imperceptible. Although a true equilibrium is not achieved and some movement is always to be found, for many practical purposes the moisture profile attained at the end of a few days after the cessation of infiltration may be regarded as sensibly stationary, and the moisture content at this stage is commonly called the field capacity. It is usually defined in some such necessarily loose terms as being the soil moisture content after infiltration when drainage has ceased, it being assumed that there is no groundwater to complicate the profile.

## 12.11 The concept of specific yield

Hydrologists are accustomed to define a property, characteristic of an aquifer, called the specific yield $Y$. This is defined as the volume of water given up by or extracted from the groundwater per unit area of water table, when the said water table is lowered by unit distance. While this concept has undoubted utility in the case of long term fluctuations of the water table, it is misleading, as has been shown by Childs (1960a), when the fluctuations are relatively rapid or when the water table is very near the surface, for then the specific yield may vary over a wide range. Such rapid fluctuations may take place during intermittent rainfall on drained land or during well pumping tests, such as those of Theis (1935), which are particularly designed to provide an estimate of the specific yield.

The analysis given here follows Gonçalvez dos Santos Jr. (1967). Let the moisture profile be described by expressing the moisture content $c$ as a function of the height $z$ measured always from the water table, wherever that may be. The height $Z$ of the water table itself is measured from a fixed datum level. Thus

$$c = c(z)$$

and is an unchanging function of $z$ for so long as the profile is of unchanging shape relative to the water table, even though it may be rising or falling with the water table itself.

A selected zone of the moisture profile may be specified by specifying the upper and lower boundaries as measured from the water table. Let the upper boundary be at $u$ and the lower at $l$ as measured from the water table at a given moment. The volume of water stored between these boundaries per unit cross-section of a vertical column is $S$ where

$$S = \int_l^u c \, dz \qquad (12.75)$$

That is to say, $S$ is a function of the moisture content $c$ and of each of the limits $l$ and $u$, and each of these three variables may be allowed to vary independently. Thus,

$$S = S(c, l, u)$$

Thus when the three variables change as time passes, $S$ may vary with time because $c$ varies with time, because $l$ varies with time, or because $u$ varies with time, in each case the other two variables not mentioned being held constant. When all three variables are allowed to change, the total change of $S$ is the sum of the changes due to alteration of the variables singly. Thus one may write,

$$\left(\frac{dS}{dl}\right)_{c,u} = \left(\partial\left[\int_l^u c \, dz\right]/\partial l\right)_{c,u} = -c_l$$

The subscripts $c,u$ indicate the variables which are being held constant during the differentation with respect to $l$ while the subscript $l$ indicates that the moisture content $c$ is that at the level $l$. If $l$ is changing with time $t$, then the change of $S$ may be related to the change of time in the form,

$$\left(\frac{dS}{dt}\right)_{c,u} = \frac{\partial S}{\partial l}\frac{dl}{dt} = -c_l\frac{dl}{dt} \qquad (12.76)$$

Similarly,

$$\left(\frac{dS}{dt}\right)_{c,l} = \frac{\partial S}{\partial u}\frac{du}{dt} = c_u\frac{du}{dt} \qquad (12.77)$$

Also,

$$\left(\frac{dS}{dt}\right)_{l,u} = \left\{\partial\left[\int_l^u c \, dz\right]/\partial t\right\}_{l,u} = \left[\int_l^u (\partial c/\partial t)dz\right]_{l,u} \qquad (12.78)$$

Then one may write,

$$\frac{dS}{dt} = \frac{dS}{dt_{l,u}} + \frac{dS}{dt_{c,u}} + \frac{dS}{dt_{c,l}}$$

or, from Equations (12.76) and (12.77),

$$\frac{dS}{dt} = \frac{dS}{dt_{l,u}} - c_l \frac{dl}{dt} + c_u \frac{du}{dt} \tag{12.79}$$

Thus in what follows $\frac{dS}{dt_{l,u}}$ , as defined in Equation (12.78), indicates the component of the rate of change of stored water between $u$ and $l$ due to the change of shape of the moisture profile as measured from the water table, the remainder, constituting the second two terms of the right-hand side of Equation (12.79), being due to the change of the boundaries.

If the boundaries of the selected portion of the profile are fixed soil levels, then both $l$ and $u$ change only as a consequence of the movement of the water table, increasing together as the water table falls away from the section, and decreasing together as the water table rises towards the section. Thus,

$$\frac{dl}{dt} = \frac{du}{dt} = -\frac{dZ}{dt}$$

Substitution of these values in Equation (12.79) yields

$$\frac{dS}{dt} = \frac{dS}{dt_{l,u}} + (c_l - c_u)\frac{dZ}{dt} \tag{12.80}$$

In particular, when the section comprises the whole of the moisture profile from the level momentarily occupied by the water table to the surface,

$$\frac{dS}{dt} = \frac{dS}{dt_{0,u}} + (c_0 - c_u)\frac{dZ}{dt} \tag{12.81}$$

where $c_0$ is the saturation water content at the water table and $c_u$ is now the moisture content at the surface.

The rate of storage must be equal to the difference between the rate of flow, $v_0$, of water at the level momentarily occupied by the water table and the rate $v_u$ at the surface, i.e.

$$\frac{dS}{dt} = v_0 - v_u \tag{12.82}$$

so that, from Equations (12.81) and (12.82),

$$v_0 - v_u = \frac{dS}{dt_{0,u}} + (c_0 - c_u)\frac{dZ}{dt} \tag{12.83}$$

In the circumstances of the definition of the specific yield, the water table s falling at the rate, say $\frac{dF}{dt}$, equal to $-\frac{dZ}{dt}$, groundwater is being lost so that the rate of downward flow is, say, $q$ where $q$ is identical with $-v_0$, and it may be reasonably supposed that the rate of loss by evaporation from the unsaturated surface, $v_u$, is negligibly small. Then by the definition of the specific yield $Y$,

$$Y\frac{dF}{dt} = q$$

or, substituting for $\dfrac{dF}{dt}$ and $q$ from the above identities,

$$Y\frac{dZ}{dt} = v_0 \tag{12.84}$$

A comparison of Equation (12.84) with Equation (12.83) shows that, with zero $v_u$,

$$Y = c_0 - c_u + \left(\frac{dS}{dt_{0,u}}\right)\Big/\left(\frac{dZ}{dt}\right)$$

$$= c_0 - c_u - \left(\frac{dS}{dt_{0,u}}\right)\Big/\left(\frac{dF}{dt}\right) \tag{12.85}$$

One can now discuss the possible variations of $Y$. Firstly, if the moisture profile is falling without change of shape, the factor $\dfrac{dS}{dt_{0,u}}$ vanishes and the specific yield takes the value,

$$Y = c_0 - c_u$$

One such circumstance in which this can occur, illustrated in Figure (12.22), arises when the water table is deep and is falling fairly slowly, so that the upper part of the profile may be taken to have the constant and uniform moisture content which is recognized as the field capacity, $c_{fc}$. Then the specific yield is

$$Y = c_0 - c_{fc} \tag{12.86}$$

and this is a constant value, characteristic of the soil type. This is the value understood by the hydrologist, and which is too often supposed to be the unique value.

Another circumstance in which the moisture profile may fall without change of shape arises when the water table is so near the surface that the surface soil itself is saturated, the suction being less than the air entry value. This is illustrated in Figure (12.23). Again $\dfrac{dS}{dt_{0,u}}$ vanishes, but in

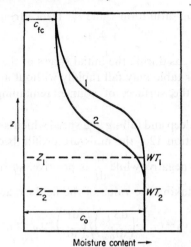

Figure 12.22 A moisture profile falling without change of shape at the same rate as the water table. The water table is deep and the upper part of the profile is at field capacity. $WT_1$ is the water table, at height $Z_1$, corresponding to profile 1, and $WT_2$, $Z_2$ corresponding to subsequent profile 2.

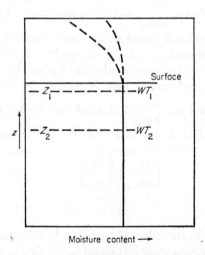

Figure 12.23 As Figure (12.22), but the water table is shallow and the profile is everywhere saturated. $WT$, $Z$ and the numerals 1 and 2 have the same significance as in Figure (12.22).

this case $c_u$ also has the saturation value $c_0$. Hence from Equation (12.85),

$$Y = 0$$

In such circumstances, as during the initial stages of the drainage of water-logged land, the water table may fall rapidly without any reduction of the groundwater at all, the surface, of course, remaining saturated under increasing suctions.

If the water table is deep and falling at a speed which is steadily decreasing, then as shown in Section 12.4 the moisture profile becomes steadily less extended. Thus $\dfrac{dS}{dt_{0,u}}$ s negative while $\dfrac{dF}{dt}$ is positive, so that from Equation (12.85), with $c$ again taking the value of the field capacity $c_{fc}$,

$$Y = c_0 - c_{fc} - \left(\frac{dS}{dt_{0,u}}\right) \bigg/ \left(\frac{dF}{dt}\right) > c_0 - c_{fc}$$

This case is illustrated in Figure (12.24).

In the next case, suppose that because of intense rainfall the moisture profile above the water table tends asymptotically to a very high value, little short of saturation, at the surface, as described in Section 12.2 and shown in Figure (12.25). On cessation of rainfall the moisture profile will at once begin to subside to the equilibrium form, or to field capacity, whichever is the greater.

During the subsidence, $\dfrac{dS}{dt_{0,u}}$ will be at first very strongly negative but will decrease in magnitude as the steady shape profile is approached. The surface moisture content $c_u$ will in the early stages be high, possibly approaching $c_0$. Hence the contributions to the specific yield as indicated in Equation (12.85) will vary in their dominance as time passes, and the contribution governed by $\dfrac{dS}{dt_{0,u}}$ will depend on whether the water table is being closely controlled. If drainage control is perfect and the water table unvarying, then

$$\left(\frac{dS}{dt_{0,u}}\right) \bigg/ \left(\frac{dZ}{dt}\right) \to \infty$$

and

$$Y \to \infty$$

no matter what may be the values of $c_0$ and $c_u$. In this case water is removed from the stock of groundwater just equal to the rate at which it arrives at the water table due to the collapse of the moisture profile, without change of water table level. The infinitely great value of the specific yield follows from the elementary definition, Equation (12.85).

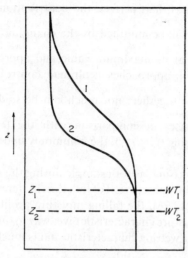

Moisture content ⟶

Figure 12.24 Moisture profile stages above a water table which is falling at decreasing speeds. *WT*, *Z* and the numerals 1 and 2 have the same significance as in Figure (12.22).

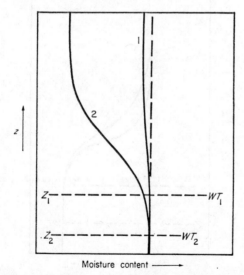

Moisture content ⟶

Figure 12.25 Moisture profile stages above a water table which is falling due to the cessation of intense infiltration. The profile collapses as it descends. *WT*, *Z* and the numerals 1 and 2 have the same significance as in Figure (12.22).

If the drainage is less than perfect, the contribution made by the term which contains $\dfrac{dS}{dt_{0,u}}$ will be modified by the changing value of $\dfrac{dZ}{dt}$, which will in general begin at its maximum value and approach zero asymptotically as the water table approaches its ultimate controlled level. Thus both $\dfrac{dS}{dt_{0,u}}$ and $\dfrac{dZ}{dt}$ decrease together; not much can be said at this stage as to how the ratio decreases in comparison with the term $(c_0 - c_u)$ which increases up to the value $(c_0 - c_{fc})$, the commonly understood value of the specific yield.

If the water table is rising, as for example during the recovery period of a well pumping test, following a period of fall, then hysteresis comes into effect. As shown in Figure (12.26), the falling moisture profile which reflects the form of the drying moisture characteristic gives way to the rising profile which reflects the form of the wetting characteristic and is much more compressed. In Equation (12.85), the factor $\dfrac{dS}{dt_{0,u}}$ is rendered negative by the increasing compression of the profile in the early stages, while $\dfrac{dZ}{dt}$ is now negative.

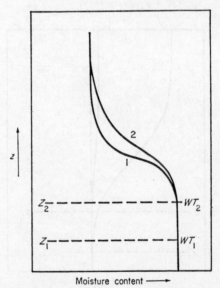

Moisture content ⟶

Figure 12.26 Moisture profile stages above a water table which has changed its direction of motion from downwards to upwards. The difference of shape is due to hysteresis. *WT*, *Z* and the numerals 1 and 2 have the same significance as in Figure (12.22).

Hence from Equation (12.85),

$$Y < c_0 - c_{fc}$$

Enough has been said to indicate the wide range of variation of which $Y$ is capable. This has been confirmed experimentally by Youngs and Smiles (1963) during well pumping tests in a sand tank, using the analysis due to Theis (1935).

Although not very apposite to the present discussion, it may be mentioned here that the concept of specific yield is applicable also to confined aquifers, where the pressure head is measured by the height of water standing in a bore hole which penetrates through the confining stratum. A reduction of pressure permits the confining stratum to compress the elastic aquifer slightly, thereby yielding water. The yield per unit change of pressure head is in this case the specific yield, and from the nature of the phenomenon the value is commonly very small. A typical specific yield of free ground-water might be about ten per cent, while the specific yield of a confined aquifer might well be from a tenth of this value down to negligibly small quantities.

## NOTES

### Note 27. The equation for the moisture profile above a moving water table

The form of the equation of continuity best suited to this case is that of Equation (11.34), namely,

$$-\partial z/\partial t = \partial[K + K(\partial H/\partial z)]/\partial c \qquad \text{(N27.1)}$$

Since it is assumed that sufficient time has elapsed for the attainment of a state such that the moisture profile moves without change of shape at the same rate, $V$, as the water table itself, $(\partial z/\partial t)$ does not in this case vary either with $c$ or with time, but is constant and equal to $V$ everywhere. Hence,

$$-V = \partial[K + K(\partial H/\partial z)]/\partial c \qquad \text{(N27.2)}$$

This equation may be integrated directly between the limits $c$ and $c_u$, where the latter is the sensibly uniform moisture content at the upper end of the profile, endowing the soil in that region with conductivity $K_u$. The result is

$$-V(c - c_u) = K - K_u + K(\mathrm{d}H/\mathrm{d}z) - K_u(\mathrm{d}H/\mathrm{d}z)_u$$

or

$$\mathrm{d}z/\mathrm{d}H = 1/[(K_u/K)\{1 + (\mathrm{d}H/\mathrm{d}z)_u\} - 1 - (V/K)(c - c_u)]$$

This may be integrated between the limits of the water table, where both $z$ and $H$ have the value zero, and any chosen value of $H$ which is to be found at the height $z$. The integrated form is

$$z = \int_0^H \mathrm{d}H/[(K_u/K)\{1 + (\mathrm{d}H/\mathrm{d}z)_u\} - 1 - (V/K)(c - c_u)] \qquad \begin{cases} \text{(N27.3)} \\ \text{(12.11)} \end{cases}$$

**Note 28. The shape of the ultimate infiltration moisture profile**

The moisture content at the surface of the profile is $C$ while that in the unaffected soil yet to be reached by the water front is $c_0$. Hence, if the profile is descending without change of shape at a velocity $V$, then each unit of time sees a distance $V$ at moisture content $c_0$ eliminated from the soil below the water front and an equal length at the greater moisture content $C$ added at the top. Hence if $S$ denotes the water stored in the profile, the rate of increase, $dS/dt$ is given by

$$dS/dt = V(C - c_0) \tag{N28.1}$$

This rate of storage may be expressed on a different basis. In a section of the profile extending from the surface to some distance below the water front, the moisture gradient is vanishingly small at both ends, since the unaffected soil is at uniform moisture content $c_0$ and the top of the profile, as has been demonstrated, and has had sufficient time to achieve a finite depth at uniform moisture content $C$. Hence at each end, the rate of flow, in accordance with Darcy's law, is given by Equation (12.41) for the appropriate uniform moisture content. Thus the rate of infiltration is $K_C$, while the rate of outflow at the bottom is $K_0$. Hence, the rate of storage in the profile as it advances is

$$dS/dt = K_C - K_0 \tag{N28.2}$$

Combining this with Equation (N28.1), one arrives at a value of $V$ given by

$$V = (K_C - K_0)/(C - c_0) \tag{N28.3}$$

The equation of continuity for this case, Equation (12.49), now takes the form,

$$d[D(dc/dz) + K]/dc = (K_C - K_0)/(C - c_0) \tag{N28.4}$$

The right-hand side contains only parameters of constant value for the particular case. Further, it is known that the moisture gradient $dc/dz$ vanished at both ends of the profile. Hence, the integration is elementary, and may be carried out between the limits $c_0$ and a chosen $c$ or between $C$ and $c$ at will. Either way the result is

$$D(dc/dz) = [K_C(c - c_0) + K_0(C - c) + K(c_0 - C)]/(C - c_0) \tag{N28.5}$$

Now let the height of the water front relative to an arbitrarily chosen datum level be $z_f$, at which point the moisture content is $c_0$ since the front has only just reached it. Integration of Equation (N28.5) between the limits $z_f$ and $z$, where the moisture content is $c$, results in the equation,

$$z - z_f = (C - c_0) \int_{c_0}^{c} D \, dc / [K_C(c - c_0) + K_0(C - c) + K(c_0 - C)] \quad \left\{ \begin{matrix} \text{(N28.6)} \\ \text{(12.50)} \end{matrix} \right.$$

**Note 29. The determination of the coefficients of Equation (12.51)**

At a sufficiently great depth the moisture content at a given moment is undisturbed at the value $c_0$, the corresponding hydraulic conductivity is $K_0$, and the moisture gradient $\partial c/\partial z$ vanishes. Hence, Equation (12.48) may be integrated between the limits $c_0$ and $c$ with the result,

$$-\int_{c_0}^{c} (\partial z/\partial t) dc = D(\partial c/\partial z) + (K - K_0) \tag{N29.1}$$

Equation (N29.1) is subject also to the condition that at the surface, where $z$ may arbitrarily be assigned the value zero, the moisture content is $C$ at all times. Substituting for $(\partial z/\partial t)$ from Equation (12.51), one has

$$-\int_{c_0}^{c}(\partial z/\partial t)\mathrm{d}c = \tfrac{1}{2}t^{-\frac{1}{2}}\int_{c_0}^{c}\lambda\ \mathrm{d}c + \int_{c_0}^{c}\mu\ \mathrm{d}c + 1\tfrac{1}{2}t^{\frac{1}{2}}\int_{c_0}^{c}v\ \mathrm{d}c + 2t\int_{c_0}^{c}\xi\ \mathrm{d}c + \ldots \quad \text{(N29.2)}$$

It follows from Equation (12.51), the assumed solution, that

$$-\partial z/\partial c = t^{\frac{1}{2}}(\mathrm{d}\lambda/\mathrm{d}c) + t(\mathrm{d}\mu/\mathrm{d}c) + t^{1\frac{1}{2}}(\mathrm{d}v/\mathrm{d}c) + t^{2}(\mathrm{d}\xi/\mathrm{d}c) + \ldots \quad \text{(N29.3)}$$

Inverting Equation (N29.3) one has, by long division

$$-\partial c/\partial z = t^{-\frac{1}{2}}/(\mathrm{d}\lambda/\mathrm{d}c) - (\mathrm{d}\mu/\mathrm{d}c)/(\mathrm{d}\lambda/\mathrm{d}c)^{2} + t^{\frac{1}{2}}[(\mathrm{d}\mu/\mathrm{d}c)^{2} +$$
$$- (\mathrm{d}v/\mathrm{d}c)(\mathrm{d}\lambda/\mathrm{d}c)]/(\mathrm{d}\lambda/\mathrm{d}c)^{3} - t[(\mathrm{d}\mu/\mathrm{d}c)^{3} +$$
$$- 2(\mathrm{d}\lambda/\mathrm{d}c)(\mathrm{d}\mu/\mathrm{d}c)(\mathrm{d}v/\mathrm{d}c) + (\mathrm{d}\lambda/\mathrm{d}c)^{2}(\mathrm{d}\xi/\mathrm{d}c)]/(\mathrm{d}\lambda/\mathrm{d}c)^{4} +$$
$$+ \ldots \quad \text{(N29.4)}$$

Hence Equation (N29.1) may be rewritten, by using this substitution for $(\mathrm{d}c/\mathrm{d}z)$, as

$$-\int_{c_0}^{c}(\partial z/\partial t)\mathrm{d}c = -Dt^{-\frac{1}{2}}/(\mathrm{d}\lambda/\mathrm{d}c) + [(K-K_0) + D(\mathrm{d}\mu/\mathrm{d}c)/(\mathrm{d}\lambda/\mathrm{d}c)^{2}] +$$
$$- Dt^{\frac{1}{2}}[(\mathrm{d}\mu/\mathrm{d}c)^{2} - (\mathrm{d}v/\mathrm{d}c)(\mathrm{d}\lambda/\mathrm{d}c)]/(\mathrm{d}\lambda/\mathrm{d}c)^{3} +$$
$$+ Dt[(\mathrm{d}\mu/\mathrm{d}c)^{3} - 2(\mathrm{d}\lambda/\mathrm{d}c)(\mathrm{d}\mu/\mathrm{d}c)(\mathrm{d}v/\mathrm{d}c) +$$
$$+ (\mathrm{d}\lambda/\mathrm{d}c)^{2}(\mathrm{d}\xi/\mathrm{d}c)]/(\mathrm{d}\lambda/\mathrm{d}c)^{4} + \ldots \quad \text{(N29.5)}$$

Equating coefficients of corresponding terms in Equations (N29.2) and (N29.5), one has

$$\tfrac{1}{2}\int_{c_0}^{c}\lambda\ \mathrm{d}c = -D(\mathrm{d}c/\mathrm{d}\lambda) \quad \text{(N29.6)}$$

$$\int_{c_0}^{c}\mu\ \mathrm{d}c = (K-K_0) + D(\mathrm{d}c/\mathrm{d}\lambda)^{2}(\mathrm{d}\mu/\mathrm{d}c) \quad \text{(N29.7)}$$

$$(1\tfrac{1}{2})\int_{c_0}^{c}v\ \mathrm{d}c = D[(\mathrm{d}c/\mathrm{d}\lambda)^{2}(\mathrm{d}v/\mathrm{d}c) - (\mathrm{d}c/\mathrm{d}\lambda)^{3}(\mathrm{d}\mu/\mathrm{d}c)^{2}] \quad \text{(N29.8)}$$

$$2\int_{c_0}^{c}\xi\ \mathrm{d}c = D[(\mathrm{d}c/\mathrm{d}\lambda)^{2}(\mathrm{d}\xi/\mathrm{d}c) - 2(\mathrm{d}c/\mathrm{d}\lambda)^{3}(\mathrm{d}\mu/\mathrm{d}c)(\mathrm{d}v/\mathrm{d}c) +$$
$$+ (\mathrm{d}c/\mathrm{d}\lambda)^{4}(\mathrm{d}\mu/\mathrm{d}c)^{3}] \quad \text{(N29.9)}$$

and so on for further unspecified coefficients.

Equation (N29.6) is identical in form with Equation (12.40), differing only in the symbols used, and the procedure for solving it to obtain a plot of $\lambda$ as a function of $c$ is the same as is described in the last paragraph of Section 12.6. As in that case, the prior requirement is a knowledge of $D$ as a function of $c$. Thus Equation (N29.6) is an auxilliary equation for computing the coefficient $\lambda$.

When $\lambda$ has been determined in this way it follows that $(\mathrm{d}c/\mathrm{d}\lambda)$ is also a known function of $c$ for the purpose of solving Equation (N29.7), so that the product $D(dc/d\lambda)^2$ may be written as the single quantity $P$, which is now a known function of $c$. Equation (N29.7) then becomes

$$\int_{c_0}^{c} \mu \, dc = P(d\mu/dc) + (K - K_0) \tag{N29.10}$$

A solution of this equation in the form of a plot of $\mu$ as a function of $c$ carries with it a knowledge of $(d\mu/dc)$ as a function of $c$, so that the product $D(dc/d\lambda)^3(d\mu/dc)^2$ is now a known function of $c$ alone and may be expressed by the symbol $Q$. Hence Equation (N29.8) may be written:

$$(1\tfrac{1}{2}) \int_{c_0}^{c} v \, dc = P(dv/dc) - Q \tag{N29.11}$$

Again, when this equation is solved to give $v$ and therefore $(dv/dc)$ as functions of $c$, then the product $D(dc/d\lambda)^3(d\mu/dc)[2(dv/dc) - (d\mu/dc)^2(dc/d\lambda)]$ is at this stage also known as a function of $c$, and is indicated hereafter by the symbol $R$. Hence Equation (N29.9) becomes

$$2 \int_{c_0}^{c} \xi \, dc = P(d\xi/dc) - R \tag{N29.12}$$

and so on. Equation (N29.10), (N29.11) and (N29.12) are auxilliary equations for determining the coefficients, $\mu$, $v$ and $\xi$, respectively.

The procedure for solving these equations is the same for all since they are all of the same general form, namely,

$$\int_{c_0}^{c} \varepsilon \, dc = U(d\varepsilon/dc) - W \tag{N29.13}$$

where $\varepsilon$ represents one or other of the various coefficients and $U$ and $W$ are the appropriate functions of $c$ indicated in the corresponding auxilliary equation. At all times the moisture content is known to be $C$ at the surface, where $z$ has the value zero, so that from Equation (12.49) each of the coefficients, $\lambda$, $\mu$, $v$, $\xi$, and so on, must also have the value zero when the moisture content has the value $C$, in order that this condition for $z$ may be satisfied. Hence, the general boundary condition for Equation (N29.13) is that $\varepsilon$ has the value zero when $c$ has the value $C$.

The procedure for integrating Equation (N29.13) follows in general that for determining $\lambda$ (or $\chi$) from Equation (12.40) as outlined in Section 12.6. One begins with the knowledge that the value of $\varepsilon$ for a moisture content $C$ is zero, that is to say, with the position of one extreme point on the curve of $\varepsilon$ versus $c$. The area under the curve of $\varepsilon$ versus $c$ within the positive quadrant of the axes is $\int_{c_0}^{C} \varepsilon \, dc$, and is not at first known, but one makes a guess at the value.

Then since $U$ and $W$ are known at the moisture content $C$, one can deduce from Equation (N29.13) the value of $(d\varepsilon/dc)$, and hence the value of $\varepsilon$ at a moisture content $C - \Delta c$, where the step $\Delta c$ is suitably small but chosen at will. Thus a second point on the curve is obtained as is also the area under the curve between the limits $C$ and $C-\Delta c$, so that the remaining area $\int_{c_0}^{C-\Delta c} \varepsilon \, dc$, is found by subtraction from the initially guessed area. Again the new values of $U$ and $W$ at the

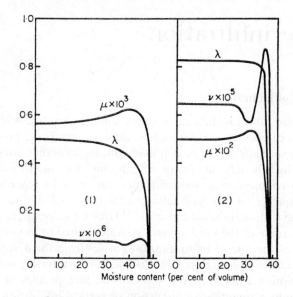

Figure N29.1 The coefficients $\lambda$, $\mu$ and $\nu$, of Philip's equation for the descending moisture profile, as functions of moisture content, in (1) slate dust and (2) grade 15 Ballotini.

moisture content $C-\Delta c$ are known so that the value of $(d\varepsilon/dc)$ at this moisture content may now be found from Equation (N29.13). Repetition of this process produces the delineation of the whole curve of $\varepsilon$ versus $c$ if the initial guess was correct. The test of this correctness is that in the last stage the subtraction of the last element of area should reduce the residual area, i.e. $\int_{c_0}^{c_0} \varepsilon \, dc$ to zero, and if this is not so, then a new start must be made with another initial guess. Much reiteration is not required, since the linearity of the equation enables one to interpolate or extrapolate from any two initial erroneous guesses in order to arrive at the correct one. For full details, including the derivation of the working finite difference equation from Equation (N29.13), the original paper should be read. Some examples of functions, such as $\mu$ and $\nu$, are shown in Figure (N29.1).

# CHAPTER 13

# Surface infiltration

## 13.1 The rate of infiltration

THE rate of infiltration is the rate at which water crosses the soil surface, insofar as this can be defined and located, and thus enters the profile from outside. It is especially not to be confused with the hydraulic conductivity of the soil, which is only one factor entering into the rate of infiltration. According to circumstances, the infiltration rate may be regarded either as the consequence of the hydraulic conductivity and of the potential gradient at the surface, in accordance with Darcy's law, or alternatively as the rate of increase of the total amount of water stored in the soil profile. A discussion of the rate of infiltration thus arises from the description of the development of moisture profiles presented in Chapter 12, and insofar as that description dealt with the vertical moisture profiles in uniform soil profiles only, a quantitative discussion of vertical infiltration must be similarly limited. Qualitative inferences as regards heterogeneous profiles may be possible subsequently.

It was shown in Section 12.7(b) and illustrated in Figure (12.11) that the sudden surface flooding of an initially relatively dry homogeneous soil profile results in an initially very great rate of infiltration, due to a very steep gradient of moisture content acting in a thin surface layer with the high conductivity and diffusivity of saturation, and that this rate steadily declines as the development of the moisture profile reduces the gradient of moisture content at the surface. When this gradient at last becomes vanishingly small, the rate of infiltration settles down to the constant rate which, in accordance with Equation (11.10), is simply equal to the saturation conductivity at the surface. If the surface moisture content is raised to and maintained at a constant value less than saturation, then the rate of infiltration settles down to the lower rate equal to the conductivity of the unsaturated surface soil. The saturated conductivity, $K_{sat}$, is the limiting value of the steady infiltration rate.

In fact the discussion of the maintenance of the surface at a constant moisture content other than saturation is highly academic, since even in the laboratory, with its possibilities of close control, such a boundary

274

condition can be maintained only with extreme difficulty, if at all. In nature, if the surface is not saturated, it is invariably the rate of infiltration which is imposed, whether by rainfall or by the rate of melting of snow cover. To be sure, the arrival of a single raindrop at the soil surface must result in momentary saturation, but this must be followed by an interval during which the free water is absorbed and redistributed to a lower moisture content before the arrival of the next drop in the same spot. This redistribution has been discussed inadequately, but as far as is possible, in Section 12.10. For the present this situation may be regarded as one of average surface moisture content less than saturation, determined by the rate of infiltration imposed by the intensity of rainfall. The consequences have been described in Section 12.7(b), and illustrated in Figure (12.13). For so long as the surface moisture content in such circumstances is less than saturation, the rate of infiltration is determined by the rate of precipitation at the surface and the discussion is trivial. However, as soon as the surface reaches saturation, if the rate of rainfall is sufficient to achieve this state, the limiting value of surface conductivity is attained and the infiltration rate is from this moment determined by the surface moisture gradient. There is nothing to distinguish this case from that of surface flooding.

The course of infiltration from a flooded surface as a function of time is usually termed the law of infiltration. The ultimate steady value of the infiltration rate is often referred to as the infiltration capacity. When the words "infiltration rate" are used briefly without elaboration and without a reference to time, the infiltration capacity is usually implied.

A number of different formulas expressing the law of infiltration have been proposed, based on either the analysis of more or less naïve models, on intuition, on frank empiricism or on genuinely ascertained physical properties of porous materials in general and of soil in particular. The aim of all such formulas is to express the rate of infiltration as a function of time elapsed from the inception of surface flooding, and, in particular, to account for the rapid decrease from initially very high values and, for uniform soils, the asymptotic approach to an ultimate constant value. In the treatment that follows here the first concern will be with such homogeneous soil profiles.

## 13.2 The infiltration law of Green and Ampt

Among the earliest of the proposed infiltration laws was that of Green and Ampt (1911). These authors prefaced their treatment with a discussion of the saturated conductivity of soil based on the capillary tube model, on the lines of Section 10.2, but their subsequent analysis is subject to this only to the extent that they suppose, consistently with such a model, that the advancing water front is a precisely defined surface at which the pressure

$H_f$, negative because a suction, is a constant characteristic of the soil. This front separates uniformly saturated soil behind it, of uniform hydraulic conductivity $K_{sat}$, from uniformly unsaturated and as yet uninfluenced soil beyond it. Such a supposition is, in fact, an assumption that the moisture characteristic is a step-shaped curve indicating the reduction of moisture content sharply from saturation to a constant degree of unsaturation at an air entry value $H_f$, and would be satisfied by a granular body in which the pore shapes and sizes were quite uniform.

To avoid the tedious repetition of negative signs as a consequence of the convention of taking the upward direction of $z$ to be positive, the depth to which the water front has penetrated at time $t$ will be indicated by $l$. For generality, it will be supposed also that water stands on the soil surface to a height $H_0$, which is therefore the hydrostatic pressure head at that level. Applying Darcy's law to the saturated column of length $l$ and unit cross-section, for which purpose one may most conveniently refer to the form given in Equation (9.5), one finds that the rate of flow, $dQ/dt$, down the column, is

$$dQ/dt = K_{sat}(H_0 + l - H_f)/l \qquad (13.1)$$

This may be written in the form,

$$dQ/dt = A + B/l \qquad (13.2)$$

where the constants $A$ and $B$ are given respectively by

$$\left. \begin{array}{l} A = K_{sat} \\ B = K_{sat}(H_0 - H_f) \end{array} \right\} \qquad (13.3)$$

Since there can be no storage of water in the saturated column other than is due to its increase of length, $dQ/dt$ must have the same value everywhere and in particular is the rate of infiltration.

Equation (13.2) evidently satisfies both of the requirements that $dQ/dt$ should be large when $l$ is small, at the onset of infiltration, and that it should tend to a constant value $K_{sat}$ after a long time when $l$ has become large. However, since $H_f$ is not a precisely definable constant for real soils, the formula does not give $dQ/dt$ in absolute terms. Insofar as the constants $A$ and $B$ of Equation (13.2) may be determined, in practice, by observations of the rate of infiltration at two known depths of penetration of the profile, to that extent the formula is to be regarded as empirical.

The total quantity of water, $Q$, penetrating in time $t$ from the initiation of flooding may be computed by integrating Equation (13.2) after an appropriate substitution for $dQ/dt$. If $\Delta f$ is the difference between the water-filled porosity before and after the passage of the water front, then

$$dQ/dt = \Delta f(dl/dt)$$

and Equation (13.2) may be written:

$$\Delta f(\mathrm{d}l/\mathrm{d}t) = (lA+B)/l \qquad (13.4)$$

At the onset of flooding both $t$ and $l$ have the value zero, hence, as shown in Note 30 at any subsequent stage,

$$t/\Delta f = [l-(B/A)\ln\{(lA+B)/B\}]/A \qquad (13.5)$$

Since the total infiltration of water in time $t$, namely $Q$, is equal to the product $l\Delta f$, the above equation may be written in the form,

$$t = [Q - C\,\ln(1+Q/C)]/A \qquad (13.6)$$

where

$$C = B\Delta f/A \qquad (13.7)$$

It has to be admitted that none but the most artificial of granular media, such as an array of spheres of uniform size, presents a moisture profile which exhibits the feature of a well-defined plane of separation, at a well-defined suction, between saturated material behind the water front and unaffected material ahead of it. Nevertheless Green and Ampt did in fact demonstrate the applicability of their formula, Equation (13.5), to the case of infiltration into a column of disturbed soil material. Yet again, Swartzendruber and Huberty (1958) used an equation of the type of Equation (13.2), regarded as wholly empirically based, and compared it with observed rates of infiltration into natural soil profiles. Agreement was generally surprisingly good for total infiltrations of up to three inches of water, the surprise being occasioned by the fact that natural profiles are often far from being uniform as is assumed when the equation is derived on physical grounds. In some cases, while the equation was fitted by the observations with fair success, negative values of the constant $A$ emerged, which cannot be reconciled with the above theory of infiltration into uniform soil profiles since negative hydraulic conductivity $K$ is implied. Since no physical basis was in fact appealed to, the anomaly may be dismissed as an example of the danger of extrapolating frankly empirical formulas beyond the range of observed validity. The constant $A$ is in fact the value of $(\mathrm{d}Q/\mathrm{d}t)$ expected at infinite depth of infiltration after infinite time.

### 13.3 The infiltration law of Horton

As an example of an intuitive law one may quote that of Horton (1940), although an essentially similar one was proposed earlier by Gardner and Widtsoe (1921), who based themselves on physical but rather obscure arguments. Many natural decay processes conform to a law that the rate of approach of the variable to an ultimate value is proportional to the amount by which that variable differs from the said ultimate value. In the present case one may specify the rate of infiltration, $dQ/dt$, as the variable, which for brevity will be indicated by $\dot{Q}$, and assume that the ultimate value is $\dot{Q}_f$. Then the intuitive law is

$$d\dot{Q}/dt = -E(\dot{Q} - \dot{Q}_f)$$

The negative sign in front of the right-hand side indicates that the rate of infiltration is *slowing down* to the ultimate value. It is not strictly necessary, since $E$ may be allowed to emerge from the analysis as a negative constant. The rate of infiltration may be taken to be a known value $\dot{Q}_i$ at an initial moment which may be taken as the arbitrary time zero, and then integration of the above equation gives

$$\ln[(\dot{Q} - \dot{Q}_f)/(\dot{Q}_i - \dot{Q}_f)] = -Et$$

or

$$\dot{Q} = \dot{Q}_f + (\dot{Q}_i - \dot{Q}_f)e^{-Et} \tag{13.8}$$

This is best regarded as an intuitively inspired empirical equation with three parameters, namely $\dot{Q}_i$, $\dot{Q}_f$ and $E$. These parameters may be determined in any particular case from any three separate observations of pairs of simultaneous values of $\dot{Q}$ and $t$. Opinion as to the applicability of this equation has markedly varied (Swartzendruber and Huberty, 1958; Philip, 1957c) and this is no doubt inevitable, since formulas which have no basis in adequate physical theory can only reflect experience.

### 13.4 The infiltration law of Kostiakov

Next may be mentioned an equation introduced by Kostiakov (1932) in the form, frankly empirical,

$$\dot{Q} = \dot{Q}_i t^{-\theta} = dQ/dt \tag{13.9}$$

where the symbols have the same meaning as in the preceding paragraph and $\theta$ is a constant. If Equation (13.9) is integrated with the boundary condition that $Q$, the total infiltration, has the value zero at the arbitrary zero of time at the onset of infiltration, one has

$$Q = \dot{Q}_i t^{(1-\theta)}/(1-\theta)$$

In order that $\dot{Q}$ should decay and not be augmented, $\theta$ must be positive, while in order that $Q$ should not be negative, since negative $Q$ is impossible, $(1-\theta)$ must be positive. Hence $\theta$ must be a positive number lying between zero and unity. It is evident from Equation (13.9) that the predicted value of $\dot{Q}$ must become zero after a sufficient time, and this is contrary to experience and to theory based on acceptable physical principles. The ultimate infiltration rate in a uniform profile is known to be equal to $K_{sat}$, as shown in Section 13.1. According to Equation (13.9), this rate should be attained in a time $T$ given by substituting $K_{sat}$ for $\dot{Q}$, namely,

$$T = [\dot{Q}_i/(K_{sat})]^{1/\theta} \qquad (13.10)$$

Hence if Equation (13.9) has any application at all, it may be expected to be valid in the initial stages of infiltration and certainly not for longer than $T$ as given by Equation (13.10). Here again there are two parameters, $\dot{Q}_i$ and $\theta$, which must be determined empirically from observation of two pairs of values of $\dot{Q}$ and $t$.

## 13.5 The infiltration law of Philip

Finally, one may quote an infiltration equation due to Philip (1957c), which he derived from the moisture profile equation, Equation (12.51). As shown in Note 31 it is

$$Q = At^{\frac{1}{2}} + Bt \qquad (13.11)$$

The constants $A$ and $B$ are of course not the same as those similarly indicated in Equation (13.2). The differential form, expressing the rate of infiltration as a function of time, is

$$dQ/dt = \tfrac{1}{2}At^{-\frac{1}{2}} + B \qquad (13.12)$$

It satisfies the requirement of providing for a very large rate of infiltration when the lapsed time is very small, and appears to provide also for a constant rate of infiltration at ultimate stages after a sufficiently great lapse of time. However, the constant $B$ as derived in Note 31 is not the surface conductivity that is required to give the ultimate infiltration rate correctly. Just as Equation (12.51) is itself a profile formula which is applicable only to the early stages of the advance of the moisture profile, so is Equation (13.12) valid only for early stages of infiltration if the constants are accepted as those derived from Equation (12.51). If, however, the infiltration formula, Equation (13.11), is regarded as an empirical formula to which one is led on physical grounds, and may therefore regard $A$ and $B$ as empirical constants to be determined by observation of the infiltration itself, then the formula does satisfy the requirements of predicting the infiltration behaviour at both extremes of lapsed time.

Equation (13.11) is more convenient than Green and Ampt's equation, since $Q$ is expressed explicitly as a function of $t$ instead of vice versa. However, the only reported test of this equation is a comparison by Philip with the result of a detailed computation of $Q$ using Equation (12.51) itself, and since this latter is the basic equation from which Equation (13.11) is derived, the test is not stringent. It has already been remarked that

Youngs (1957) compared experimentally observed profiles with those calculated from Equation (12.51) and found good agreement for the materials used; one may therefore infer that good agreement should be found between observed infiltration rates and those calculated from Equation (13.11). Such a comparison has never, however, been reported.

## 13.6 The rate of infiltration into a composite profile

Consider a composite profile of two layers separated by a sharply defined junction. A steady state of flow $v$ may be maintained provided that $v$ is less than the saturated conductivity of the less permeable layer, and, as shown in Section 12.7(a), the moisture contents at points sufficiently remote from the junction will be uniform at $c_U$ in the upper layer and $c_L$ in the lower layer, where $c_U$ and $c_L$ are the moisture contents which confer hydraulic conductivities $K_U$ and $K_L$ respectively in these two layers of different materials, such that

$$v = -K_U = -K_L \tag{13.13}$$

In the neighbourhood of the junction, the moisture profile is determined by the necessity of continuity of the hydrostatic pressure, and is calculable by the methods described in Section 12.5, the profile in the upper layer being in accordance with Equation (12.8), subject to a suction at the base (the junction) determined by the moisture content $c_L$ and the moisture characteristic of the lower layer.

A sudden and maintained increase of the surface moisture content to $C_U$ initiates the effects described in Section 12.7(b) to (d) for so long as the advancing water front remains in the upper layer, and in particular the rate of infiltration begins at a new high rate and settles down to an ultimate rate $K'_U$ appropriate to the new and higher moisture content $C_U$. All this occurs as though the lower layer were non-existent, naturally, if the junction is not reached by the water front. When the junction is reached, the subsequent events depend upon whether the rate of flow $K'_U$ can be accommodated in the lower layer with a uniform moisture profile at moisture content $C_L$ and conductivity $K'_L$, that is to say, whether the maximum value of $K'_L$, the saturation value, is not demanded. If this is so, then a new steady state is ultimately set up with

$$v' = -K'_U = -K'_L$$

with a moisture content $C_U$ in the upper layer, except near the junction, and $C_L$ in the lower layer. Hence the infiltration rate is unaffected by the presence of the lower layer.

If the stated conditions prevail even when the surface is maintained at saturation, i.e. when the saturated conductivity of the upper layer is less than that of the lower layer, the infiltration capacity is quite unaffected by the presence of the lower layer, and is in fact simply the infiltration capacity of the upper layer by itself.

The situation changes when $C_L$ reaches saturation while $C_U$ is still less than saturation, for the rate of flow is then limited by $K'_L$ while $K'_U$ is still capable of increase. Thus a sudden saturation of the surface inaugurates infiltration in accordance with the infiltration law of the upper material, with a decreasing asymptotic approach to the infiltration capacity $_U K_{sat}$ the saturation conductivity of the upper layer. When, however, the water front reaches the junction, the limiting conductivity $_L K_{sat}$ exerts its influence. Since in such a case the upper layer is already saturated when the water front reaches the junction, with the exception of a very short zone in the immediate neighbourhood, it has no available capacity for water storage. Hence the limitation of infiltration exerted by the lower layer is experienced very quickly after the water front arrives at the junction.

A rigorous analysis of this situation is not yet possible. The development of the moisture profile during horizontal diffusion into a heterogeneous soil profile was described in Section 12.8, and was limited to very restricted specifications of conductivity variation and moisture dependence. Progress may, however, be made by appeal to the approximate methods of Green and Ampt (1911). Confidence in this approach is encouraged, even for composite profiles, from the circumstance that, as described in Section 13.2, it is adequately confirmed by agreement with observed results in uniform profiles, while, as mentioned in Section 13.5, Philip's infiltration equation is derived from his profile equation which has been tested against experiment by Youngs. Hence by inference, it may be taken that Green and Ampt's results are compatible with the more rigorously deduced formula of Philip, which is itself not capable of extension to composite layers.

## 13.7 Extension of Green and Ampt's analysis for infiltration into a composite profile

The very general analysis to be presented here is a development of some calculations for some particular cases of a similar kind communicated to the author by Philip. The following results are arrived at in Note 32.

Let the depth of penetration of the water front, $l$, be expressed as a fraction of the depth $L$, over which the hydraulic conductivity is specified. Thus in a profile of two well-defined layers let the saturated conductivity be $K_U$ throughout the upper layer to a depth $L$, and $K_L$ at depths beyond $L$. The concept of Green and Ampt requires the specification of a negative pressure $_f H_U$ at the water front when the latter is located in the upper layer and $_f H_L$ when it is in the lower. For generality, let it be supposed that the land surface is flooded to a depth $H_0$ at all times, although in practice $H_0$ will often be vanishingly small. Then the rate of infiltration $dQ/dt$ is given in accordance with Section 13.2 and Note 32 by

$$dQ/dt = A_U + B_U/(l/L) \tag{13.14a}$$

$$\text{for } 0 < (l/L) < 1$$

$$dQ/dt = A_U(B_L/A_L + l/L)/[1 + (A_U/A_L)(l/L - 1)] \tag{13.14b}$$

$$\text{for } (l/L) > 1$$

where

$$\left.\begin{array}{l} A_U = K_U \\ A_L = K_L \\ B_U = K_U(H_0 - {}_fH_U)/L \\ B_L = K_L(H_0 - {}_fH_L)/L \end{array}\right\} \tag{13.15}$$

In conformity with Section 13.2 these equations are written in terms of the constants $A$ and $B$ because these are in fact the experimentally determined parameters of the infiltration equations for each of the two layers taken separately.

The integrated forms of the equations of (13.14) may be found by the procedure of Note 32, after expressing $dQ/dt$ in the form $\Delta f\, dl/dt$, where $\Delta f$ is the change of water content occasioned by the passage of the water front. The results are

$$t = (L/A_U)\Delta f_U[l/L - (B_U/A_U)\ln\{1 + (A_U/B_U)l/L\}] \tag{13.16a}$$

$$\text{for } 0 < (l/L) < 1$$

$$t - t_L = (L/A_L)\Delta f_L[(A_L/A_U - 1 - B_L/A_L)\ln\{(l/L + B_L/A_L)/(1 + B_L/A_L)\} + l/L - 1] \tag{13.16b}$$

$$\text{for } (l/L) > 1$$

where $t_L$ is the time at which the water front meets the junction.

As in Equation (13.6), these equations may be written to give the observable total infiltration $Q$ instead of the penetration $l$ which is not readily observable in the field.

Thus,

$$Q = l\Delta f_U$$

$$\text{for } 0 < (l/L) < 1$$

$$Q = L\Delta f_U + (l - L)\Delta f_L$$

$$\text{for } (l/L) > 1$$

whence, from Equation (13.16a),

$$t = [Q - C\ln(1 + Q/C)]/A_U \tag{13.17}$$

$$\text{for } 0 > (l/L) > 1$$

This is naturally the same as Equation (13.6) since it is precisely the same case, with the exception that

$$C = B_U L \Delta f_U / A_U$$

instead of as given in Equation (13.7), because lengths of penetration are here expressed as fractions of $L$.

Equation (13.16b) takes the form,

$$t - t_L = [(Q - Q_L) + \{C'' - C'\} \ln\{1 + (Q - Q_L)/C'\}]/A_L \quad (13.18)$$

$$\text{for } (l/L) > 1 \text{ or } Q > Q_L$$

In this equation the constants $C'$ and $C''$ are contractions given by

$$C' = L \Delta f_L (1 + B_L/A_L)$$

$$C'' = L \Delta f_L A_L / A_U$$

In Equation (13.18) $Q_L$ is the total volume of water infiltrated up to the time $t_L$ when the water front has reached the junction between the two layers.

An experimental test of the equations of (13.16) is shown in Figure (13.1), by courtesy of M. Bybordi.

If the soil profile is not discontinuous at a sharply defined junction, but presents instead a continuous distribution of hydraulic conductivity and air entry pressure over the depth range $L$, then Darcy's law may be expressed for any point in the saturated part of the column in the form,

$$dQ/dt = -K \, d\phi/dl$$

When the water front is at depth $l$, integration between the limits 0 and $l$ gives

$$dQ/dt \int_0^l dl/K = \phi_0 - \phi_l = H_0 - H_f + l$$

where the conductivity $K$ and the air entry pressure $H_f$ are functions of $l$. Again emphasizing that $K$ and $H_f$ are specified in any practical case only over a range $L$, one may write this equation,

$$dQ/dt \int_0^{l/L} d(l/L)/K = (H_0 - H_f)/L + l/L \quad (13.19)$$

$$\text{for } 0 < (l/L) < 1$$

When the forms of $K$ and $H_f$ as functions of $l/K$ are known, the integral may be evaluated, by graphical or numerical methods if by no other way, and $dQ/dt$ may thus be calculated at any required value of $l/L$. For example, if $K$ decreases linearly from $K_0$ at the surface to $K_L$ at depth $L$, a situation which is expressed by the equation,

$$K = K_0 - (K_0 - K_L)(l/L)$$

K*

then substitution of this value in Equation (13.19) followed by integration
yields

$$dQ/dt = \frac{(K_0 - K_L)\{l/L + (H_0 - H_f)/L\}}{\ln\{K_0/[K_0 - (K_0 - K_L)l/L]\}}$$   (13.20)

$$\text{for } 0 < (l/L) < 1$$

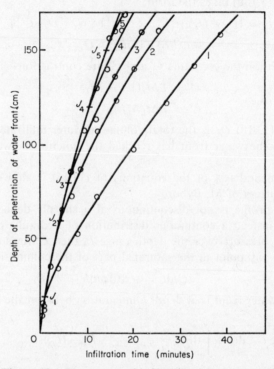

Figure 13.1 The course of infiltration into a composite
column, the upper part being of coarser material than
the lower. The full lines are theoretical curves and the
circles are experimental points. *J* represents a junction
position between the two materials, and the numbers
against the curves correspond to the appropriate junction positions.

Finally, one may invert the argument in order to express the hydraulic
conductivity at a specified depth in terms of the observed law of infiltration
at the surface. Thus if one writes the variable ratio $l/L$ of Equation (13.19)
as $\lambda$, one has

$$dQ/dt \int_0^\lambda (1/K)d\lambda = \lambda + (H_0 - H_f)/L$$

Differentation with respect to $\lambda$ yields

$$1/K = d\,[\{\lambda+(H_0-H_f)/L\}/(dQ/dt)]/d\lambda \qquad (13.21)$$

where $K$ is the hydraulic conductivity at depth $\lambda$ at which the differential coefficient is measured. This expression is of particular interest since it offers a prospect of being able to study the conductivity profile merely from observations of the surface infiltration. To do this, one needs to be able to assess the depth of penetration $\lambda$ from a measurement of the total infiltration $Q$, and the deficit from saturation $\Delta f$, as well as to have a knowledge of the air entry pressure $H_f$ down the profile. While these are more or less static profile properties and do not seem to offer any insuperable difficulties, the method has not yet been developed.

Another point of interest of Equation (13.21) is that it would seem that, within certain limits, it should be possible to determine the conductivity profile required to produce any specified form of infiltration law. In particular, it is of interest to determine the profile required to produce the infiltration law of Green and Ampt, namely

$$dQ/dt = A'+B'/\lambda \qquad (13.22)$$

$$\text{for } 0<\lambda<1$$

where $A'$ and $B'$ are freely chosen and not necessarily related by the equations of (13.15), which are appropriate to the special case of the uniform soil profile.

Substitution of Equation (13.22) in Equation (13.21) followed by differentiation with respect to $\lambda$ results in

$$1/K = \frac{(A'+B'/\lambda)(1+dC/d\lambda)+(\lambda+C)(B'/\lambda^2)}{(A'+B'/\lambda)^2} \qquad (13.23)$$

where

$$C = (H_0-H_f)/L \qquad (13.24)$$

This equation places restrictions upon the possible values of the constants $A'$ and $B'$. At the surface, $K$ has the value $K_0$ so that, from Equation (13.23),

$$\underset{\lambda\to 0}{Lt}\ 1/K = C/B' = 1/K_0 \qquad (13.25)$$

Since $H_f$ is inherently negative, $C$ is inherently positive from Equation (13.24); $K_0$ also is inherently positive, hence from Equation (13.25) the constant $B'$ of the infiltration law must be positive.

At the other extreme of the range of penetration, one has, from Equation (13.23),

$$\underset{\lambda\to 1}{Lt}\ 1/K = \frac{(A'+B')(1+dC/d\lambda)+B'(1+C)}{(A'+B')^2} \qquad (13.26)$$

If the infiltration law is taken to hold over an infinite range of depth, it follows from Equation (13.24) that at infinite $L$ the value of $C$ for unit value of $\lambda$ is zero. Since with infinitely great $L$ the value of $\lambda$ remains negligibly small for substantial but finite penetrations $l$ of the water front, the value of $B'$, from Equation (13.22), must also be negligibly small, since otherwise the infiltration rate would remain infinitely great during the stage of the substantial penetration. This point is illustrated by the equations of (13.15) where $B_U$ is necessarily small for very large $L$. Thus when $L$ tends to infinity, Equation (13.26) becomes

$$\underset{\lambda \to 1}{Lt}\ 1/K = (1 + dC/d\lambda)/A' \tag{13.27}$$

Since the profiles with which this section is concerned are those in which $K$ decreases with increasing depth, the inference is that the effective pore sizes are becoming smaller and the air entry suctions are increasing with depth, so that $dC/d\lambda$ is inherently positive if it is not zero, and since the conductivity also is inherently positive, it follows from Equation (13.27) that $A'$ may be negligibly small but may not have a negative value. It will in fact take the value zero if $K$ itself tends to zero at great depth.

When, however, the infiltration law is specified as in Equation (13.22) over a limited range of depth only, Equation (13.26) applies with finite non-zero values of $B'$ and $C$, and, with positive values of $B'$ and of $C$ as well as of $dC/d\lambda$, it is evident that $A'$ may take negative values up to an ascertainable limit while remaining compatible with positive values of $K$.

By the use of Equation (13.23), it is possible to determine the dependence of $K$ upon $\lambda$ which provides an infiltration law of the type specified in Equation (13.22), with any values of $A'$ and $B'$ which are not in conflict with Equation (13.26), even though such values may include zero or slightly negative values of $A'$. The observed and calculated infiltration laws for such profiles are shown in Figure (13.2), kindly communicated by M. Bybordi. Were such an infiltration law to be interpreted in accordance with Section 13.2, Equations (13.2) and (13.3), appropriate to uniform profiles, the conclusion would be drawn that the negative value of $A'$ indicates a negative hydraulic conductivity, which is impossible. Such negative values have, in fact, been observed in the field by Swartzendruber and Huberty (1958). The fallacy of so interpreting such an observation is indicated quite clearly in the Equations of (13.3), since a negative value of $K$ also demands a negative value of $B'$. Hence a negative $A'$ while $B'$ remains positive is evidence of the non-uniformity of the profile and the necessity of interpreting the infiltration law in accordance with Equation (13.23).

That Equation (13.23) is compatible with a uniform profile is readily demonstrated by substituting a constant $K$ instead of $A'$ and $K(H_0 - H_f)/L$

instead of $B'$ in Equation (13.23), these values of $A'$ and $B'$ being appropriate to the uniform profile as expressed in the equations of (13.3). Since the profile is uniform, the value of $dC/d\lambda$ is zero. The substitution reduces the right-hand side of Equation (13.23) to $1/K$ in agreement with the left-hand side.

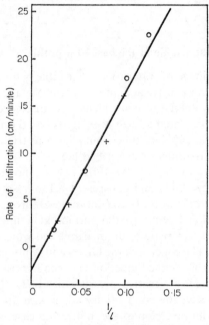

Figure 13.2 The rate of infiltration into a composite column as a function of the inverse of the penetration of the water front. The column has been made up to give a straight line law with a negative intercept on the infiltration axis. Crosses are experimental points and circles are theoretical points computed from Green and Ampt's equation.

Yet another influence on the course of infiltration needs to be mentioned, namely the change with time of the hydraulic conductivity of a given horizon which may take place as a result of the modification of soil structure. This is particularly likely to happen either as a result of very violent rainstorms or of irrigation of a violent kind, especially when the irrigation water used has an appreciable proportion of dissolved sodium ions which are conducive to poor stability of aggregation, as discussed in Section 4.11. The common experience is that the incidence of water at the surface is conducive to a reduction of the surface conductivity. The implication is that, in accordance with the infiltration laws discussed in

Section 13.6, the less affected lower horizons remain capable of accepting water at the reduced rate transmitted to them by the more affected upper horizons. Hence the change of the infiltration rate with time and with the incidental increase of the depth of penetration is almost entirely a reflection of the rate of change of hydraulic conductivity of the affected surface horizon.

### 13.8 Seepage to a water table at the base of a profile

Although the drainage of water to a water table is not a case of infiltration, nevertheless it may be treated approximately by the method of Green and Ampt and may therefore conveniently be discussed in this chapter. Youngs (1960) has presented a form of analysis based on the flow of water in uniform cylindrical capillary tubes, but such a restriction is unnecessary and a more general treatment is described here.

The circumstances envisaged are those where the moisture profile is located above a water table and consists of a lower saturated portion of length $z$, i.e. of height $z$ relative to the datum level through the water table, and an upper portion from $z$ to the surface at a uniform unsaturated moisture content characteristic of drained soil and which may be regarded as the field capacity. The concept of the Green and Ampt method is adopted, namely that the sharply defined junction between the two parts of the profile is characterized by a negative pressure head $H_f$. The soil profile is uniform, that is to say, the conductivity when saturated is the same, $K$, everywhere. This is an inversion of the infiltration case, with the saturated column underneath the unsaturated. The difference between saturation and field capacity is expressed by the symbol $\Delta f$, as in Section 13.2.

At an initial moment from which the seepage is timed, the water front is observed to be at a height $z_0$, higher than the level to which it will ultimately settle. The problem is to determine the rate of fall of the water front and the consequent seepage to the water table.

Darcy's law is in this case written:

$$dQ/dt = K(H_f + z)/z \qquad (13.28)$$

where $dQ/dt$ is the rate of flow down the saturated part of the profile, per unit cross-sectional area, and is therefore the rate of seepage to the water table at the base. Also,

$$dQ/dt = -\Delta f \, dz/dt \qquad (13.29)$$

the negative sign being occasioned from the circumstances that the rate of seepage is positive when the water front is falling, i.e. when $dz/dt$ is negative. Thus, from Equations (13.28) and (13.29),

$$dz/dt = -(K/\Delta f)(H_f + z)/z \qquad (13.30)$$

or

$$-(K/\Delta f)dt = z\,dz/(H_f + z)$$

This is similar in form to Equation (13.4) and may be integrated in a similar manner, as described in Note 30. In this case the subsidiary transformation which renders the integration simple is the substitution of $\xi$ for $H_f + z$. The final result is, for integration between the limits $z_0$ and $z$,

$$Kt/\Delta f = z_0 - z + H_f \ln[(H_f + z)/(H_f + z_0)] \qquad (13.31)$$

It is evident from this solution that after an infinite lapse of time $z$ assumes a value $z_u$ given by

$$z_u + H_f = 0 \qquad (13.32)$$

Since $H_f$ is negative $z_u$ is of course positive. This value of $z_u$ is in keeping with the known fact that the ultimate static moisture profile reflects the moisture characteristic, with the suction numerically equal to the height, so that the moisture content is sharply reduced from saturation at a height equal to the suction at the water front.

Since the parameters and variables of Equation (13.31) are not all readily identified in practice, the solution cannot be tested experimentally until it is expressed in terms of observable quantities. Let $Q$ be the quantity of water drained from the profile at the expiry of time $t$ when the water front has fallen to a height $z$; $Q_\infty$ the quantity ultimately drained when movement has sensibly ceased and the water front is at a height $z_u$; $Q_m$ the quantity which would have drained ultimately had $H_f$ been negligibly small; and $F_0$ the value of $dQ/dt$ at the initial moment when the water front is at height $z_0$. Then from Equation (13.28),

$$F_0 = K(H_f + z_0)/z_0 \qquad (13.33)$$

It follows from the above definitions, with the help of Equation (13.32), that

$$Q = \Delta f(z_0 - z) \qquad (13.34)$$

$$Q_\infty = \Delta f(z_0 - z_u) = \Delta f(z_0 + H_f) \qquad (13.35)$$

$$Q_m = \Delta f\, z_0 \qquad (13.36)$$

From these equations the following are readily derived. From Equation (13.34), by mere rearrangement,

$$z_0 - z = Q/\Delta f \qquad (13.37)$$

From Equations (13.35) and (13.36),

$$H_f = (Q_\infty - Q_m)/\Delta f \qquad (13.38)$$

From Equations (13.33), (13.35) and (13.36),

$$K = F_0 Q_m / Q_\infty \qquad (13.39)$$

Lastly, from Equations (13.34) and (13.35),

$$(H_f + z)/(H_f + z_0) = 1 - Q/Q_\infty \qquad (13.40)$$

Substitution of Equations (13.37) to (13.40) in Equation (13.31) yields,

$$[(Q_\infty/Q_m) - 1]\ln(1 - Q/Q_\infty) + Q/Q_m = F_0 t/Q_\infty \qquad (13.41)$$

All the quantities in this equation are measurable with the exception of $Q_m$, and even this may be estimated approximately if $H_f$ is estimated from Equation (13.32) by a visual estimate of $z_u$, the height of the static water front. The flux $F_0$ is simply the slope at the initial moment of the plot of $Q$ as a function of time. The uncertainty as to $Q_m$ may be eliminated for the early stages of experimental observation by expanding $\ln(1 - Q/Q_\infty)$ in the form,

$$\ln(1 - Q/Q_\infty) = -Q/Q_\infty - (Q/Q_\infty)^2/2 - (Q/Q_\infty)^3/3 - \ldots$$

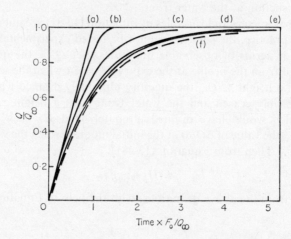

curve (a), $\dfrac{Q_\infty}{Q_m} = 1 \cdot 0$; curve (b), $\dfrac{Q_\infty}{Q_m} = 0 \cdot 9$; curve (c),

$\dfrac{Q_\infty}{Q_m} = 0 \cdot 5$; curve (d), $\dfrac{Q_\infty}{Q_m} = 0 \cdot 1$; curve (e), $\dfrac{Q_\infty}{Q_m} = 0$;

curve (f), experimental

Figure 13.3 The rate of percolation $Q$ from a saturated column to a water table as a function of time. $Q_\infty$ is the initial drainable content; $Q_m$ is the initial total content; and $F_0$ is the initial rate of percolation.

Substitution in Equation (13.41) except for the term $-\ln(1-Q/Q_\infty)$ itself results in the elimination of the term $Q/Q_m$ to give

$$\ln(1-Q/Q_\infty)+\tfrac{1}{2}Q^2/(Q_mQ_\infty)+(1/3)Q^3/(Q_mQ_\infty{}^2)+$$
$$+(1/4)Q^4/(Q_mQ_\infty{}^3)+\ldots = -F_0t/Q_\infty \qquad (13.42)$$

If all terms except the first on the left-hand side may be neglected, this equation takes the simpler form,

$$1-Q/Q_\infty = \mathrm{e}^{-F_0t/Q_\infty} \qquad (13.43)$$

By comparing the expanded form of $\ln(1-Q/Q_\infty)$ above with the sum of the neglected terms, it will be seen that in the former the factor $Q/Q_\infty$ takes the place of $Q/Q_m$ in Equation (13.42) wherever it occurs, so that the condition that the stated terms may be neglected is that

$$Q/Q_m \ll Q/Q_\infty$$

or that $Q_\infty/Q_m$ is negligibly small in comparison with unity.

All the quantities in Equation (13.43) are readily determined in an experiment in which the quantity of liquid drained from a column of porous material is measured as time elapses. Such experiments have been reported by Youngs (1960) using various granular materials and a variety of different liquids. The comparison between observed and computed relationships is shown in Figure (13.3). The computed curves are drawn from the complete Equation (13.42) with various assumed values of the ratio $Q_\infty/Q_m$. Of these curves, that for vanishing $Q_\infty/Q_m$ corresponds to Equation (13.43).

## NOTES

### Note 30. The total infiltration by Green and Ampt's method
The differential equation which expresses the rate of advance of the water front, Equation (13.4) is

$$dt/dl = \Delta f\, l/(lA+B) \qquad (N30.1)$$

where $\Delta f$ is the difference between water contents before and after the passage of the water front.

Make the transformation,

$$\lambda = lA+B$$

so that

$$\left. \begin{aligned} l &= (\lambda-B)/A \\ dl &= d\lambda/A \end{aligned} \right\} \qquad (N30.2)$$

Then Equation (N30.1) takes the form

$$dt/d\lambda = \Delta f(\lambda-B)/A^2\lambda$$
$$= \Delta f(1/A^2 - B/A^2\lambda)$$

Hence by an elementary integration,

$$t = \Delta f[\lambda/A^2 - (B/A^2)\ln\lambda + \text{constant}]$$

Substituting back the appropriate value of $\lambda$ in terms of $l$ from the equations of (N30.2), one has

$$t = \Delta f[(lA+B)/A^2 - (B/A^2)\ln(lA+B) + \text{constant}] \qquad \text{(N30.3)}$$

Since both $t$ and $l$ vanish together one has a value for the constant of integration given by

$$\text{constant} = (B/A^2)\ln(B - B/A^2)$$

whence finally Equation (N30.3) reduces to

$$t = \Delta f[l - (B/A)\ln\{(lA+B)/B\}]/A \qquad \left\{ \begin{array}{l} \text{(N30.4)} \\ \text{(13.5)} \end{array} \right.$$

### Note 31. The rate of infiltration by Philip's equation

Equation (12.51) for the moisture profile during infiltration is, writing $l$ instead of $-z$ so that $l$ is the depth from the surface,

$$l = \lambda t^{\frac{1}{2}} + \mu t + v t^{1\frac{1}{2}} + \xi t^2 + \ldots \qquad \text{(N31.1)}$$

Hence the total amount of water stored in the profile, from the surface at moisture content $C$ to a depth $L$ which is located in the zone at uniform moisture content $c_0$ to which the water front has not yet penetrated, is $S_L$ where

$$S_L = \int_0^C l\,dc = Lc_0 + \int_{c_0}^C l\,dc$$

In this case the upper limit of moisture content, $C$, is generally saturation, but the analysis is quite general. By the use of Equation (N31.1) one may write this in the form,

$$S_L = Lc_0 + t^{\frac{1}{2}}\int_{c_0}^C \lambda\,dc + t\int_{c_0}^C \mu\,dc + t^{1\frac{1}{2}}\int_{c_0}^C v\,dc + t^2\int_{c_0}^C \xi\,dc + \ldots$$

The rate of increase of the stored water is therefore

$$dS_L/dt = \frac{1}{2}t^{-\frac{1}{2}}\int_{c_0}^C \lambda\,dc + \int_{c_0}^C \mu\,dc + 1\frac{1}{2}t^{\frac{1}{2}}\int_{c_0}^C v\,dc + 2t\int_{c_0}^C \xi\,dt + \ldots \qquad \text{(N31.2)}$$

since the quantities represented by the integrals are functions of $c$ only and are independent of $t$.

The rate of storage is given in another form as the excess of the infiltration rate, $dQ/dt$, over the outflow rate at the lower boundary $L$, where the moisture gradient vanishes and the rate of flow is, as described in Section 12.7(a), Equation (12.41), equal to $K_0$ downward, where $K_0$ is the hydraulic conductivity at moisture content $c_0$. Thus,

$$dS_L/dt = dQ/dt - K_0 \qquad \text{(N31.3)}$$

For sufficiently small time intervals following the inception of surface flooding, all the terms on the right-hand side of Equation (N31.2) except the first two may be neglected, since not only do the increasing powers of $t$ operate to diminish these terms but also the coefficients become rapidly smaller with successive terms. Hence with this modification of Equation (N31.2) and combination with Equation (N31.3), one obtains

$$dQ/dt = K_0 + \tfrac{1}{2}t^{-\frac{1}{2}}\int_{c_0}^{C}\lambda\,dc + \int_{c_0}^{C}\mu\,dc$$

which may be written:

$$dQ/dt = \tfrac{1}{2}At^{-\frac{1}{2}} + B \qquad \left\{\begin{matrix}\text{(N31.4)}\\ \text{(13.12)}\end{matrix}\right.$$

The integrated form is given directly as

$$Q = At^{\frac{1}{2}} + Bt \qquad \left\{\begin{matrix}\text{(N31.5)}\\ \text{(13.11)}\end{matrix}\right.$$

where

$$A = \int_{c_0}^{C}\lambda\,dc$$

$$B = K_0 + \int_{c_0}^{C}\mu\,dc$$

The constants $A$ and $B$ are soil properties which depend on the moisture limits $c_0$ and $C$.

### Note 32. The calculation of the rate of infiltration into a composite profile

Consider a profile which consists of a layer of depth $L$ and conductivity $K_U$ resting on a lower layer of conductivity $K_L$. The hydrostatic pressure at the water front, in accordance with the hypothesis of Green and Ampt, is $_fH_U$ when the water front is located in the upper layer and $_fH_L$ when it is in the lower.

Then as shown in Section 13.2, Equations (13.2) and (13.3),

$$dQ/dt = A_U + B_U/(l/L) \qquad \left\{\begin{matrix}\text{(N32.1)}\\ \text{(13.14a)}\end{matrix}\right.$$

$$\text{for } 0 < (l/L) < 1$$

where

$$A_U = K_U$$

$$B_U = K_U(H_0 - {}_fH_U)/L$$

These equations apply when the water front is still in the upper layer and the effective profile is therefore uniform.

When the water front is in the lower layer, the length of effectively conducting upper column is $L$, while that of the lower column is $l-L$, where $l$ is the total penetration from the surface. Applying Darcy's law to the two parts of the column separately, one has

$$dQ/dt = K_U(H_0 - H_J + L)/L$$

$$dQ/dt = K_L(H_J - {}_fH_L + l - L)/(l - L)$$

where $H_J$ is the unknown pressure head at the junction of the two layers. $H_J$ may be eliminated between these two equations to give

$$H_0 - {}_fH_L + l = (dQ/dt)[L/K_U + (l-L)/K_L]$$

This may be rearranged to give

$$dQ/dt = \frac{A_U(B_L/A_L + l/L)}{1 + (A_U/A_L)(l/L - 1)} \qquad \begin{cases} \text{(N32.2)} \\ \text{(13.14b)} \end{cases}$$

where

$$A_L = K_L$$

$$B_L = K_L(H_0 - {}_fH_L)/L$$

$A_L$ and $B_L$ may be determined by an infiltration experiment using the material of the lower layer by itself.

Since Equation (N32.1) is effectively the same equation as Equation (13.2), differing only in the specification of the constant $B$, it may be integrated in the same way, as described in Note 30, to give essentially the same equation as Equation (13.5), namely,

$$t = (L/A_U)\Delta f_U[l/L - (B_U/A_U)\ln\{1 + (A_U/B_U)l/L\}] \qquad \begin{cases} \text{(N32.3)} \\ \text{(13.16a)} \end{cases}$$

where $\Delta f_U$ is the increase of moisture content suffered by the upper layer at the passage of the water front.

Equation (N32.2) may be integrated in a similar fashion. It is written in the form,

$$dt = \frac{\Delta f_L d\left[1 + (A_U/A_L)(l/L - 1)\right]}{A_U(B_L/A_L + l/L)} \qquad \text{(N32.4)}$$

where $\Delta f_L$ is the moisture content increase due to the passage of the water front in the lower layer.

One now transforms the variable $l$ by the equation,

$$B_L/A_L + l/L = \lambda \qquad \text{(N32.5)}$$

so that

$$dl/L = d\lambda \qquad \text{(N32.6)}$$

With these substitutions Equation (N32.4) takes the form,

$$dt = (L\Delta f_L \, d\lambda/A_L)[(A_L/A_U - 1 - B_L/A_L)/\lambda + 1]$$

Upon integration between the limits $L$ and $l$, corresponding to $t_L$ and $t$, one arrives at the solution,

$$t - t_L = (L/A_L)\Delta f_L[(A_L/A_U - 1 - B_L/A_L)\ln\lambda + \lambda]$$

Substituting back from $\lambda$ to $l$ from Equation (N32.5), one arrives at

$$t - t_L = (L/A_L)\Delta f_L[(A_L/A_U - 1 - B_L/A_L)\ln\frac{l/L + B_L/A_L}{1 + B_L/A_L} + l/L - 1] \qquad \begin{cases} \text{(N32.7)} \\ \text{(13.16b)} \end{cases}$$

CHAPTER 14

# The flow of groundwater

## 14.1 The nature of groundwater flow

IT has been noted in Chapter 12 that the study of the development of moisture profiles in unsaturated soil is largely an analysis of water movement in the vertical direction. This is because changes of moisture profile are primarily a consequence of changes of rates of precipitation or of evaporation at the soil surface; because such changes are relatively uniform over considerable areas; and because unsaturated soils have relatively low hydraulic conductivity so that quite large suction gradients, beyond what are commonly observed in the horizontal direction, are required to produce appreciable movements of water.

On the other hand, groundwater, which is by definition beneath the water table and therefore at positive hydrostatic pressure, saturates the soil to give it its maximum possible hydraulic conductivity, and at the same time this uniformity of saturation removes all possibility of large suction gradients contributing predominantly to the potential gradient. Hence the dominant water moving force is gravity, and the pore water therefore moves much as does free water in bulk. Its passage is hindered in this direction by the intervention of impermeable barriers, in that direction by intercepting channels, and in yet another direction by unsaturated soil of low hydraulic conductivity above the water table. In short, the boundary conditions operate to guide the flow of the water in many different directions.

The distinction between the movement of water in the soil and in the saturated zone beneath the water table lies firstly in the distinction between media of variable moisture content and hydraulic conductivity, both in space and time, and media of constant and relatively uniform moisture content and conductivity; and secondly in the characteristic difference between the boundary conditions. This does not mean that the study of these two different regions is a study of inherently unconnected matters, since clearly the percolation of water down a profile to a water table may well provide, by contributing to groundwater, a boundary condition for groundwater movement; while groundwater movement, by changing the height of the water table, may well cause a change of moisture profile. It is simply that the analysis of the two kinds of movements are best simplified in the elementary stages by considering them separately since the methods of analysis are rather distinct.

The equation of continuity for the unsaturated state, discussed at length in Chapter 11, is complicated by the variability of moisture content and hydraulic conductivity (and indeed this variability is at the root of the analysis), but this complication is compensated by the simplicity of uni-directional movement. The complication of groundwater study due to movement in any of all possible directions is compensated by the simplicity inherent in constant and uniform moisture content and conductivity.

The equation of continuity appropriate to these circumstances is, in the most general form, Laplace's equation, Equation (11.22), namely,

$$\partial^2\phi/\partial x^2 + \partial^2\phi/\partial y^2 + \partial^2\phi/\partial z^2 = 0$$

As explained in Section 11.3, the more complicated equation which describes water movement in soils with anisotropic conductivity is readily converted into Laplace's equation by a simple distortion of the coordinates, so that a separate treatment of anisotropic problems is not called for.

## 14.2 Circumstances of groundwater movement

Most situations which cause groundwater movement are more or less complicated mixtures of separately recognizable factors, and it is simplest to idealize matters so that these factors can be described and analysed singly. Such factors are specified in terms of the particular boundary conditions that characterize them. These boundary conditions define the sources of the water supplies that maintain the groundwater, the sinks to which the groundwater drains, and the impermeable boundaries that limit the zone in which flow can take place. These boundary conditions may be listed and described as follows.

Where water enters or leaves the soil by flow through the bed of a water-filled channel, the surface which separates the soil from the water in the channel is an equipotential. For if the surface of the free water is at a height $Z$, then the height of any other point in the channel water at a depth $d$ below the surface is $Z-d$, while the hydrostatic pressure head $H$ at that depth is $d$. Hence the potential at such a point is

$$\phi = z + H = Z - d + d = Z$$

Thus the whole cross-section of the channel water, including the wetted perimeter, is an equipotential whose potential is that of the surface of the free water. A buried drain full of water to the roof, but with no back-pressure, i.e. with zero pressure at the roof, is similarly an equipotential labelled by the potential at the roof.

Where groundwater is intersected by the soil surface elsewhere than at a drainage channel, where it can be in contact with free water as above, it wells out over a surface of seepage. It then runs down such a seepage surface as a thin sheet on its way to the nearest surface drain. Since the flowing sheet has only negligible thickness and yet, being flat, has negligible curvature, the pressure head is neither positive nor negative, but is simply negligibly small. Hence the potential $\phi$ is given simply by the height component $Z$. The shape and position of the surface of seepage is known, since it is the known land surface, but the extent over which seepage takes place is not initially known. The total rate of flow across the surface of seepage and any drainage channel equipotential with which it may be continuous is equal to the total input at the input boundaries in the steady state. However, the partition between the surface of seepage and the drainage channel is not initially known.

At a water table also, the potential component due to hydrostatic pressure vanishes, and again $\phi$ and $Z$ are equal, where as before $Z$ is the value of $z$ at the zero pressure surface. At the water table, the distribution of flow is known if this surface is stationary, since it is, or may reasonably be assumed to be in many cases, identical with the rate of infiltration at the land surface above. If the water table is moving, however, the distribution of flow has a component due to the change of stored groundwater, and since neither the shape nor position of the water table are initially known, neither can be changes of such position nor any quantities dependent upon such changes.

An impermeable boundary is one across which no flow can take place, so that it coincides with the direction of flow and therefore with the direction of the potential gradient. Similarly, a surface about which the pattern of flow is symmetrical is necessarily plane and defines the direction of flow and of the potential gradient. The way in which these boundary conditions may be assigned to the various parts of the complete boundary defines the particular problem.

Thus the effects of local precipitation by itself may be studied by supposing that foreign water is excluded by basing the conducting soil layer on a truly horizontal impermeable bed and by postulating continuity of uniformity of soil properties and precipitation to infinity in the horizontal plane. The distribution of sources and of natural or artificial sinks (e.g. spring lines and drains) introduces the features of localized control of the flow of water.

The movement of foreign water may be idealized by considering the movement of groundwater in a uniformly sloping impermeable bed from an upstream source to a downstream sink in the complete absence of local precipitation. Again this may be characterized by a distribution of natural or artificial drains. Artesian water may be idealized as a special case of foreign water with a natural flow direction vertically upwards, from a deep source at high potential, to the surface or to a natural or artificial system of sinks at or nearer to the surface.

In some cases the source and the sink may be interchanged to provide a new problem. Thus the flow of artesian water to a surface drain channel may be reversed to give a problem of the seepage of water from a surface canal to a deep water table, and the flow of surface precipitated rainfall to a drain is the reverse of the flow of irrigation water from a soil pipeline to the general soil surface where evaporation or transpiration provides the sink.

Lastly, it is worth mentioning one specific case of a combination of foreign water and local precipitation. In localities near the sea coast, and particularly in islands and long narrow isthmuses, there is frequently a body of locally precipitated fresh water superimposed on a body of salt foreign water for which the sea is the source. Such a fresh water body commonly has a lenticular shape and is known as a Ghyben-Herzberg lens, from the originators of the study of this problem, namely Ghyben and Drabbe (1888-89) and Herzberg (1901).

As an example of a problem of local precipitation, one may consider the flow of groundwater to a system of long parallel drains at uniform depth, spaced uniformly and relatively closely in soil of uniform hydraulic conductivity resting on a level impermeable bed of infinite extent. The water table is to be regarded as stationary, the rate of drainage just equalling the rate of recharge by rainfall at a steady rate $q$. The situation is shown in

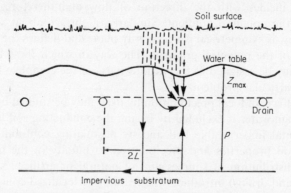

Figure 14.1 Cross-section of groundwater with local rainfall drained by a system of uniformly spaced tile lines. Some streamlines are sketched.

Figure (14.1), in which the separation of neighbouring drains is $2L$ and the height of the water table above the drain level is $Z_{max}$ at its highest point, midway between the drains. The impermeable bed is at depth $p$ below the drain level.

Because of the low ratio between the separation of the drains and their length, or distance to the nearest cross drain, the problem is two-dimensional with the directions of flow restricted to planes perpendicular to the drain axes. The streamlines from the water table to the drains are of the form sketched.

The problem is essentially to locate the water table and to draw the flow-net within the boundaries; and to generalize the results by relating the water table height to the drain spacing, the rate of recharge, the hydraulic conductivity of the soil, the diameter of the drain, and the depth of the impermeable bed, when all these parameters are varied. As subsequent modifications one may relax the simplifying but restricting assumptions, such as that the upper boundary is indeed the water table, by taking into account that the effective zone of flow is extended by the capillary fringe, and even by exploring the whole zone between the soil surface and the impermeable bed; that the rainfall is steady and the water table stationary, by examining the consequences of interrupting or intensifying the recharge rate; and that the problem is two-dimensional, by examining situations in which the drainage boundaries are at finite distances in all directions.

Simplified geometries of the kind described are advisable for the establishment of principles, but many complicated natural situations may be approximated to one or other of such simplified boundaries and may be illuminated by the analysis.

As an example of the flow of foreign water one may examine the perturbation of the flow-net and water table when the natural flow of the groundwater is intercepted by a drain as shown in Figure (14.2). Here the slope of the impermeable bed is taken to be uniform at the angle $\theta$ and the extent to be infinite in both upstream and downstream directions. There is no surface precipitation and one does not enquire into the nature of the upstream ultimate source or the fate of the groundwater downstream. The immediate object is to establish the shape of the water table as perturbed by the drain, to calculate the rate of flow of water drawn off by the drain, and to relate these to the pertinent parameters of the problem.

The special case of artesian foreign water is depicted in Figure (14.3), in which an aquifer at depth $p$ below drain level and in which the hydrostatic pressure is $P$, is in contact with the overlying soil, of uniform conductivity $K$, at a horizontal boundary of infinite extent. The drain parameters are labelled as in the case of local surface precipitation. The solution required here is the location of the water table, the delineation of the contained flow-net, the rate of flow of artesian groundwater to the drains, and the relation of these quantities to the depth and pressure of the aquifer, the soil conductivity, and the design of the drainage system.

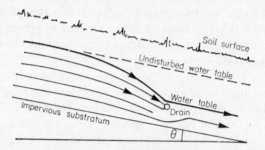

Figure 14.2 Cross-section of groundwater on a sloping bed with foreign water drained by a tile line. Some streamlines are sketched.

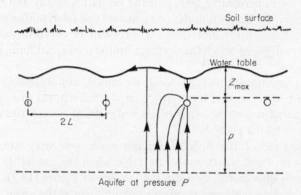

Figure 14.3 Cross-section of groundwater with artesian inflow drained by a system of uniformly spaced tile lines. Some streamlines are sketched.

The special case of the Ghyben-Herzberg lens is illustrated in Figure (14.4), which shows a two-dimensional case of an isthmus separating two zones of sea water at different levels. A zone of salt groundwater is continuous with the external sea and there is a flow from one side to the other. Superimposed on the salt groundwater is a body of fresh groundwater due to local surface precipitation or rainfall, located under a water table and draining to the edges. Thus there is an overall flow pattern of the kind indicated by the sketched streamlines. The flow of fresh water is confined between the water table and the depressed boundary between the fresh and salt water, while the flow of salt water is of a quite different kind without recharge. The object of the analysis is to discover the locations of the water table and of the boundary between the fresh and salt water bodies, and secondarily to calculate the rate of throughflow of the salt water. The relevant parameters which determine the solution are the rate of rainfall, the conductivity of the soil, the dimensions of the cross-section, the levels of the sea, and the densities of the fresh and salt water.

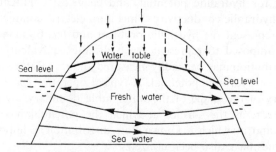

Figure 14.4 The Ghyben-Herzberg lens. Fresh groundwater fed by rainfall is supported on saline water fed by
the sea. Some streamlines are sketched.

## 14.3 Methods of analysis

The direct solution of Laplace's equation with initially specified boundary conditions, that is to say for a specified particular case, is generally not feasible, and recourse must be made to various expedients.

Firstly a considerable group of problems are two-dimensional, flow taking place in the plane containing the vertical or $z$ direction and a horizontal direction, say $x$. It may then be established that a very general class of functions of the variables $x$ and $z$ satisfies Laplace's equation, giving an appropriate potential distribution, so that one may choose freely any one function of such a class, and by examining the resulting potential distribution, determine the boundary conditions within which it is confined. That is to say, one assumes a solution and discovers the particular problem that one happens to have been analysing. Such chance solutions are useful more often than one might expect, and in any case frequently give an insight into other problems of greater inherent interest.

Secondly, while it is difficult to discover a potential distribution which both satisfies Laplace's equation and fits specified boundaries, it is by no means difficult to test whether a suggested solution does in fact do so. Thus one may begin by guessing a potential distribution, guided by the known values of potential or flow rate at the boundaries, and by systematic test and correction of errors, arrive in due course at a sufficiently correct potential distribution. The best known of such reiterative processes of numerical solution is Southwell's (1946) method of relaxation of restraints.

Thirdly, one may set up an analogue of the hydraulic problem, using laws of conduction which are formally analogous to Darcy's law. For example, the law of conduction of electricity is formally identical with Darcy's law, with electric current taking the place of flowing water, voltage substituting for hydraulic potential, and electrical conductivity playing the part of hydraulic conductivity. Thus if an electric conductor is shaped to scale to represent the hydraulic problem and has boundary potentials or currents imposed to represent the hydraulic equivalents, the interior voltage distribution may be readily measured and reveals the corresponding distribution of hydraulic potential. A typical operation by this method is the discovery of the shape of the water table boundary, by a process of trial and error, as the surface at which the potential has no pressure component and at which a known flow distribution is imposed. Similar heat analogues have also been utilized.

Fourthly, if the shape of the boundaries of the zone of flow is such that it is a reasonable assumption that the direction of flow can nowhere depart markedly from the horizontal, then one may use a greatly simplified mathematical analysis. Since in such a case the equipotentials must be vertical, and are labelled by the height at which they intersect the water table, the potential gradient is simply the slope of the water table immediately above the given point. This, the Dupuit-Forchheimer approximation (Dupuit, 1863; Forchheimer, 1914), yields equations from which the shape of the water table may be computed directly.

Lastly, it is possible to treat certain highly idealized cases by rigorous mathematical analysis in which the variables are transformed in more or less complicated stages. The solutions which emerge from the idealized problem contain information which guides one to their application to much more general situations. Before proceeding with a detailed discussion of methods of analysis, it is necessary to become familiar with the significance of the imaginary operator $i$.

### 14.4 The imaginary number i

The operator $i$ is defined by the equation,

$$i^2 = -1$$

Any real number that one can conceive, either positive or negative, has a positive square. Hence $i$, or $(-1)^{\frac{1}{2}}$, is not a conceivable number. It is therefore referred to as imaginary, somewhat oddly, since it cannot be imagined. The product of $i$ with a real number, say $z$, is also evidently imaginary. A number which is the sum of a real term $x$ and an imaginary term $iz$ is called in this context a complex number. Thus, if

$$\rho = x + iz$$

$\rho$ is a complex number with a real part $x$ and an imaginary part $iz$.

Since a real number cannot be equated to an imaginary number, if two complex numbers are equal, it follows that the real part of the one equals the real part of the other, and similarly with the imaginary parts.

The algebra of complex quantities is carried out in good faith with the symbols representing the complex quantities in just the same way as in the algebra of real quantities. An equation which represents the solution of a problem is always expressible as two separate equations from the property of the separate equality of the real and imaginary parts, so that a single complex equation provides a pair of simultaneous equations for determining two unknown variables. This is the first important feature of the operator. The second is that, as shown in Note 33, the exponential function $e^{i\theta}$ is expressible as a complex circular function by the equation,

$$e^{i\theta} = \cos\theta + i\sin\theta \tag{14.1}$$

from which it follows that

$$e^{-i\theta} = \cos\theta - i\sin\theta \tag{14.2}$$

A special case is

$$\left. \begin{array}{l} e^{i\pi} = -1 \\ i\pi = \ln(-1) \end{array} \right\} \tag{14.3}$$

This relationship enables one to adopt a convention that a complex number may be represented in a vector-like diagram as the sum of a real number, represented by a displacement on the real axis $x$, and an imaginary component, represented by a displacement on the perpendicular $z$ axis. This is shown in Figure (14.5), in which the point $(x,z)$ is at a distance $r$ from the origin 0, and the radius $r$ makes the angle $\theta$ with the $x$ axis.

If $r$ were a vector, one would write

$$r = \mathbf{i}x + \mathbf{j}z$$

where $\mathbf{i}$ and $\mathbf{j}$ are unit vectors in the directions of $x$ and $z$ respectively, and

$$x = r\cos\theta$$
$$z = r\sin\theta$$

One writes similarly, according to the convention of the representation of complex numbers,

$$\rho = x + iz \tag{14.4}$$
$$= r\cos\theta + ir\sin\theta$$

and, from Equation (14.1),

$$\rho = re^{i\theta} \tag{14.5}$$

Thus the specification of the complex number $\rho$ in terms of the radius vector $r$ and the argument $\theta$ in accordance with Equation (14.5) is entirely equivalent to the Cartesian expression, Equation (14.4). The representation of the complex number in this vector-like way is called the Argand diagram.

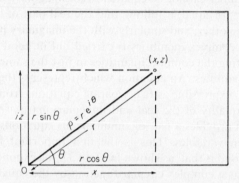

Figure 14.5 The representation of a complex number in a vector-like diagram. The Argand diagram.

## 14.5 Conjugate functions as solutions of Laplace's equation

Laplace's equation as it frequently occurs in the theory of groundwater is often restricted to two dimensions, say the horizontal and vertical directions, $x$ and $z$, to take the form,

$$\partial^2\phi/\partial x^2 + \partial^2\phi/\partial z^2 = 0 \tag{14.6}$$

Every point in the $x,z$ plane is ranged over by an independent variation of $x$ and $z$, and at each point there is an appropriate value of the potential $\phi$. The expression of $\phi$ as a function of the varying $x$ and $z$, that is to say, the distribution of $\phi$ over the plane, constitutes the solution of the equation.

As is shown in Note 34, any function whatever of the complex variable $\rho$, or $x+iz$, is a solution of Equation (14.6). Moreover, such a function, say $W(\rho)$, must in general itself be complex and have a real part, say $\phi$, and an imaginary part, say $\psi$, each of which is a function of $x$ and $z$; these functions are separately solutions of Equation (14.5). Thus to summarize, if

$$W(\rho) = \phi + i\psi$$

where

$$\rho = x + iz$$

then

$$\partial^2 W/\partial x^2 + \partial^2 W/\partial z^2 = 0$$

$$\partial^2 \phi/\partial x^2 + \partial^2 \phi/\partial z^2 = 0$$

$$\partial^2 \psi/\partial x^2 + \partial^2 \psi/\partial z^2 = 0$$

The two sets of solutions, $\phi$ and $\psi$, are related in the following way. The distribution of $\phi$ may be indicated in the diagram of the $x,z$ plane by a third axis, perpendicular to the plane, so that $\phi$ describes a surface lying above the plane. This surface may be indicated on the plane of the diagram by contours or lines joining points of equal $\phi$, the complete surface being covered by a family of such contours at stated intervals or increments of $\phi$. Such contours are called equipotentials. The distribution of $\psi$ may be indicated similarly by a family of curves of equal $\psi$ at increments of $\psi$. Then it is found, as shown in Note 34, that each curve of the $\phi$ family cuts each curve of the $\psi$ family at right angles. Such families of curves are said to be orthogonal.

As Equation (14.6) is developed in Section 11.3, the variable $\phi$ is the hydraulic potential and the contours of $\phi$ are equipotentials. As discussed in Section 9.1, the gradient of potential, in the direction in which the water is urged, is everywhere perpendicular to the equipotentials, so that the contours of equal $\psi$ depict everywhere the direction of the force on the moving water. Hence the contours of equal $\psi$ are called streamlines and $\psi$ itself is called the stream function. The pair of functions together are called conjugate functions. The network of meshes formed by the two orthogonal families of equipotentials and streamlines is called a flow-net. If the intervals between neighbouring equipotentials and streamlines are sufficiently small, the elementary meshes of the flow-net are rectangles, and are frequently referred to as such even when the meshes are somewhat larger and the sides are more or less curvilinear.

Figure (14.6) depicts a portion of a flow-net. The zone between any pair

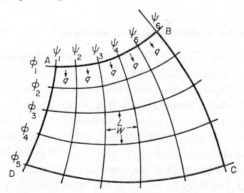

Figure 14.6 The flow-net inside specified boundaries.

of neighbouring streamlines is called a streamtube, and at every cross-section of a streamtube the total rate of flow remains the same, since a streamline which begins between the two limiting streamlines stays between them. Let the increments of potential between any pair of neighbouring equipotentials, $\phi_1 - \phi_2$, $\phi_2 - \phi_3$, and so on, be the same, namely $\Delta\phi$, and furthermore let the streamlines, $\psi_1$, $\psi_2$, $\psi_3$, and so on, be so chosen as to provide the same rates of flow, $q$, in all the streamtubes. Then on applying Darcy's law to any of the mesh elements, of width $W$ and length $L$, one has,

$$q = K\Delta\phi(W/L)$$

It follows that in a conductor of uniform $K$, the ratio of $W$ to $L$, that is to say the shape, of each of the elementary rectangular meshes is the same, since $q$ and $\Delta\phi$ are the same for all.

There is no reason why $\phi$ should have been chosen to represent potential, since $\psi$ also satisfies Laplace's equation and could be the potential. In that case $\phi$ would be the stream function. Whether $\phi$ is the potential and $\psi$ the stream function or vice versa depends upon the boundary conditions. Thus in Figure (14.6), if the boundaries shown are the actual boundaries of the conductor, then the boundaries $AB$ and $CD$ are lines of constant $\phi$ while the boundaries $BC$ and $DA$ are lines of constant $\psi$. Thus if $AB$ and $CD$ are inflow and outflow faces maintained at constant potential, then $\phi$ is the potential function and $\psi$ is the stream function, and in particular $BC$ and $AD$ are bounding streamlines. But if $BC$ and $AD$ are, in practice, the constant potential inflow and outflow faces, then $\psi$ is the potential function and $\phi$ is the stream function, and $AB$ and $CD$ are the bounding streamlines. The flow-net is in each case precisely the same, with the functions of $\phi$ and $\psi$ reversed. One case is called the inversion of the other, and problems may sometimes be solved more easily by inverting the boundary conditions in this way.

## 14.6 Some particular cases of conjugate functions

Since any function of $(x + iz)$ is a solution of Laplace's equation, it will provide a flow-net which will agree with some boundary or other consisting of portions with specified potential functions or stream functions. Hence one may boldly select a function at will and plot the potential and stream functions in order to discover possible boundaries which the function fits. That is to say, one chooses a solution and discovers the problem to which that solution is pertinent. Each such solution will provide two problems, one in which the function $\phi$ is the potential and $\psi$ the stream function, and one in which the roles of $\phi$ and $\psi$ are interchanged. The method will be illustrated by reference to some examples. For a further treatment, the reader is referred to Muskat (1937) and to Polubarinova-Kochina (1962).

*Case (a)*

As a first example choose horizontal coordinates, $x$ and $y$, rather than $x$ and the vertical coordinate $z$; and examine the chosen function $W$ where

$$W = \phi + i\psi = A \ln(x+iy)$$
$$= A \ln(re^{i\theta})$$
$$= A \ln r + iA\theta$$

where $A$ is a constant.

Since the real and imaginary parts of the two sides must separately be equal, this yields the result,

$$\phi = A \ln r \qquad (14.7)$$
$$\psi = A\theta \qquad (14.8)$$

Hence if $\phi$ is chosen to be the potential and $\psi$ the stream function, the equipotential $\phi_R$ is the circle of radius $R$ given by the equation,

$$R = e^{\phi_R/A}$$

The streamline $\psi_\Omega$ is the straight line radiating from the origin given by

$$\theta = \psi_\Omega/A$$

The solution is illustrated in Figure (14.7).

Thus the equipotentials are a set of co-axial cylinders, since the

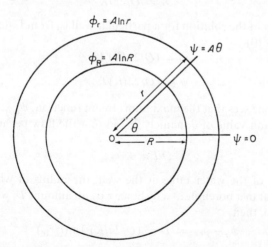

Figure 14.7 Concentric circular equipotentials and radial streamlines. The flow of water to a well.

L

supposition is that there is no variation in the vertical $z$ direction, and the streamlines are in planes radiating from the common axis. This is therefore the solution for the flow of water to a cylindrical well bored into a confined aquifer, confined because the condition is that the flow must be everywhere horizontal, and this would not be the case in groundwater beneath a water table, which would be drawn into a cone of depression near the well.

The value of the constant $A$ may be derived in terms of the properties of the soil and the size and the functioning of the well. Thus from Equation (14.7),

$$d\phi/dr = A/r$$

and, by Darcy's law,

$$v = -K\,d\phi/dr$$
$$= -KA/r \tag{14.9}$$

If the well is being pumped at a steady rate of $Q$ units of volume per unit time, this is the rate of flow towards the well at all distances from it, i.e. in the negative direction of $r$, across the cylindrical surface of radius $r$ concentric with the well and with length $l$, where $l$ is the thickness of the confined aquifer which is supposed to be completely penetrated by the well. Hence,

$$Q = -2\pi r l v$$

and, from Equation (14.9),

$$Q = 2\pi l A K$$

Hence the constant $A$ is given by

$$A = Q/2\pi l K \tag{14.10}$$

The final form of the solution for a particular well is, from Equations (14.7), (14.8) and (14.10),

$$\phi = (Q/2\pi l K)\ln r \tag{14.11}$$

$$\psi = (Q/2\pi l K)\theta \tag{14.12}$$

From this one sees that the maximum stream function, $\psi_{max}$, corresponds to the maximum value of $\theta$, namely $2\pi$, so that the flow per unit length of well is

$$Q/l = K\psi_{max} \tag{14.13}$$

If the level of the water table at the well, the radius of which is $r_W$, is lower than that in a borehole at a distance $r$ by an amount $D$, which is called the drawdown, then

$$\phi_r - \phi_{rW} = D = (Q/2\pi l K)\ln(r/r_W) \tag{14.14}$$

This formula is the basis of Thiem's method of measuring $K$ by pumping a well to a stage which approximates to a steady state (Thiem, 1906).

Although the method is strictly valid only for a well in a confined aquifer, nevertheless it is commonly applied to the pumping of groundwater under a free water table. The consequent error is not serious if the drawdown, or in this case the cone of depression, is small compared with the effective depth of the well, but its uncritical use without an examination of its suitability in given circumstances is to be deprecated.

The solution could equally well be applied by assigning $\psi$ to the potential and $\phi$ to the stream function. The equipotentials would then be planes radiating from the axis of concentric cylindrical surfaces which contain the streamlines. This would be the description of the flow between the plane radial faces of a segment of a cylindrical annular conductor.

## Case (b)

In the solutions that now follow, the water table plays an important part. Since the potential $\phi$ is the sum of height $z$ and pressure head $H$, the latter of which vanishes at the water table, the water table condition is recognized by the equality of $\phi$ and $z$, which is then written $Z$. Now examine the solution,

$$W = \phi + i\psi = (x + iz)e^{-i\theta} \sin \theta + h \cos \theta$$

where $h$ and $\theta$ are constants, the physical significance of which remains to be learned. After expanding the exponential term to the form (cos $\theta$ + $i$ sin $\theta$) and equating the real and imaginary parts separately, one has the pair of equations,

$$\phi = (x \cos \theta + z \sin \theta)\sin \theta + h \cos \theta \qquad (14.15)$$

$$\psi = (z \cos \theta - x \sin \theta)\sin \theta \qquad (14.16)$$

The streamline $\psi_\Omega$ is, from Equation (14.16), the straight line of slope $\theta$ given by

$$z = x \tan \theta + \psi_\Omega/(\sin \theta \cos \theta) \qquad (14.17)$$

where the intercept on the $z$ axis is $\psi_\Omega/(\sin \theta \cos \theta)$. Hence all the streamlines are parallel to each other and make intercepts of increasing magnitude according to the value of the labelling stream function. The zero stream function for vanishing $\psi$ lies along the impermeable bed which is the straight line of slope $\theta$ passing through the origin. The problem is thus that of the flow of groundwater down an impermeable substratum. It is illustrated in Figure (14.8).

The water table emerges as the line where the pressure is zero, so that, from the definition of potential, $\phi_{WT}$ is simply $Z$, where $Z$ is the water table height and $\phi_{WT}$ is the potential at the water table. Then from Equation (14.15),

$$\phi_{WT} = Z = (x \cos \theta + Z \sin \theta)\sin \theta + h \cos \theta$$

whence

$$Z = x \tan \theta + h/\cos \theta \qquad (14.18)$$

Since this also is a straight line parallel to the impermeable bed with slope $\theta$, it is a streamline, with an intercept on the $z$ axis equal to $h/\cos \theta$. If it is supposed that there is no capillary fringe, the water table is the limiting upper surface of the zone of flow where the stream function is $\psi_{max}$, and the thickness of the flow zone, measured perpendicular to the streamlines, is $h$. By comparing Equation (14.17) with Equation (14.18), one finds that

$$\psi_{max}/(\sin \theta \cos \theta) = h/\cos \theta$$

or

$$\psi_{max} = h \sin \theta \qquad (14.19)$$

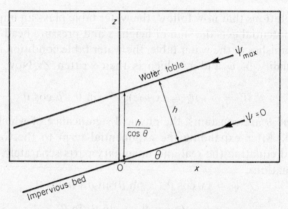

Figure 14.8 The uniform flow of water between a uniformly sloping bed and the parallel water table.

Let $Q$ be the rate of flow per unit width of flow zone, i.e. measured in the direction perpendicular to the plane of $x$ and $z$. Then by Darcy's law

$$Q = -hK \, \text{grad} \, \phi \qquad (14.20)$$

where the negative sign as usual indicates that the flow is downhill, the potential increasing positively upwards. The magnitude of grad $\phi$, $|\text{grad} \, \phi|$, is

$$|\text{grad} \, \phi| = \{(\partial \phi/\partial x)^2 + (\partial \phi/\partial z)^2\}^{\frac{1}{2}}$$

The partial differential coefficients may be derived from Equation (14.15), whence

$$|\text{grad} \, \phi| = \{(\cos \theta \sin \theta)^2 + \sin^4 \theta\}^{\frac{1}{2}}$$
$$= \sin \theta \qquad (14.21)$$

Hence from Equations (14.20) and (14.21),

$$Q = -hK \sin \theta$$
$$= -K\psi_{max} \tag{14.22}$$

*Case (c)*

Now consider the parabolic form,

$$W^2 = (\phi + i\psi)^2 = A(x + iz)$$

or

$$\phi^2 - \psi^2 + 2i\phi\psi = A(x + iz)$$

Equating the real and imaginary parts, one finds,

$$x = (\phi^2 - \psi^2)/A \tag{14.23}$$
$$z = 2\phi\psi/A \tag{14.24}$$

The zero streamline (for which $\psi$ vanishes), which marks the impermeable bed, is evidently the line

$$z = 0; \quad x = \phi^2/A$$

that is to say, it is the positive axis of $x$.

The water table, for which the pressure vanishes so that $Z$ and $\phi$ in Equation (14.24) are synonymous, corresponds to

$$\left.\begin{array}{l} \phi_{WT} = Z \\ \psi = A/2 = \psi_{max} \end{array}\right\} \tag{14.25}$$

and since this is constant the water table is also a streamline, limiting the zone of flow. The shape of this surface is obtained by substituting these values of $\psi_{max}$ and $\phi$ from Equation (14.25) in Equation (14.23), with the result,

$$Z^2 - Ax - A^2/4 = 0$$

This is a parabola whose axis of symmetry is the $x$ axis and with focus at the origin, as shown in Figure (14.9). It cuts the $x$ axis at $-A/4$.

The condition of vanishing $z$ which defines the $x$ axis can also be satisfied by the corresponding forms of Equation (14.24) and (14.35), namely,

$$\phi = 0 \tag{14.26}$$
$$x = -\psi^2/A \tag{14.27}$$

That is to say, the negative axis of $x$ is a zero equipotential along which the stream function increases to its maximum at the water table intersection at $-A/4$. The total potential as well as the height component of potential both vanish together along this axis, and it follows that the pressure component must also vanish. Thus the negative axis of $x$ marks a surface of seepage at which the water emerges at zero pressure.

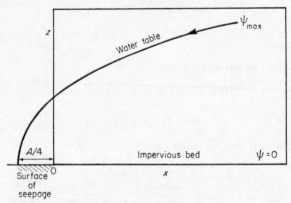

Figure 14.9 The flow of foreign water on a horizontal
bed to an adjoining horizontal surface of seepage.

Hence the solution describes the problem of the flow of foreign ground-water along a horizontal bed from an upstream source to a horizontal overhanging surface of seepage as shown in Figure (14.9). Such a sink might be a rubble-filled trench as shown.

The rate of flow $Q$ per unit width of flow zone, measured perpendicular to the plane of the diagram, may be determined as follows. At the surface of seepage, which is an equipotential, the potential gradient is vertical. Thus by differentiating Equation (14.24) with respect to $z$,

$$\text{grad } \phi = \partial\phi/\partial z = A/(2\psi) \tag{14.28}$$

The potential gradient, and therefore the rate of flow, thus decreases as $\psi$ increases in the negative direction of $x$, and the total flow must be determined by integrating the contribution $dQ$ from elementary strips $dx$ of the surface of seepage. By Darcy's law, and using Equation (14.28),

$$\begin{aligned} dQ &= -K\,dx\,\text{grad }\phi \\ &= -KA\,dx/(2\psi) \end{aligned} \tag{14.29}$$

But along the surface of seepage, Equation (14.27) gives

$$dx/d\psi = -2\psi/A$$

Hence, substituting this value of $dx$ in Equation (14.29),

$$dQ = K\,d\psi$$

and

$$Q = \int_0^{\psi_{max}} K \, d\psi$$

$$= K\psi_{max} \tag{14.30}$$

In terms of the constant $A$, from Equation (14.25),

$$Q = KA/2$$

The constant $A$ of the equation thus turns out to be $2Q/K$ and the final form of the solution, in terms of imposed parameters, to be

$$W^2 = (\phi + i\psi)^2 = (2Q/K)(x + iz)$$

*Case (d)*

The next solution for examination is

$$x + iz = -A \, e^{\pi W/B} + iW + B/2$$

with

$$W = \phi + i\psi$$

Expansion of this expression with the help of Equation (14.1) yields

$$x + iz = -A \, e^{\pi\phi/B}\{\cos(\pi\psi/B) + i \sin(\pi\psi/B)\} + i\phi - \psi + B/2$$

whence

$$x = -A \, e^{\pi\phi/B}\cos(\pi\psi/B) - \psi + B/2 \tag{14.31}$$

$$z = -A \, e^{\pi\phi/B}\sin(\pi\psi/B) + \phi \tag{14.32}$$

The zero streamline, namely at vanishing $\psi$, is seen to satisfy the equations,

$$z = \phi = Z \tag{14.33}$$

$$x = -A \, e^{\pi Z/B} + B/2 \tag{14.34}$$

The first of these equations indicates that the zero streamline is also a water table, since the pressure component of potential is zero. The second gives the streamline shape, shown in Figure (14.10), from which it appears that

$$\left. \begin{array}{l} x = B/2 - A \text{ at } Z = 0 \\ x \to B/2 \text{ as } Z \to -\infty \end{array} \right\} \tag{14.35}$$

Again from Equation (14.34), $x$ vanishes at

$$Z = (B/\pi)\ln(B/2A)$$

but it will be shown that this region of the solution has little application.

From Equation (14.32) it is evident that a second stream function, namely $\psi = B$, also gives the water table condition,

$$z = \phi = Z$$

and since the two water tables must contain between them the whole zone of flow, this second stream function must be the maximum, $\psi_{max}$. Thus

$$\psi_{max} = B \tag{14.36}$$

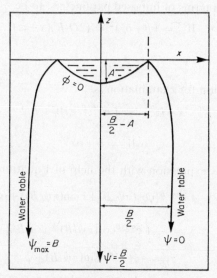

Figure 14.10 The flow of water from the bed of a canal to a deep water table.

Then the shape of this streamline-cum-water table is given from Equation (14.31) by

$$x = A\, e^{\pi Z/B} - B/2 \tag{14.37}$$

This is clearly symmetrical about the $z$ axis with the zero streamline, Equation (14.34), with

$$x = A - B/2 \text{ at } Z = 0$$
$$x \to -B/2 \text{ as } Z \to -\infty \tag{14.38}$$

As one would expect from this symmetry, the $z$ axis is given, as shown by substitution in Equation (14.31), by

$$\psi = B/2$$
$$x = 0$$

The shape of the zero equipotential surface, i.e. at vanishing $\phi$, is, from Equations (14.31) and (14.32),

$$x = -A \cos(\pi\psi/B) - \psi + B/2 \qquad (14.39)$$

$$z = -A \sin(\pi\psi/B) \qquad (14.40)$$

with

$$0 < \psi < B$$

The limits, obtained by substituting the limiting values of $\psi$, are therefore the points

$$_0x_1 = -A + B/2; z = 0$$
$$_0x_2 = A - B/2; z = 0$$

and these are the previously obtained intersections of the bounding streamlines with the $x$ axis. The width of the zero equipotential at vanishing $z$ is thus $C$, where

$$C = {_0x_1} - {_0x_2} = B - 2A \qquad (14.41)$$

The intersection of the zero equipotential with the $z$ axis is obtained by substituting $B/2$ for $\psi$ in Equations (14.39) and (14.40) with the result,

$$x = 0; z = -A$$

The constant $A$ is thus the maximum depth of the zero equipotential surface. The remainder of the equipotential is shaped as shown in Figure (14.10). It could be, for example, the shape of the bed of a canal containing water to the brim. The zone of flow in the upward direction would not exist in such a system and the flow-net for positive $z$ need not be discussed further.

It therefore emerges that the flow-net corresponds to the seepage of water from an unlined canal, of width $C$ and maximum depth $A$, to a water table so far below it as to be irrelevant to the discussion. The seepage band begins at the top with a width $C$ and increases at great depth to a width $B$ such that, from Equation (14.41),

$$B = C + 2A \qquad (14.42)$$

The rate of seepage $Q$, per unit length of canal, may be calculated from the flow at great depth where the streamlines and therefore the gradient of potential are vertical. In this region the equipotentials are horizontal, so that if $z$ is constant on any given equipotential, so is the pressure. At the intersection with the bounding streamline the pressure is known to be zero, since this is a water table, hence the pressure is everywhere zero in this region.

Hence,

$$\phi = z$$
$$\text{grad } \phi = \partial\phi/\partial z = 1$$

L*

By Darcy's law,

$$Q = -KB \operatorname{grad} \phi = -KB$$

Thus from Equation (14.42),

$$Q = -K(C+2A)$$

or from Equation (14.36),

$$Q = -K\psi_{max} \tag{14.43}$$

The negative sign indicates as usual that the movement is downward when the gradient of potential is upward.

In all of these cases other than (a), which is for a confined aquifer, the assumption that a water table bounds the zone of flow is artificial, because the negative pressure or suction just outside the flow zone thus revealed does not immediately reduce the hydraulic conductivity there to zero, except in the case of very coarse granular materials which become unsaturated at very low suctions indeed. For example, the seepage from a canal described in Case (d) might very well approximate to the truth for seepage from an irrigation feeder in a very coarse sand, and for this reason the furrow method of irrigation is not suited to such land. The seepage would be very largely accounted for by losses to the general deep ground-water with very little sideways spread to crop rooting zones. In soils of more usual texture there would be a capillary fringe of appreciable thickness extending the width of the flow zone, and beyond this there would be a zone of lateral moisture profile development as discussed in Section 12.4, or more precisely, the movement would be a process of two-dimensional diffusion of which the above description is a rough approximation.

## 14.7 Steady and transient stages of the water table

In the examples (b), (c) and (d) of Section 14.6 the water table is also a bounding streamline. The boundaries are divided into parts which are equipotentials and other parts which are streamlines, namely, either the impermeable bed or the water table. Thus the operation of Darcy's law accommodates the flow between the input and output equipotentials within a flow-net which has no tendency to expand, since the one free surface, the water table, has at no point any component of flow perpendicular to it. The water table is therefore stationary and the rate of flow will remain unchanged for as long as the potential of the inflow surface remains constant. The flow-net is said to be that of a steady or stationary state.

If, however, the boundary conditions are changed, usually by a change of the input equipotential by a rise of level of water in the supply canal, the solution will naturally be changed to provide a new flow-net corresponding to a new and changed stationary state. The changed dimensions of the boundaries and the rates of flow are given by the solutions in the changed circumstances. The stages by which the flow-net changes from the steady state, appropriate to the initial boundary conditions, to that appropriate to the changed conditions are referred to as non-steady or transient stages, on which the analyses of the previous section throw no light.

The water table of the drainage situation shown in Figure (14.1) is not a streamline, for a rate of precipitation is transmitted across it. Nevertheless, if it is regarded as an imposed surface of uniform zero pressure, then its potential is also imposed and it is a potential-conditioned inflow surface, the drain being the outflow equipotential. The resulting flow-net is determined by these boundaries together with the limiting streamlines located by the conditions of symmetry, so that the rate of flow is determined. In particular, the distribution of the flow across the water table is determined. If the distribution of the flow at the water table so determined is maintained by an imposed input from outside, as for example, by precipitation, the water table will remain stationary; otherwise it will move. If the precipitation is at a rate less than that determined by the water table potential, then a contribution must be made by the groundwater itself by a fall of the water table; or if the rate of precipitation is in excess of that which is determined by the water table position, then there will be groundwater storage and the water table will rise. In each case, the water table will eventually come to rest in a position such that the rate of arrival just matches the rate of flow from the water table to the drain, in accordance with the potential at the water table as determined by its shape and position.

At each transient stage between a change of the boundary conditions and the ultimate stationary state, there will be a distribution of flow appropriate to the water table stage, such that if it were matched by the input distribution, that stage would be a steady state. Thus each transient stage may be regarded as a momentary stationary state appropriate to that particular distribution of flow across the water table.

The next four chapters will be concerned with stationary states as a preliminary to tracing the course of transient stages with the passage of time.

## NOTES

**Note 33. The relationship between the exponential and circular functions**
    The definition of the exponential function $e^{\theta}$ is

$$e^{\theta} = 1 + \theta + \theta^2/\underline{|2} + \theta^3/\underline{|3} + \theta^4/\underline{|4} + \dots \tag{N33.1}$$

where

$$\underline{|n} = (1)(2)(3)(4)\dots\dots(n-1)(n)$$

One may expand $\sin \theta$ into a series of powers of $\theta$ in the form,

$$\sin \theta = A + B\theta + C\theta^2 + D\theta^3 + E\theta^4 + \ldots \qquad \text{(N33.2)}$$

where the coefficients, $A$, $B$, $C$, have to be determined. By differentiation of Equation (N33.2),

$$d(\sin \theta)/d\theta = \cos \theta = B + 2C\theta + 3D\theta^2 + 4E\theta^3 + \ldots$$

When $\theta$ vanishes, $\cos \theta$ takes the value unity, hence,

$$1 = B$$

Again, by further differentiation with respect to $\theta$,

$$d^2(\sin \theta)/d\theta^2 = -\sin \theta = 2C + (3)(2)D\theta + (4)(3)E\theta^2 + \ldots$$

When $\theta$ vanishes, so does $\sin \theta$, hence,

$$C = 0$$

Repeating the differentiation yields

$$d^3(\sin \theta)/d\theta^3 = -\cos \theta = (3)(2)\,D + (4)(3)(2)E\theta + \ldots$$

At vanishing $\theta$

$$-1 = (1)(2)(3)\,D$$

or

$$D = -1/\underline{3}$$

Repetition of the process gives the required coefficients, which, substituted in Equation (N33.2), give the final formula,

$$\sin \theta = \theta - \theta^3/\underline{3} + \theta^5/\underline{5} - \ldots \qquad \text{(N33.3)}$$

The same procedure applied to the function $\cos \theta$ results in the formula,

$$\cos \theta = 1 - \theta^2/\underline{2} + \theta^4/\underline{4} - \theta^6/\underline{6} + \ldots \qquad \text{(N33.4)}$$

Since

$$i^2 = -1$$
$$i^4 = +1$$
$$i^6 = -1$$

and so on, these formulas, Equations (N33.3) and (N33.4), may be rewritten, so as to eliminate all the negative signs. Thus,

$$i \sin \theta = i\theta + i^3\theta^3/\underline{3} + i^5\theta^5/\underline{5} + \ldots \qquad \text{(N33.5)}$$

$$\cos \theta = 1 + i^2\theta^2/\underline{2} + 1^4\theta^4/\underline{4} + \ldots \qquad \text{(N33.6)}$$

Also,

$$e^{i\theta} = 1 + i\theta + (i\theta)^2/\underline{2} + (i\theta)^3/\underline{3} + \ldots \qquad \text{(N33.7)}$$

Comparison of these formulas shows that the sum of Equations (N33.5) and (N33.6) equals Equation (N33.7). Thus,

$$e^{i\theta} = \cos \theta + i \sin \theta \qquad \left\{ \begin{array}{l} \text{(N33.8)} \\ \text{(14.1)} \end{array} \right.$$

**Note 34. Conjugate function relationships**
Consider the function $W(\rho)$, where

$$\rho = x + iz \tag{N34.1}$$

In general it has the real and imaginary part $\phi$ and $i\psi$, so that

$$W(\rho) = \phi + i\psi \tag{N34.2}$$

Hence on differentiating $W$ partially with respect to $x$, and using Equation (N34.1),

$$\partial W/\partial x = (\mathrm{d}W/\mathrm{d}\rho)(\partial\rho/\partial x) = \mathrm{d}W/\mathrm{d}\rho \tag{N34.3}$$

$$\partial^2 W/\partial x^2 = (\mathrm{d}^2 W/\partial\rho^2)(\partial\rho/\partial x) = \mathrm{d}^2 W/\mathrm{d}\rho^2 \tag{N34.4}$$

Similarly,

$$\partial W/\partial z = (\mathrm{d}W/\mathrm{d}\rho)(\partial\rho/\partial z) = i\,\mathrm{d}W/\mathrm{d}\rho \tag{N34.5}$$

$$\partial^2 W/\partial z^2 = i^2\,\mathrm{d}^2 W/\mathrm{d}\rho^2 = -\mathrm{d}^2 W/\mathrm{d}\rho^2 \tag{N34.6}$$

Hence from the addition of Equations (N34.4) and (N34.6),

$$\partial^2 W/\partial x^2 + \partial^2 W/\partial z^2 = 0$$

This is Laplace's equation, which $W(\rho)$ therefore satisfies and of which it is therefore a solution. $W(\rho)$ is any function of $\rho$, since in the above analysis it was not specified.

Furthermore, from Equation (N34.2),

$$\partial^2 W/\partial x^2 + \partial^2 W/\partial z^2 = \partial^2\phi/\partial x^2 + \partial^2\phi/\partial z^2 + i(\partial^2\psi/\partial x^2 + \partial^2\psi/\partial z^2) = 0$$

Hence the real and imaginary parts of the equation must separately vanish, yielding the two equations,

$$\partial^2\phi/\partial x^2 + \partial^2\phi/\partial z^2 = 0$$
$$\partial^2\psi/\partial x^2 + \partial^2\psi/\partial z^2 = 0$$

Thus $\phi$ and $\psi$ separately satisfy Laplace's equation and each is a solution of that equation.

Next, from Equations (N34.3) and (N34.2),

$$\partial\phi/\partial x + i\,\partial\psi/\partial x = \mathrm{d}W/\mathrm{d}\rho$$

or

$$i\,\partial\phi/\partial x - \partial\psi/\partial x = i\,\mathrm{d}W/\mathrm{d}\rho \tag{N34.7}$$

Similarly, from Equations (N34.5) and (N34.2),

$$\partial\phi/\partial z + i\,\partial\psi/\partial z = i\,\mathrm{d}W/\mathrm{d}\rho \tag{N34.8}$$

Hence from Equations (34.7) and (34.8),

$$i\,\partial\phi/\partial x - \partial\psi/\partial x = \partial\phi/\partial z + i\,\partial\psi/\partial z$$

Equating the real and imaginary parts separately, one arrives at

$$\partial\phi/\partial x = \partial\psi/\partial z \tag{N34.9}$$

$$\partial\psi/\partial x = -\partial\phi/\partial z \tag{N34.10}$$

Now $\partial\phi/\partial x$ is the component of the gradient of $\phi$ in the $x$ direction, so that

$$\partial\phi/\partial x = \cos\theta \text{ grad } \phi$$

and similarly,

$$\partial\phi/\partial z = \sin\theta \text{ grad } \phi$$

where $\theta$ is the angle which the direction of grad $\phi$ makes with the $x$ axis.

In a similar manner, it may be seen that

$$\partial\psi/\partial x = \cos\theta' \text{ grad } \psi$$

$$\partial\psi/\partial z = \sin\theta' \text{ grad } \psi$$

where $\theta'$ is the angle between the $x$ axis and the direction of grad $\psi$. Substitution of these values in Equations (N34.9) and (N34.10) yields

$$\cos\theta \text{ grad } \phi = \sin\theta' \text{ grad } \psi \qquad (N34.11)$$

$$\sin\theta \text{ grad } \phi = -\cos\theta' \text{ grad } \psi \qquad (N34.12)$$

Hence by division of Equation (N34.12) by Equation (N34.11),

$$\tan\theta = -\cot\theta' = \tan(\theta'-\pi/2)$$

Hence,

$$\theta'-\theta = \pi/2$$

This indicates that the gradient of $\phi$ is at right angles to the gradient of $\psi$ at the point of intersection; and since the gradient of each of the functions is at right angles to the contours of constant value of that function, it follows that the contours of constant $\phi$ are at right angles to the contours of constant $\psi$ at the point of intersection.

Since $\phi$ satisfies Laplace's equation, it is a possible solution for the potential function, in which case the contours of $\phi$ are equipotentials and therefore the contours of $\psi$ must be streamlines, so that $\psi$ is the stream function. But, quite equally, $\psi$ may be the potential function since it also satisfies Laplace's equation, in which case $\phi$ is the stream function.

CHAPTER 15

# The flow of groundwater: approximate solutions

## 15.1 Numerical solutions of Laplace's equation by successive approximation

IT is evidently a practicable proposition to test whether a proposed potential distribution, however derived, is in fact in accordance with Laplace's equation, since it is possible to deduce the second differentials from the distribution and to test whether

$$\partial^2 \phi/\partial x^2 + \partial^2 \phi/\partial y^2 + \partial^2 \phi/\partial z^2 = 0 \tag{15.1}$$

The process of amending the distribution if it is found not to satisfy Laplace's equation requires further discussion. While the appropriate procedure is quite possible in three dimensions, particularly with the aid of modern digital computers, it has been commonly restricted to two-dimensional problems and will be so limited in this presentation.

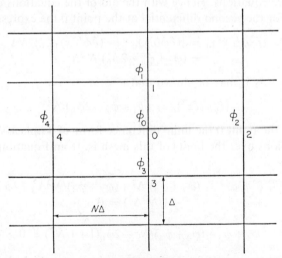

Figure 15.1 A grid of rectangular mesh showing a single relaxation pattern.

321

The distribution of potential may be presented in the form of a geometrical grid of squares, rectangles, or equilateral triangles, drawn within the boundaries of the problem space and with the prevailing values of potential marked at the grid intersections. Laplace's equation must first be written in terms of the finite differences of potential between neighbouring mesh points. To do this, one supposes that the gradient of potential is approximately uniform between such neighbouring points and it follows therefore that these distances must not be too great. Since potentials may vary more sharply in some parts of the system than in others, the mesh must be finer in those parts than in the others. Devices are available for fitting such finer meshes into the more general pattern of coarser mesh.

In Figure (15.1) is shown a unit of rectangular mesh pattern showing one point at potential $\phi_0$ surrounded by the four neighbouring points at potentials respectively, $\phi_1$, $\phi_2$, $\phi_3$ and $\phi_4$. Let the vertical spacing, namely between the points at potentials $\phi_3$ and $\phi_0$ and between those at $\phi_0$ and $\phi_1$, be $\Delta$; and let the horizontal spacing be $N\Delta$ where $N$ is a constant numerical factor. Then one may substitute for the differential coefficients, the approximations

$$\left.\begin{array}{l}(\partial\phi/\partial x)_{4\to 0} = (\phi_0-\phi_4)/N\Delta \\ (\partial\phi/\partial x)_{0\to 2} = (\phi_2-\phi_0)/N\Delta\end{array}\right\} \tag{15.2}$$

Even though the potential may not increase strictly linearly, it is nevertheless a good approximation to assign to the mid-point of an interval a gradient of potential equal to the mean gradient over the interval expressed by the above equations. Hence with the aid of the equations of (15.2) one may write for the second differential at the point 0 the expression,

$$\begin{aligned}(\partial^2\phi/\partial x^2)_0 &= \{(\partial\phi/\partial x)_{0\to 2}-(\partial\phi/\partial x)_{4\to 0}\}/N\Delta \\ &= (\phi_2+\phi_4-2\phi_0)/N^2\Delta^2\end{aligned} \tag{15.3}$$

In a similar way it may be shown that

$$(\partial^2\phi/\partial z^2)_0 = (\phi_1+\phi_3-2\phi_0)/\Delta^2 \tag{15.4}$$

Hence at the point 0 the finite difference form of Laplace's equation in two dimensions over the limits of this mesh is, from Equations (15.3) and (15.4),

$$\partial^2\phi/\partial x^2+\partial^2\phi/\partial z^2 = (\phi_1+\phi_3)/\Delta^2+(\phi_2+\phi_4)/N^2\Delta^2-2\phi_0(1/\Delta^2+ \\ +1/N^2\Delta^2) = 0$$

or

$$\phi_1+\phi_3+(\phi_2+\phi_4)/N^2-2\phi_0(1+1/N^2) = 0 \tag{15.5}$$

By far the commonest form of mesh, and one to which the rest of this discussion will for the most part be restricted, is the square mesh with $N$

having the value unity. The finite difference form of Laplace's equation is then

$$\phi_1 + \phi_2 + \phi_3 + \phi_4 - 4\phi_0 = 0 \tag{15.6}$$

This is, of course, just a statement that the potential at a mesh point is the arithmetic mean of the potentials at the four neighbouring mesh points. This equation provides the basis of the method of Southwell (1946) for solving Laplace's equation by successive approximation and which he called the method of relaxation of restraints or, in brief, relaxation.

In any given problem, the potential will be known along certain specified boundaries, for it is this specification which describes the particular problem. Within this boundary a mesh network is drawn and potentials are assigned to the mesh points by guesswork, with the guidance only of the boundary potentials. At a point such as 0 in Figure (15.1) the potential so assigned will be $\phi'_0$ instead of the true value $\phi_0$ and likewise with the four surrounding potentials. Since these potentials so guessed can only fortuitously satisfy Laplace's equation, in general it will be found that, for a square mesh network

$$\phi'_1 + \phi'_2 + \phi'_3 + \phi'_4 - 4\phi'_0 = R \tag{15.7}$$

where $R$ is called the residual. The object of the procedure is to reduce all the residuals to zero by a systematic adjustment of potentials. To do this one notes, dropping subscripts and denoting guessed potentials by $\phi$, that

$$\phi_1 + \phi_2 + \phi_3 + \phi_4 - 4(\phi_0 + \delta\phi) = R - 4\delta\phi$$

That is to say, if one increases the potential at a point by a specified amount, the residual at that point is decreased by four times that amount. Further,

$$(\phi_1 + \delta\phi) + \phi_2 + \phi_3 + \phi_4 - 4\phi_0 = R + \delta\phi$$

Thus an increase of potential by a specified amount at a mesh point increases the residual by the same amount at each of the points which surround it, for $\phi_1$ is common to four surrounding unit meshes.

The procedure therefore is as follows. One calculates the residual at each of the mesh points and begins by eliminating the largest. This may be done by increasing the potential at that point by one-quarter of the residual if the latter is positive or decreasing it by that amount when the residual is negative. Since the chosen grid point is at the same time the central point of one relaxation pattern and a peripheral point of each of the four surrounding unit patterns, the residuals at each of the four surrounding points must in the first case be increased, or in the second place decreased, by an amount equal to this change of potential. These changes of residual at the surrounding points may make them smaller or larger, but in any case, these changes will be relatively small, so that there is an overall improvement of the potential distribution. One then transfers attention to the largest remaining residual and so on. Eventually, in this way all of the residuals are reduced to acceptably small values. The result is a potential distribution which satisfies Laplace's equation within acceptable limits, at the same time fitting the boundary conditions, so that it is the solution of the problem.

The practised computer can introduce many variations of technique to hasten the work, but for these modifications the reader is referred to the standard texts on this subject. Certain modifications must, however, be mentioned briefly here since they are of basic importance. The derivations of the formulas given may be found in Note 35.

When a relaxation pattern is not symmetrical with four arms of equal length, as for example when a problem boundary lies at an angle with the grid lines and intersects them at varying distances from grid points, the equation for the residual must be reconsidered. If one arm only is cut short, as shown in Figure (15.2), so that $n$ is the fraction which remains within the boundary, and the guessed potential at the end of this arm is $\phi_1$, the equation for the residual is

$$R = \phi_1/n + \phi_2 + \phi_3 + \phi_4 - \phi_0(3 + 1/n) \qquad (15.8)$$

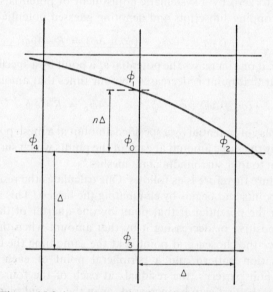

Figure 15.2 A square mesh relaxation pattern with arms of unequal length due to the intersection with a boundary.

For each truncated arm the appropriate term is similarly modified. Thus if arms 1 and 2 are incomplete, the included fractions being $n_1$ and $n_2$, one would have the formula,

$$R = \phi_1/n_1 + \phi_2/n_2 + \phi_3 + \phi_4 - \phi_0(2 + 1/n_1 + 1/n_2)$$

Thus the relaxation requires the following procedure. If the potential at a point is changed by the amount $+\delta\phi$, the residual at that point is changed by $-\delta\phi$ for each of the complete arms which radiate from it to neighbouring grid points, and by $-\delta\phi/n$ for each truncated arm of which only the fraction $n$ lies within the problem boundary. The residual at each neighbouring grid point within the boundary is changed by $+\delta\phi$. At a neighbouring point on a boundary where the potential is imposed there is no change of residual, since such a boundary point is not relaxed.

Now consider a boundary which is a streamline coinciding with one of the grid lines. Such a boundary might be, for example, an impermeable bed. The grid is extended by one row of meshes into a hypothetical space beyond the boundary, and the additional row of grid points so obtained is treated as a mirror image of the row lying one mesh length inside the boundary. Every change of potential and of residual at one of these points in the problem space is repeated exactly at the corresponding point in the image space.

In general, at uncontrolled boundaries of this kind the problem is to add flow-nets in the hypothetical neighbouring space in such a way as to leave the problem flow-net unchanged, but forming a part of an extended flow-net in unrestricted space, to which the relaxation formulas may be applied. An example is shown in Figure (15.3).

A boundary may occur between two zones of different hydraulic conductivities, say $K_1$ and $K_2$ respectively. Figure (15.4) shows such a boundary coinciding with a grid line running parallel to the $x$ axis, so that $\phi_1$ lies in the medium of conductivity $K_1$, $\phi_3$ lies in that of conductivity $K_2$, and $\phi_2$, $\phi_4$ and $\phi_0$ lie on the boundary. The equation for the residual here is

$$R = \phi_1\{2K_1/(K_1+K_2)\} + \phi_2 + \phi_3\{2K_2/(K_1+K_2)\} + \phi_4 - 4\phi_0 \quad (15.9)$$

Thus if a point on such a boundary is relaxed by a change of potential $+\delta\phi$, the residual at that point is changed by $-4\delta\phi$, and the residuals at each of the neighbouring grid points are changed by $+\delta\phi$. If a grid point in the first row on that side of the boundary which has conductivity $K_1$ has its potential changed by $+\delta\phi$, the residual at that point is changed by $-4\delta\phi$ and the residuals at all neighbouring grid points except on the boundary are changed by $+\delta\phi$. At the neighbouring point on the boundary the residual is changed by $+2K_1\delta\phi/(K_1+K_2)$. Similarly, if one changes by $+\delta\phi$ the potential of a point in the first grid row on the side at

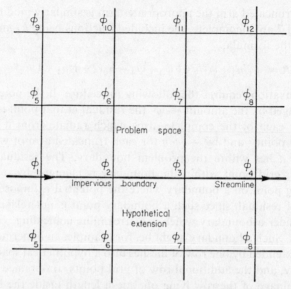

Figure 15.3 The extension of the relaxation mesh into a hypothetical adjoining space by reflection in an impervious boundary.

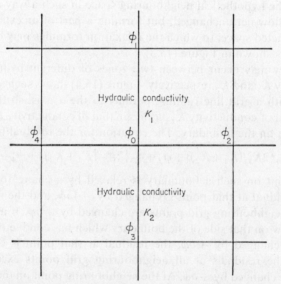

Figure 15.4 The relaxation pattern divided between adjoining spaces of different hydraulic conductivities.

conductivity $K_2$, the residuals are changed by $-4\delta\phi$ at that point, by $+\delta\phi$ at all surrounding points except on the boundary, and by $+2K_2\delta\phi/(K_1+K_2)$ at the point on the boundary.

It sometimes happens that the potential is known initially over only a very small proportion of the boundary, which makes it difficult to guess reasonably the first trial potential distribution and also to control the relaxation. In such cases it frequently happens that a relatively large part of the boundary consists of one or other of two bounding streamlines, that is to say, they are boundaries at which the stream function is imposed and known. Since, as has been shown in Section 14.5, the stream function, equally with the potential function, obeys Laplace's equation, the distribution of stream function may be found by relaxation just as readily as can the potential function, and in the circumstances described this will prove to be the more convenient procedure. When either the potential function or the stream function has been derived throughout the region within the boundaries, the contours of that function may be drawn and the contours of the remaining function obtained by drawing in the orthogonal family of curves, thus completing the flow-net.

As an illustration of the method, one may consider the case of drainage of locally precipitated rainfall by a system of parallel equidistant drains at uniform depth laid in soil of uniform hydraulic conductivity resting on an impermeable bed, as described in Section 14.2. By reference to Figure (14.1) it will be seen that with the exception of drains very near to the edge of the drained land, the symmetry of the flow-net requires that the vertical line through the centre of the drain section shall constitute a pair of stream-lines, one from the water table down to the drain and the other from the impermeable bed up to the drain; and similarly that the vertical line midway between drains shall also constitute a streamline from the water table down to the impermeable bed. The impermeable bed must itself be a bounding streamline. The net result is that the section shown in Figure (15.5) constitutes a unit which, together with its mirror image in the medial plane, provides the flow-net pattern between two drains. The repetition of this pattern across the drainage section provides the complete flow-net of the drained land. It is therefore sufficient to solve the problem of determining the water table and flow-net in the half section depicted in Figure (15.5).

Let it be supposed that the drain, $D$, is just full of water so that its periphery is an equipotential, and let it be defined to be the zero equi-potential, so that its roof is the origin of coordinates. The water table, $PQ$, where $P$ is vertically over the drain, is known to have zero pressure, so that when it is located, the boundary condition is known to be that the potential is simply the height component. At this boundary, if it be assumed that the flux due to rainfall at the rate $q$ is uniformly distributed, the stream function increases uniformly from zero at $P$ to $qL$ at $Q$, where $L$ is half the distance separating the drains. The line $PD$ is the zero bounding streamline, and the line $QRSD$ is the maximum bounding streamline. Hence, if one chooses to solve this problem by relaxing the potential function, the only parts of the boundary at known potential are $QP$ and the drain perimeter $D$. If, however, one chooses to relax the stream function, the whole of the boundary is at a known value of stream function with the exception of the very small drain perimeter. Hence it is natural to choose the latter procedure.

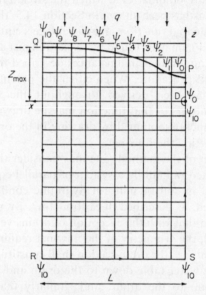

Figure 15.5 Cross-section of a drainage system as shown
in Figure (14.1), set out for solution by relaxation of the
stream function.

There are two elements of guesswork, since the water table location as well as the distribution of stream function is initially unknown. One therefore begins with a chosen position for $Q$ at height $Z_{max}$ above the drain and draws in a trial water table curve $QP$ as a first approximation. Next one draws in the boundaries $PD$, $DS$, $SR$, and $RQ$ to scale, to represent the known boundaries of the problem. The impermeable bed is at depth $p$ below the drain. On such a diagram the drain will almost certainly be so small as to be represented by a point, and through this point one draws the $x$ axis. This, together with $PDS$ as the $z$ axis, defines the directions of the grid lines of the square mesh network into which one divides the interior space. This might conveniently divide the water table into, say, ten parts.

Starting at $P$ one labels the water table divisions with successively increasing values of stream function from 0 to, say, 1000 in steps of 100, ending at $Q$. All mesh points at the boundary $PD$ excepting $D$ itself, are labelled zero, and all points on the rest of the boundary, $QRSD$, again excepting $D$, are labelled with the number 1000. At $D$ itself, where all streamlines converge, the stream function is assigned the average value, namely 500 in this case. At each of the interior mesh points a value of stream function is assigned by personal judgment, which will improve remarkably with experience. One then improves all these interior stream functions in turn by relaxation as described. It will, of course, be found that certain residuals which are liquidated at one stage of the relaxation will reappear later due to correction of the stream function at neighbouring points, but eventually the stream function everywhere will be found to satisfy Laplace's equation within the prescribed limits, which in this case might well be one or two units.

The next stage is to draw a set of, say, nine interior streamlines at intervals of stream function of 100 units. This is done in the usual way by plotting the stream function along each of the grid lines of the network and by interpolating on these lines the points at which one finds the selected stream functions. The points of constant stream function are then joined smoothly to give the required streamline. The flow-net is then completed by drawing a family of equipotentials such that wherever an equipotential crosses a streamline, it does so at right angles. Neighbouring equipotentials are so spaced that each mesh has the same ratio of width to length, that is to say, it is a curvilinear rectangle. It may conveniently be "square". The equipotentials which pass through the two ends of the trial water table can be labelled since the respective potentials are simply the heights of these two points, the pressure component of potential being zero at the water table. The intermediate equipotentials therefore may also be labelled, since they are at equal intervals of potential between these two limits, as described in the general discussion of the flow-net in Section 14.5. Hence these labelled potentials indicate the heights at which the respective equipotentials should intersect the water table, and this locus may therefore be drawn in and the flow-net extended to it, if necessary. In general it will be found that this emergent "water table" is not the same as that which was drawn as the trial guess. Neither of course is it the true required water table, since the incident rainfall is not distributed uniformly over it. One therefore uses it as a second trial for a repetition of the relaxation procedure. Eventually, after perhaps three or four repetitions, one arrives at a coincidence between the trial and the emergent water tables at which simultaneously the rainfall is uniformly distributed and the potential equals the height above the datum drain level. This then is the true water table.

From the nature of this procedure one also has the complete flow-net within the boundaries, from which one can estimate the relationship between the rainfall rate $q$ and the hydraulic conductivity $K$, which together produce this flow pattern. If one considers any one of the curvilinear rectangles, one sees that it is bounded by two labelled equipotentials such that the difference of potential $\delta\phi$ is given by

$$\delta\phi = \delta Z$$

where $\delta Z$ is the vertical height between the intersections of the equipotentials with the water table. This is the same whichever pair of equipotentials is in question. The remaining boundaries of the rectangle are neighbouring streamlines which intercept the water table with a horizontal separation $\delta L$, and this also is the same whichever pair of streamlines is in question. The rate $\delta Q$ of flow of water through the rectangle is thus equal to the rate at which rainfall arrives at the catchment area $\delta L$, that is,

$$\delta Q = q\delta L$$

Now let the width and length of the rectangle be respectively $X$ and $Y$. Application of Darcy's law to this element yields the result,

$$q\delta L = KX\delta Z/Y$$

or

$$q/K = (X/Y)(\delta Z/\delta L) \qquad (15.10)$$

The ratios $(X/Y)$ and $(\delta Z/\delta L)$ are known when the flow-net has been revealed by the relaxation process, the former ratio by measuring a sufficient number of elementary rectangles of the flow-net to provide an acceptable average. Hence the value of $q/K$ which is responsible for this particular flow-net, namely the solution of this particular problem, is identified.

It may further be noted that the same solution may be interpreted at any scale desired, since the proportional enlargement of all linear dimensions leaves unchanged the ratios $(X/Y)$ and $(\delta Z/\delta L)$. In particular, the crucial ratio $Z_{max}/L$ is always the same for the same value of $q/K$.

The solution of a sufficient number of particular cases of this kind permits the construction of a set of curves presenting the relationship between $Z_{max}/L$ and the significant parameters $q/K$ and $p/L$. In principle the drain radius $r$ should emerge from the solution as the zero equipotential, so that $Z_{max}/L$ could also be expressed as a function of $r/L$, but in practice the accuracy which the method permits in the very small region of sharply varying potential near the origin is inadequate for the purpose. The whole question of the influence of the drain size must be left for discussion in a later section.

**15.2 Procedure by analogy**

One of the elements of initial guesswork which are characteristic of the relaxation method may be eliminated by substituting for the porous material, instead of a meshwork of points on paper, another conducting body in which may be constructed an analogue of the hydraulic problem. The commonest of such methods is that of the electric analogue, and in this section the detailed discussion of such procedures will be restricted to electrical methods. However, heat analogues also are practicable.

The law of conduction of electricity in ohmic conductors is

$$I = CA \text{ grad } V$$

and this is formally similar to Darcy's law, with $I$, the strength of the electric current, playing the part of the rate of flow of water in the hydraulic conductor; $C$, the electrical conductivity, simulating the hydraulic conductivity $K$; and the voltage, $V$, substituting for the hydraulic potential or head, $\phi$. In both cases $A$ is the area of cross-section of the conductor. Hence if one can discover in the electrical system an analogue of the boundary conditions in the hydraulic problem, one may construct to scale an analogue or model of the problem and from it derive the required elements of the solution. Again one may illustrate the method by reference to the problem that was studied in Section 15.1.

Since this problem is two-dimensional, the appropriate electric conductor is also two-dimensional, namely a conducting sheet, the conductivity of which is sufficiently low to enable one to impose equipotentials where necessary by sealing to the sheet copper electrodes maintained at a suitable voltage. Such sheets are commonly and conveniently of graphited paper, which may be obtained commercially. Alternatively, appropriately shaped shallow containers of solutions of electrolytes may be used.

No difficulty arises in the case of the boundaries which coincide with the bounding streamlines $PD$ and $QRSD$ of Figure (15.5). A cut along a streamline affects the system not at all, since there is no tendency for the current to flow from the region on one side of the streamline to that on the other. Hence the sheet conductor is simply cut to shape to follow these boundaries to any convenient scale.

The drain perimeter is the zero equipotential, hence the appropriate analogue is a copper conductor shaped and placed to scale to represent the drain and maintained at a constant voltage which is conveniently the zero of reference.

The position and shape of the water table is not known initially, and in fact this information is a main part of the required solution of the problem. The boundary conditions here are that the distribution of inflow is prescribed, and that the hydrostatic pressure is zero. The latter condition specifies that the hydraulic potential is simply the height of the water table point relative to the adopted datum level, which is conveniently the drain. Again the distribution of inflow will be taken to be uniform, as for rainfall, but this is not a necessary restriction. It is evident that the distribution of inflow of water may be simulated by a similar distribution of electric current led in by suitable electrodes, but the location of the boundary at which it enters must be guessed in the first place. The analogue eliminates the element of guesswork as to the interior potential distribution, but the characteristic successive approximations to the true water table remains. The test of the adequacy of the trial water table is the extent to which the voltage at that boundary is proportional to the height, as is required by the analogy to the hydraulic potential boundary condition.

The known boundary conditions of analogue shape, drain electrode potential and current distribution over the guessed water table having been duly imposed, the interior potential distribution is ascertained by exploration with a probe electrode connected to a potentiometer or to an electronic voltmeter, or some such voltage measuring device which does not itself upset the voltage distribution. The measurements are conveniently made at a regular network of interior points, from which equipotentials may be interpolated if required and from which in particular one may plot the locus of points at which the voltage is proportional to the height above the drain level. The constant of proportionality is known from the circumstance that the trial water table is drawn from a chosen height at the mid-point, and the voltage at this point is known from the measurements.

If the locus of points at which the voltage is proportional to the height does not coincide with the initially chosen boundary over which the current inflow condition is satisfied, then neither this chosen boundary nor the discovered locus is the water table analogue, since the two conditions must be satisfied simultaneously. Hence the discovered locus is chosen as the second trial boundary, the conducting sheet is recut to this shape to represent the trial water table, the current input distribution is readjusted to this new boundary, and the interior potential distribution again observed. From this a second locus of points emerges which satisfies the potential boundary condition at the water table. A few repetitions of this procedure suffice to determine a boundary at which the two boundary conditions of potential and inflow distribution are simultaneously satisfied, and this boundary on the analogue simulates the water table in the hydraulic problem.

At this stage, the known potential distribution enables the experimenter to draw the complete flow-net, and from this the procedure to determine the rainfall and conductivity parameters which are responsible for the flow-net are the same as was described in Section 15.1.

Although, if a sufficient number of solutions are available, it may then be possible to derive some general empirical laws of the relationships between the relevant parameters, the chief use of these methods is to solve particular specified problems. This is not the place to describe in detail an exhaustive list of such solutions, but such a list would contain items dealing with local and foreign drained groundwater (Childs 1943b, 1946), and with idealized cases of the Ghyben-Herzberg lens (Childs, 1950). It is, however, permissible to mention certain matters of principle which have been illuminated by the method of analogues. It is common practice in the field to measure the height of the water table by observing the level of water in a borehole. This is satisfactory when the groundwater is stationary or stagnant, but when it is moving in any direction other than the horizontal, the borehole itself perturbs the flow of groundwater by intersecting neighbouring equipotential surfaces and providing a short-circuiting path of low resistance. The result is to produce a drawdown at the borehole so that the water level observed is not that which obtains in the general vicinity. The magnitude of this error has been studied by electrical analogue methods (Childs, 1945a).

Secondly, it is usual to suppose that the water table is the boundary between the groundwater zone, where the flow is in uniform saturated material, and an upper zone where the flow is vertical. In fact, it is known that there is a capillary fringe of essentially saturated material lying above the water table. The analogue of the upper boundary of the fringe, which is more nearly the boundary between the two zones of flow, may be specified, since this boundary may be idealized as the surface at which the suction is known, since it is the suction at which, as shown by the moisture characteristic, appreciable water begins to be lost. Thus at the water table the potential obeys the law,

$$\phi = z$$

where $z$ is the height relative to the datum level, but at the boundary of the capillary fringe,

$$\phi = z + H_{cf}$$

where $H_{cf}$ is the negative pressure head which characterizes that boundary. Hence while the analogue of the water table condition is

$$V = Bz$$

where $V$ is the voltage at the water table analogue and $B$ is a constant, the analogy to the capillary fringe boundary is

$$V = Bz + E$$

where $E$ is a constant which represents the constant air entry pressure $H_{cf}$

of the soil. The electric analogue procedure may be carried out as readily with the fringe boundary condition as with that proper to the water table (Childs, 1945a), and it has been observed that there is very little difference between the fringe boundary thus determined and that which results from determining the water table boundary by ignoring the effects of the capillary fringe and superimposing on that boundary a layer of thickness approximately equal to the static thickness of the capillary fringe, namely $H_{cf}$. This would not be true for rates of infiltration approaching in intensity the saturation conductivity of the soil, as may be seen very easily by computing the moisture profile by the methods of Section 12.2, but in any case this would be a condition where effective drainage of groundwater is almost impossible so that the discussion becomes somewhat academic.

One may also study the errors introduced by the simplifying assumption that the flow in the upper unsaturated material is truly vertical, by treating the whole zone from the surface to the impermeable bed as one single region for which the analogue must be found (Childs, 1945b). At any point in this zone the potential is simulated by the voltage of the analogue, and since the height above datum is directly represented in the analogue, the difference of potential and height, which represents the hydrostatic pressure component, is determinable from the analogue. Thus if the moisture characteristic is known, it becomes possible to determine from the analogue the moisture content at the relevant points in the hydraulic flow-net. The moisture content in turn determines the hydraulic conductivity as a fraction of that of the saturated soil, and therefore in the analogue the electric conductivity at this point must bear this same ratio to the electric conductivity which represents the saturated zone. In principle it is possible to achieve this condition by an additional process of trial and error by differential painting of the sheet conductor with a suspension of colloidal graphite, but in fact only highly idealized types of moisture characteristic are practicably amenable to treatment.

Nevertheless, enough has been done in this direction to show that the error introduced by the arbitrary divorce between the groundwater zone and the unsaturated zone is usually small. This is for two reasons. Firstly, the refraction of the streamlines at the boundary between the zones tends in any case to produce streamlines which are more nearly vertical in the unsaturated soil than in the groundwater, because, as described in Note 35 and illustrated in Figure (15.6), the refraction is toward the normal in the region of lower conductivity. Secondly, unless the zone of unsaturated soil is very thick compared with the separation of the drains, there simply is not sufficient depth to the water table for the deviation from the vertical direction to result in a distribution of flow at the water table which is very markedly different from that at the surface, and it is this identity of rates of flow that must be assumed in order to proceed with the simplified analysis. If the unsaturated zone is indeed of great thickness, then there is no call for drainage and the problem does not in any case arise.

While the conducting sheet analogue is very easily understood and is very flexible in dealing with problems of a great variety of geometric boundaries, there are circumstances in which it has notable disadvantages, principally when the soil to be simulated is not uniform but has a distribution of conductivity between wide limits. As has been said, it is not easy to simulate such soil quantitatively by differential painting of the analogue.

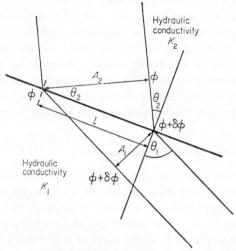

Figure 15.6 The refraction of streamlines on crossing the boundary between regions of different conductivity.

A modification of the method to deal with such circumstances is best regarded as a hybrid between the analogue and the relaxation methods. The conducting region is simulated by a grid meshwork of points between which are connected resistors which may be either fixed or variable, as may be desired, but which in any case may be changed or varied at the will of the computer much more readily than can a sheet conductor. The construction and programming of this type of analogue computer is a specialized subject and will not be described further here. The reader may be referred to Karplus (1958). The obvious field for the application of this technique is the study of the problem, referred to above, of the flow of water through the whole zone between the surface and the impermeable bed, and some exploratory work of this kind has been reported by Bouwer (1959).

Among analogues should be listed hydraulic models, since although the medium used in such models may well be a porous material, such as a uniform coarse sand, it has a physical structure so different from that of soil that confusion may be introduced into discussion if the analogue is simply referred to as a scale model. The only requirement of such an analogue is that it should have a known hydraulic conductivity, so that the ratio of flux to conductivity may be directly comparable with the measured ratio in the soil of the full-scale hydraulic problem. It does not matter that the conductivity of the soil may be in the main a structural phenomenon and that of the analogue may be a consequence of texture. It matters only that in each the movement of water should obey Darcy's law and that in each the conductivity be measurable and known.

A very particular case of hydraulic analogue is the Hele-Shaw model in which the cross-section of the two-dimensional hydraulic problem is replaced by the confined space between two narrowly separated parallel glass plates, and the fluid is usually one of enhanced viscosity. Since it has already been demonstrated that in two dimensions the solution of a problem does not depend on the scale of the model or on the scale at which the results are interpreted, further discussion is not necessary.

### 15.3 The Dupuit-Forchheimer approximation

It sometimes happens that the boundaries of a problem restrict the direction of flow of the groundwater to an approximately horizontal direction almost everywhere. For example, when there is a thin layer of permeable material of considerable extent, resting on an underlying horizontal impermeable bed, at the edges of which there is drainage, then the streamlines originating at any points which are not quite near the drainage boundary must be contained between the horizontal bed and the water table closely overhead, and are therefore constrained in approximately horizontal directions. It is only in the close vicinity of the drainage boundary that appreciable divergence from the horizontal can occur. This is a sufficient condition for the adoption of the approximation proposed by Dupuit (1863) and elaborated by Forchheimer (1914).

It follows at once from the orthogonal property of streamlines and equipotentials that if the former are approximately horizontal, then the latter must be approximately vertical, provided that the material is isotropic. At the intersection of an equipotential with the water table, the hydrostatic pressure is zero and the potential is simply the height $Z$ of the water table at that point. Hence the whole of the equipotential is at potential $Z$, and therefore the gradient of potential at all depths immediately underneath a given point on the water table is simply the gradient or slope of the water table itself. Darcy's law may therefore be written in terms of this easily observed gradient. The result is usually an equation which is amenable to elementary integration to give the shape of the water table and its height at the highest point related to the soil hydraulic conductivity and the rate of input of water at whatever input surfaces may be specified in the particular problem. The errors due to the approximation are naturally dependent upon the extent to which the supposition of horizontal streaming reflects the true facts. Usually the errors of shape of the water table are greater than the errors of relationship between the water table maximum height and the ratio between the hydraulic conductivity and the rate of flow of water.

The use of this approximation is exemplified by reference to some particular problems. In the first case the full derivation of the solution is presented, but in the subsequent problems only the solutions are presented in the text. The derivations of these solutions may be found in Note 36.

### (a) Foreign water on a horizontal bed

This case is illustrated in Figure (15.7). It is characterized by flow through

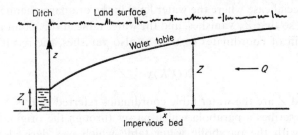

Figure 15.7 Dupuit-Forchheimer diagram. Cross-section of groundwater with foreign water draining to a ditch, which penetrates to an impervious bed.

the conducting body based on a horizontal impermeable bed, the source being at a great distance upstream. Such a source might be a canal dug to the full depth of the conducting layer, but, in fact, if sufficiently distant, it would hardly matter if it were dug to the full depth or not, so far as the solution over the zone shown is concerned. The sink or drain is a channel dug to the full depth to prevent it being by-passed by flow beneath it. There is no local addition to the flow by percolation to the water table from above.

Let the datum level, with respect to which heights $z$ are measured, be the impermeable bed; and let the horizontal direction in the plane of the flow-net be the direction of the $x$ axis. The thickness of the lamina perpendicular to the plane of the flow-net is taken to be unity, so that the area, across which liquid flows at a section where the water table height is $Z$, is itself simply $Z$. The gradient of potential at this cross-section is, in accordance with the Dupuit-Forchheimer assumption, $dZ/dx$. Hence Darcy's law as applied at this section takes the form,

$$Q = -vZ = KZ\, dZ/dx$$

where $Q$, the approximately horizontal rate of flow, is the same at all sections, i.e. for all values of $x$, provided that a steady state of flow has been attained.

This may be integrated directly to give

$$2(Q/K)(x_2 - x_1) = (Z_2{}^2 - Z_1{}^2) \qquad (15.11)$$

where $Z_1$ and $Z_2$ are the water table heights at distances $x_1$ and $x_2$ respectively. For example, $Z_2$ might be the depth of water in an upstream channel dug to the full depth of the bed, and $Z_1$ the depth of water in a similar drainage trench downstream, the outflow surface of seepage being ignored. The factor $(x_2 - x_1)$ then represents the distance between the inflow and outflow channels. Or again, this section might represent the flow between the upstream and downstream faces of an earth dam with vertical faces.

In the special case where the water level in the drainage channel is on the drain floor, so that $Z_1$ vanishes, and the water table at this point is taken to be the origin of coordinates so that $x_1$ also vanishes, the result takes the form,

$$2(Q/K)x = Z^2 \qquad (15.12)$$

where $x$ and $Z$ are the water table coordinates referred to this origin. This equation describes a parabola which passes through the origin. It may be compared with the parabolic water table which was derived in similar circumstances in Section 14.6 by appeal to conjugate function analysis. This more rigorous treatment gave the result,

$$Z^2 = 2(Q/K)x + (2Q/K)^2/4 \qquad (15.13)$$

This is precisely the same parabola displaced along the $x$ axis by the distance $Q/2K$, this distance being occupied by a draining surface of seepage. Thus it appears that quite fortuitously the errors, due on the one hand to assuming a flow limited to the horizontal direction and on the other to ignoring the surface of seepage, neutralize each other. This feature will be observed again in other examples.

### (b) Local precipitation with parallel drains

This case is illustrated in Figure (15.8), in which is shown a series of equidistant parallel drain channels with separation $2L$, dug to the full depth of the impermeable bed. It is a consequence of the symmetry of the system that the vertical planes through the drain channels and midway between neighbouring channels provide streamline boundaries, so that precipitation at the water table on one side of the medial plane proceeds wholly to the drain on that side. Again $Z$ will be measured from the impermeable bed and $x$ from the drain channel.

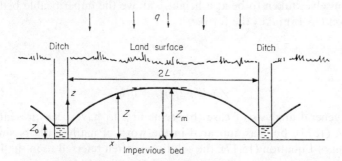

Figure 15.8 Dupuit-Forchheimer diagram. Cross-section of groundwater with local rainfall draining to parallel ditches penetrating to the impervious bed.

The equation which describes the water table is, as shown in Note 36,

$$(Z^2 - Z_0^2)/(2Lx - x^2) = q/K \tag{15.14}$$

Here $q$ is the rate of steady precipitation at the water table from above, expressed as a velocity in the same units as $K$, as for example, metres per day. $Z_0$ represents both the level of water in the drainage channel and the height of the water table at this point, since the surface of seepage is neglected.

Equation (15.14) is the equation of an ellipse with its centre at the intersection of the medial plane with the impermeable bed, and at this section the water table attains its maximum height $Z_m$ given by Equation (15.14) when $x$ is assigned the value $L$. Thus

$$(Z_m^2 - Z_0^2)/L^2 = q/K \tag{15.15}$$

In the special case when the water level in the drainage channel is kept down to bed level, so that $Z_0$ vanishes, this result reduces to the oft-quoted form,

$$Z_m^2/L^2 = q/K \tag{15.16}$$

In this last case the water emerges from the soil along a line, and the drainage channel could be a pipeline without vitiating the basis of the Dupuit-Forchheimer method. The result is, in fact, commonly applied to drainage systems which employ sub-surface drains in the form of parallel pipelines.

M

If the drain does not penetrate to the impermeable bed, there is an enhanced departure of the streamlines from the strictly horizontal direction in the neighbourhood of the drain, with a consequent increase in the error due to the adoption of the Dupuit-Forchheimer assumption. In this case, if one measures the water table height $Z'$ from the level of the drains, which are themselves taken to be at a height $p$ above the impermeable bed, then Equation (15.15) takes the form,

$$Z_m = Z' + p$$

$$Z_0 = p$$

$$Z'(Z' + 2p)/L = q/K \tag{15.17}$$

The general analysis of case (b) seems first to have been presented by Colding (1873), but has appeared in the work of many authors since. In the form of Equation (15.17), the solution is often referred to in the U.S.A. as the Donnan equation (Aronovici and Donnan, 1946).

### (c) The Ghyben-Herzberg lens

The circumstances to be treated here are shown in an idealized form in Figure (15.9). A permeable isthmus or ridge of land with vertical faces

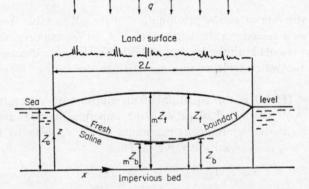

Figure 15.9 Dupuit-Forchheimer diagram. Cross-section of an idealized Ghyben-Herzberg lens.

separated by the distance $2L$ is based on a horizontal impermeable bed. The depth of sea water on each side is $Z_0$ and the rate of rainfall over the surface is $q$. The medial plane is a plane of symmetry. The boundary with the sea is taken to be the origin of $x$. The density of the sea water is taken as $\rho_s$ and that of fresh water as $\rho_f$. The hydraulic conductivity may be taken as uniform at the value $K$, irrespective of whether one refers to the fresh or salt zones, since the salt content does not markedly affect the viscosity of water upon which the conductivity depends. If the salinity causes structure breakdown, there must be a profound effect, of incalculable magnitude, on the conductivity, and the analysis is completely inapplicable.

As indicated, the flow of fresh water, which arrives at the water table and drains at the edges to the sea, takes place between a raised water table and a boundary between the fresh and salt water bodies which is depressed below sea level. Hence, one is required to find the shape and location of both of these boundaries as functions of the intensity of precipitation $q$ and the hydraulic conductivity $K$ of the soil. If $Z_f$ is the height of the water table and $Z_b$ is the height of the boundary between the fresh and salt water, both measured relative to the level of the impermeable bed, then, provided that the boundary is not depressed so deeply as to touch the impermeable bed, the required expressions are, as is shown in Note 36,

$$(Z_f - Z_0)^2/(2Lx - x^2) = (q/K)(1 - \rho_f/\rho_s) \qquad (15.18)$$

$$(Z_0 - Z_b)^2/(2Lx - x^2) = (q/K)(\rho_f/\rho_s)/(\rho_s/\rho_f - 1) \qquad (15.19)$$

The highest point of the water table, $_mZ_f$, occurs at the mid-point where $x$ takes the value $L$, so that, from Equation (15.18),

$$(_mZ_f - Z_0)^2/L^2 = (q/K)(1 - \rho_f/\rho_s) \qquad (15.20)$$

Similarly the lowest point of the salt-fresh boundary, $_mZ_b$, is obtained from Equation (15.19) by equating $x$ to $L$. The result is

$$(Z_0 - _mZ_b)^2/L^2 = (q/K)(\rho_f/\rho_s)/(\rho_s/\rho_f - 1) \qquad (15.21)$$

It follows from Equations (15.18) and (15.19) that, as is derived more basically as Equation (N36.7) of Note 36,

$$(Z_f - Z_0)/(Z_0 - Z_b) = \rho_s/\rho_f - 1 \qquad (15.22)$$

If one takes a commonly observed value for $\rho_s$, namely about $1 \cdot 025$, then Equations (15.20) and (15.21) give the values,

$$_mZ_f - Z_0 = (0 \cdot 025 \, q/K)^{\frac{1}{2}}$$

$$Z_0 - _mZ_b = (39 \, q/K)^{\frac{1}{2}}$$

while Equation (15.22) reduces to the numerical form,

$$(Z_f - Z_0)/(Z_0 - Z_b) = 0 \cdot 025$$

The boundary between the salt and fresh groundwater cannot fall below the impermeable bed, so that Equations (15.20) and (15.21) must fail at all values of $q/K$ which are large enough to demand negative values of $_mZ_b$ to satisfy Equation (15.21). For such excessive values of $q/K$, Equations (15.18) and (15.19) still apply for values of $x$ up to the value, $x_e$, at which $Z_b$ vanishes, as determined by Equation (15.19). This value $x_e$ may be determined by solving Equation (15.19) for $x$ with zero $Z_b$. The value of $Z_f$ at this point is, from Equation (15.22), $_eZ_f$ where

$$_eZ_f = Z_0 \rho_s / \rho_f \qquad (15.23)$$

For greater values of $x$ where the fresh groundwater rests directly on the impermeable bed, Equation (15.14) gives the value of $Z_f$ when the origin is suitably shifted by substituting $L - x_e$ for $L$ and $x - x_e$ for $x$. After a little algebra, the result is

$$Z_f^2 - _eZ_f^2 = (q/K)\{(x - x_e)(2L - x - x_e)\} \qquad (15.24)$$

The two limits $Z$ and $Z_0$ of Equation (15.14) are here the two limits $Z_f$ and $_eZ_f$. At the highest point where $x$ assumes the value $L$, $Z_f$ has the maximum value $_mZ_f$ given by

$$_mZ_f^2 - _eZ_f^2 = (q/K)(L - x_e)^2 \qquad (15.25)$$

Since $_eZ_f$ is known from Equation (15.23), $Z_f$ for values of $x$ in excess of $x_e$ follows from Equation (15.24), and $_mZ_f$ at the mid-point of the isthmus from Equation (15.25).

## 15.4 An exact solution for comparison with the Dupuit-Forchheimer approximation

If flow takes place between two parallel vertical plane surfaces which penetrate to an impermeable floor, and the potential distribution is known over these boundary surfaces, it is possible to determine the rate of flow by an exact theory described by Polubarinova-Kochina (1962) for uniform conductivity, and extended by Youngs (1965, 1966) to take into account non-uniform conductivity. Only the former will be described here, since the principles of the method are made plain without the additional mathematical complication necessitated by the more general case.

If the component of the velocity of flow in the horizontal $x$ direction is $v_x$ equal to $-K(\partial\phi/\partial x)$, then the total rate of flow across a vertical section of unit width from the floor to the water table at height $Z$ is $Q$, where

$$Q = -K \int_0^Z (\partial\phi/\partial x) \mathrm{d}z \qquad (15.26)$$

The subsidiary function $J$ is now introduced where

$$J = \int_0^Z \phi \, \mathrm{d}z \qquad (15.27)$$

so that

$$dJ/dx = \int_0^Z (\partial\phi/\partial x)dz + \phi_Z(dZ/dx)$$

Since $\phi$ and $Z$ are identical at the water table, where the pressure vanishes, this equation may be written, with the help of Equation (15.26),

$$dJ/dx = -Q/K + Z(dZ/dx)$$

whence, by direct integration between the limits $x_1$ and $x_2$,

$$J_2 - J_1 = -(1/K)\int_{x_1}^{x_2} Q\, dx + (Z_2{}^2 - Z_1{}^2)/2$$

This in turn may be written, with the help of Equation (15.27) and the use of a little algebra, in the form,

$$(1/K)\int_{x_1}^{x_2} Q\, dx = \int_0^{Z_1}(\phi_1 - Z_1)dz - \int_0^{Z_2}(\phi_2 - Z_2)dz + (Z_1{}^2 - Z_2{}^2)/2 \quad (15.28$$

This general expression may be applied to the special case of the flow of foreign water on a horizontal bed to a vertical ditch face if it be assumed that at a sufficient distance upstream of this face the equipotentials are vertical plane surfaces. This case is identical with the flow of water between the vertical faces of an earth dam or barrage. For the purpose of applying Equation (15.28), the downstream face is identified with the surface at $x_2$ and the upstream equipotential with the surface at $x_1$, and this latter may conveniently be taken as the origin of $x$. At this point let the water table height be $Z_m$. If the upstream boundary is the vertical face of a dam, $Z_m$ is also the height of the water at this upstream face. At the downstream or ditch face, distant $L$ from the upstream equipotential, the level of the external water is taken to be $Z_w$ and the height of the water table just inside the face is $Z_L$. This implies the existence of a surface of seepage in the downstream face between the levels $Z_w$ and $Z_L$, at which water emerges at a pressure which is limited to zero by the circumstance of its free escape.

Equation (15.28) may now be applied. At the upstream equipotential at vanishing $x$, the potential $\phi_1$ is uniform at the value $Z_m$, which is also the value of $Z_1$. Hence the first of the terms on the right-hand side vanishes. At $L$, the value of $x_2$, the potential $\phi_2$ is uniform at the value $Z_w$ up to the height $Z_w$, but between $Z_w$ and $Z_L$, which corresponds to $Z_2$, the potential is simply the height $z$. Hence Equation (15.28) takes the form,

$$(1/K)\int_0^L Q\, dx = (Z_m{}^2 - Z_L{}^2)/2 - \int_0^{Z_w}(Z_w - Z_L)dz - \int_{Z_w}^{Z_L}(z - Z_L)dz$$

$$= (Z_m{}^2 - Z_w{}^2)/2 \qquad (15.29)$$

In this example $Q$ receives no accretion between the two boundaries and is constant, hence finally,

$$(Q/K) = (Z_m^2 - Z_w^2)/2L$$

Precisely the same result is given by the Dupuit-Forchheimer approximation, Equation (15.11), in which the equipotentials are assumed to be everywhere vertical and where, in consequence, it is implied that there is no surface of seepage at the downstream face. Hence it appears that the errors that may be inherent in these two assumptions fortuitously cancel each other.

The treatment may now be extended to drainage of locally precipitated rainfall by a system of equidistant parallel ditches of separation $2L$. By symmetry, the medial plane between a pair of drains is a streamline which divides the flow-net into two mirror image catchments, and this plane may conveniently be the origin of $x$. The total flow passing the section at $x$ is thus the total rate of precipitation on the catchment area between the medial plane and the vertical plane at $x$. Thus,

$$Q = qx \tag{15.30}$$

The conditions at the ditch face, which determine the second integral of the right-hand side of Equation (15.28), are the same as before, but the potential distribution over the medial plane is not known. However, this distribution is known to lie between limits. At the one extreme, the rate of vertical flow everywhere in this plane could be zero so that the plane would be an equipotential as in the case of foreign water. Then the whole of the right-hand side of Equation (15.28) is the same as before, and the equation becomes, using Equation (15.30),

$$(1/K)\int_0^L qx \, dx = (Z_m^2 - Z_w^2)/2$$

or

$$(q/K) = (Z_m^2 - Z_w^2)/L^2 \tag{15.31}$$

Again it is to be noted that this is identical with the Dupuit-Forchheimer formula for this case, Equation (15.15), which is thus seen to be one limit of the true solution.

At the opposite extreme, it may be supposed that the vertical component of the flow rate is the same everywhere as the rate imposed at the surface, namely $-q$. Hence an application of Darcy's law in the $z$ direction in this plane gives

$$-q = -K(d\phi_1/dz)$$

It is known that the potential at the water table is identical with the value of $z$ here, namely $Z_m$, so that direct integration gives

$$\phi_1 = Z_m + (q/K)(z - Z_m) \qquad (15.32)$$

The first integral on the right-hand side of Equation (15.28) now no longer vanishes, but otherwise the solution is the same as Equation (15.31). Hence substituting Equations (15.30) and (15.32) in Equation (15.28) and integrating, one has

$$L^2(q/K) = Z_m^2(1 - q/K) - Z_w^2 \qquad (15.33)$$

Thus from the two limits expressed by Equations (15.31) and (15.33), it is evident that the true value of $Z_m$, the highest point on the water table, lies between limits expressed in the form,

$$q/K + Z_w^2/L^2 < Z_m^2/L^2 < (q/K + Z_w^2/L^2)/(1 - q/K) \qquad (15.34)$$

Thus there is no more than a ten per cent difference between these limits when the value of $q/K$ is as great as $0\cdot1$. In extending the analysis to include both vertical and horizontal stratification of conductivity, Youngs (1965, 1966) has presented sample curves to show that in most cases the difference between the limits is marginal, and, in a quoted particular case, a comparison with the experimental observations using an electric analogue showed that the assumption of the persistence of the vertical flow rate $q$ down the medial section gave results nearer the truth than those derived from the Dupuit-Forchheimer convention. One important result of Youngs's analysis is that the simple Dupuit-Forchheimer type of calculation has been demonstrated to be useful even when the conductivity is not uniform.

A limitation of this form of exact analysis is that the shape of the water table between the bounding planes is not discovered, but usually this is not of great importance. The height at the highest point is the critical item of the solution, and this is given.

## NOTES

**Note 35. Some relaxation formulas**
*(a) Incomplete meshes*

Where a boundary of a problem crosses the grid lines of the relaxation mesh, there must be some incomplete relaxation patterns, such as those shown in Figure (15.2). In this example the interval between each of three of the mesh points and the central point is the complete mesh side $\Delta$, while the fourth, at which the potential is $\phi_1$, is at a distance $n\Delta$ from the central point, where $n$ is a fraction.

If one assumes that the potential increases linearly from $\phi_0$ at the central point to $\phi_1$ and would continue to do so along a hypothetical extension of this line until it reached the value $\phi_1'$ at the complete distance $\Delta$, then one would have a complete relaxation pattern with potentials $\phi_0$ at the central point and $\phi_1'$, $\phi_2$, $\phi_3$, and $\phi_4$ at the four surrounding points. To this pattern one would justifiably apply Equation (15.6) with the result

$$\phi_1' + \phi_2 + \phi_3 + \phi_4 - 4\phi_0 = 0 \qquad \text{(N35.1)}$$

By extrapolating the linear potential increase between $\phi_0$ and $\phi_1$ to the full interval $\Delta$, one derives the expression,

$$(\phi_1 - \phi_0)/(\phi_1' - \phi_0) = n$$

Substitution for $(\phi_1' - \phi_0)$ from this result in Equation (N35.1) yields

$$(\phi_1 - \phi_0)/n + \phi_2 + \phi_3 + \phi_4 - 3\phi_0 = 0$$

It follows that the equation for the residual of incorrectly guessed potentials is

$$\phi_1/n + \phi_2 + \phi_3 + \phi_4 - \phi_0(3 + 1/n) = R \qquad \left\{\begin{matrix} \text{(N35.2)} \\ \text{(15.8)} \end{matrix}\right.$$

Each incomplete side requires a correction of this form. Thus a change of potential of $+\delta\phi$ at a mesh point requires a change of residual at that point of $-\delta\phi$ for each complete mesh side radiating from it, and $-\delta\phi/n$ for each mesh side of which only the fraction $n$ is within the boundary. At each surrounding point which is a grid point, the change of residual is $+\delta\phi$, while if the boundary is at an imposed potential, the point on the incomplete side is not relaxed at all. If the boundary potential is not imposed, then this is a special case of the relaxation of boundary points and must be dealt with separately.

### (b) Relaxation at a streamline boundary

The finite difference equivalent to Laplace's equation cannot be derived for a point on a boundary such as an impermeable bed, since there is no conducting medium on one side in which a potential gradient may be calculated. Such a boundary is a streamline, and the key to this problem is to construct a hypothetical system in an unimpeded extended conductor which would provide exactly the same flow-net in that portion of this free space which lies within the true problem boundaries.

If one were to construct a flow-net which is the mirror image of the problem, the mirror being the streamline boundary in question as shown in Figure (15.3), then the problem and its mirror image could be joined together along the mirror boundary without upsetting the flow-net in either half, since the potential distribution along both sides of the boundary would be the same before contact. One would then have the required extended flow-net to which the relaxation procedure could be applied. To each grid point on the problem side of the boundary there would exist a corresponding grid point in the mirror image at the same potential, since by symmetry it would lie on the same equipotential. Thus each time a given problem grid point is relaxed by a change of potential, precisely the same change of potential and of consequent residuals must be made in the mirror space. In practice, one need only draw one row of grid points in the mirror space, since one needs this space only as an aid to calculate the changes of residual due to changes of potential at the boundary points.

### (c) Relaxation at a boundary between regions of different conductivity

When a streamline crosses a boundary to pass from a zone of one hydraulic conductivity to a zone of different conductivity, it becomes refracted, as shown in Figure (15.6), and again the derivation of the finite difference equivalent to Laplace's equation is not simple.

Let the angle between the normal to the boundary and the streamline in the zone of conductivity $K_1$ be $\theta_1$, and the angle in zone $K_2$ be $\theta_2$. Let the intersection on the boundary made by the streamtube between two neighbouring streamlines be $l$, and the rate of flow within this streamtube be $Q$. Then the widths of the streamtubes, $A_1$ and $A_2$, in the zones indicated by the subscripts are respectively,

$$A_1 = l \cos \theta_1$$

$$A_2 = l \cos \theta_2$$

Let the potential difference between the two ends of the element $l$ of the boundary be $\delta\phi$. Then as shown in Figure (15.6) this must also be the increase of potential along a path $l \sin \theta_1$ measured along the streamline in the zone of conductivity $K_1$, so that the potential gradient in this zone, at the boundary, is

$$(\text{grad } \phi)_1 = \delta\phi/(l \sin \theta_1)$$

Similarly, in the second zone,

$$(\text{grad } \phi)_2 = \delta\phi/(l \sin \theta_2)$$

An application of Darcy's law in each of the zones may now be made, with the result,

$$Q = -K_1(l \cos \theta_1)\delta\phi/(l \sin \theta_1)$$

and

$$Q = -K_2(l \cos \theta_2)\delta\phi/(l \sin \theta_2)$$

Hence,

$$K_1/K_2 = \tan \theta_1/\tan \theta_2 \tag{N35.3}$$

Thus if $K_1$ is the higher of the two conductivities, then $\theta_1$ is the larger of the two angles.

In Figure (15.4) is shown a relaxation unit with the potentials $\phi_1$ in the medium with conductivity $K_1$ and $\phi_3$ in the medium with conductivity $K_2$. The central point at potential $\phi_0$ and the remaining two points at potentials $\phi_2$ and $\phi_4$ lie on the boundary between the two zones.

At any point on the boundary the potential is common to the two zones, so that $(\partial\phi/\partial x)$, the component of the gradient of potential which lies along the boundary, is the same on each side of the boundary. Hence the components of the velocity of flow along the boundary in the two media, $_xv_1$ and $_xv_2$, are

$$_xv_1 = -K_1(\partial\phi/\partial x)$$

$$_xv_2 = -K_2(\partial\phi/\partial x)$$

Hence,

$$_xv_1/_xv_2 = K_1/K_2 \tag{N35.4}$$

The components of flow velocity, $_zv_1$ and $_zv_2$, normal to the boundary on the two sides and immediately adjacent to it, are related to the tangential components, $_xv_1$ and $_xv_2$, by the equations,

$$_xv_1/_zv_1 = \tan \theta_1$$

$$_xv_2/_zv_2 = \tan \theta_2$$

M*

whence, using Equations (N35.3) and (N35.4), one has the result,

$$_zv_1/_zv_2 = 1 \qquad \text{(N35.5)}$$

This result may be seen to follow directly from the fact that the normal components of flow represent the rate at which water crosses the boundary from one medium to the other, and since the boundary cannot store water this rate must be the same on the two sides.

A further application of Darcy's law yields the result,

$$_zv_1 = -K_1(\partial\phi/\partial z)_1$$
$$_zv_2 = -K_2(\partial\phi/\partial z)_2$$

so that, from Equation (N35.5)

$$(\partial\phi/\partial z)_1/(\partial\phi/\partial z)_2 = K_2/K_1 \qquad \text{(N35.6)}$$

Now had the medium been continuous with conductivity $K_1$, the flow-net would have continued without refraction and would have provided a potential $\phi_3'$ instead of $\phi_3$. The gradient of potential at the boundary would have been continuous at the value observed in the medium $K_1$, so that

$$(\partial\phi/\partial z)_1 = (\phi_1-\phi_3')/2\Delta \qquad \text{(N35.7)}$$

The finite difference equation appropriate to a uniform relaxation pattern is now applicable, giving

$$\phi_1+\phi_2+\phi_3'+\phi_4-4\phi_0 = 0 \qquad \text{(N35.8)}$$

In similar fashion, had the medium been continuous across the boundary, but with conductivity $K_2$, the flow-net would have been continuous without refraction and would have provided a potential $\phi_1'$ instead of the observed $\phi_1$, such that

$$(\partial\phi/\partial z)_2 = (\phi_1'-\phi_3)/2\Delta \qquad \text{(N35.9)}$$

The finite difference equation in this relaxation pattern is

$$\phi_1'+\phi_2+\phi_3+\phi_4-4\phi_0 = 0 \qquad \text{(N35.10)}$$

and from Equations (N35.7), (N35.9), and (N35.6),

$$(\phi_1-\phi_3')/(\phi_1'-\phi_3) = (\partial\phi/\partial z)_1/(\partial\phi/\partial z)_2$$
$$= K_2/K_1 \qquad \text{(N35.11)}$$

Substitution in Equation (N35.11) of $\phi_1'$ from Equation (N35.10) and $\phi_3'$ from Equation (N35.8), with some rearrangement, gives the final relationship,

$$\phi_1[2K_1/(K_1+K_2)]+\phi_2+\phi_3[2K_2/(K_1+K_2)]+\phi_4-4\phi_0 = 0$$

It follows as before, that when the potentials are incorrectly guessed and fail to satisfy the Laplace equation, there will be a residual $R$ given by

$$R = \phi_1[2K_1/(K_1+K_2)]+\phi_2+\phi_3[2K_2/(K_1+K_2)]+\phi_4-4\phi_0 \qquad \left\{ \begin{array}{l} \text{(N35.12)} \\ \text{(15.9)} \end{array} \right.$$

## Note 36. Dupuit-Forchheimer analyses
### (a) Drainage of local precipitation
By symmetry, the rain falling to one side of the plane midway between the two

drains of Figure (15.8) flows wholly to the drain on that side. Hence the rate of flow across a vertical plane distance $x$ from the drain is equal to the rate of rainfall on the surface beyond $x$ up to the distance $L$, since $L$ is the distance to the medial plane. Hence, per unit thickness of conductor, the rate of flow across the plane at $x$ is $q(L-x)$ where $q$ is the rate of rainfall.

According to the Dupuit-Forchheimer approximation, the vertical plane at $x$ is an equipotential, and the potential gradient here is $dZ/dx$ where $Z$ is the height of the water table. $Z$ is also the height of the cross-section and therefore the area of the cross-section if the thickness is unity. Hence, an application of Darcy's law yields the equation,

$$q(L-x) = KZ \, dZ/dx$$

This equation integrates simply between the limits of distance zero and $x$ and the corresponding limits of water table height $Z_0$ and $Z$. The result is

$$(q/K)(Lx-x^2/2) = (Z^2-Z_0)^2/2$$

where $Z_0$ is both the level of water in the drain and the water table height at that point. A little rearrangement gives

$$(Z^2-Z_0^2)/(2Lx-x^2) = q/K \qquad \left\{ \begin{array}{l} \text{(N36.1)} \\ \text{(15.14)} \end{array} \right.$$

### (b) The Ghyben-Herzberg lens

A somewhat similar situation is shown in Figure (15.9), where again by symmetry one needs to treat only the half section between the drain, in this case the vertical face in contact with the sea, and the medial section at distance $L$. Since, however, there are two bodies of groundwater with different densities, namely sea water with density $\rho_s$ and fresh water with density $\rho$ , one must define two different potential functions, one for each zone. Thus Equation (9.3) is appropriately written:

$$\phi_s = z + P/g\rho_s \qquad \text{(N36.2)}$$

$$\phi_f = z + P/g\rho_f \qquad \text{(N36.3)}$$

At the distance $x$ from the sea drain the water table is at a height $Z_f$, while the height of the boundary between the salt and fresh groundwater is $Z_b$, and the pressure $P_b$ at the interboundary is common to both sides. The Dupuit-Forchheimer condition requires the vertical plane through $x$ to be the equipotential $\phi_f$ in the fresh water zone and equipotential $\phi_s$ in the saline zone. Since the pressure vanishes at the water table and takes the value $P_b$ at the interboundary, Equation (N36.3) may be expressed in either of the two ways,

$$\phi_f = Z_f$$

$$\phi_f = Z_b + P_b/g\rho_f$$

Hence,

$$Z_f - Z_b = P_b/g\rho_f \qquad \text{(N36.4)}$$

Similarly, Equation (N36.2) becomes

$$\phi_s = Z_b + P_b/g\rho_s \qquad \text{(N36.5)}$$

From the symmetry of the case, there is no flow of sea water from one side of the isthmus to the other, since the depth of sea, $Z_0$, is the same at both sea boundaries. Hence the whole of the saline groundwater is at uniform potential, and since at the sea boundary this potential is $Z_0$, this must be the potential $\phi_s$ throughout. Hence Equation (N36.5) becomes

$$Z_0 - Z_b = P_b/g\rho_s \qquad \text{(N36.6)}$$

From Equations (N36.4) and (N36.6), one has

$$(Z_f - Z_b)/(Z_0 - Z_b) = \rho_s/\rho_f$$

from which follow the two expressions,

$$(Z_f - Z_0)/(Z_0 - Z_b) = \rho_s/\rho_f - 1 \qquad \left\{ \begin{matrix} \text{(N36.7)} \\ \text{(15.22)} \end{matrix} \right.$$

$$(Z_f - Z_0)/(Z_f - Z_b) = 1 - \rho_f/\rho_s \qquad \text{(N36.8)}$$

An application of Darcy's law to the flow of fresh water, derived from the surface rainfall at the rate $q$, past the cross-section at $x$ of the fresh groundwater body, follows reasoning similar to that of the preceding case, and leads to the equation,

$$q(L - x) = K(Z_f - Z_b)\mathrm{d}Z_f/\mathrm{d}x \qquad \text{(N36.9)}$$

Substitution for $(Z_f - Z_b)$ in Equation (N36.9) from Equation (N36.8), followed by integration between the limits of $x$ from zero to $x$, corresponding to the limits of $Z_f$ from $Z_0$ to $Z_f$, results in the solution,

$$(Z_f - Z_0)^2/(2Lx - x^2) = (q/K)(1 - \rho_f/\rho_s) \qquad \left\{ \begin{matrix} \text{(N36.10)} \\ \text{(15.18)} \end{matrix} \right.$$

Equation (N36.7) may now be used in conjunction with Equation (N36.10) to derive the complementary expression for $(Z_0 - Z_b)^2$, namely

$$(Z_0 - Z_b)^2/(2Lx - x^2) = q/K(\rho_f/\rho_s)/(\rho_s/\rho_f - 1) \qquad \left\{ \begin{matrix} \text{(N36.11)} \\ \text{(15.19)} \end{matrix} \right.$$

# Conformal transformation and the hodograph; the effect of variation of depth of an impermeable sub-stratum

## 16.1 The nature of a conformal transformation

WHEN one quantity, say $\lambda$, called the dependent variable, is expressed by means of an equation as a function of another quantity, say $\mu$, the independent variable, it may also be expressed as a function of yet a third quantity, say $v$, if the relationship between $\mu$ and $v$ is known. The change of equation, from one expressing the relationship between $\lambda$ and $\mu$ to one relating $\lambda$ to $v$, is called a transformation, and the object of such a transformation is always to produce an equation which can be solved. Several examples have already been encountered, as for example in Section 12.6, Equation (12.26).

As has been seen in Chapter 14, the basic equation in the theory of groundwater movement is Laplace's equation and the solution of two-dimensional problems is often an equation which relates one complex quantity to another. Thus,

$$W(\rho) = \phi + i\psi \tag{16.1}$$

where

$$\rho - x + iz \tag{16.2}$$

When $W$ has been specified, this solution has been shown to permit the equipotentials and streamlines to be plotted in the diagram defined by the axes $x$ and $z$, a diagram conventionally referred to as the $\rho$ plane. The examples described in Chapter 14 showed how one can discover the boundary conditions (the particular problem) when the solution is specified, but usually one is required to find a solution to a specified set of boundary conditions. That is to say, one can initially plot the boundaries in the $\rho$ plane and one is required to find the solution in the form of a plot of the streamlines and equipotentials. These boundaries may be complicated and,

351

indeed, not completely known. For example, the shape of the water table of a drainage problem is not initially known.

Changing the complex variable from $\rho$ to, say $\sigma$, another complex quantity, where

$$\sigma = f(\rho) = \zeta + i\xi \tag{16.3}$$

results in another diagram with axes $\zeta$ and $\xi$ which can be calculated when $\rho$, that is to say $x$ and $z$, have been specified, so that for each point in the $\rho$ plane there is a corresponding point in the $\sigma$ plane. Thus the problem boundaries may be plotted in the $\sigma$ plane, and will in general be quite different from those in the $\rho$ plane. The transformation will have been useful if the transformed boundaries are simpler than those in the $\rho$ plane. The object of such transformations is to find ultimately a boundary situation in which the problem is soluble.

A transformation of the kind expressed in Equation (16.3) has the effect of leaving unaltered the shape of any small element of boundary, changing only its size and orientation. Thus a small square in the $\rho$ plane remains a square in the $\sigma$ plane, but may be of different size and may be turned into a different position. This is proved in Note 37. Hence the flow-net in the $\rho$ plane between source and sink remains a flow-net of the same *mesh* shape in the $\sigma$ plane and therefore still satisfies Darcy's law with unaltered $K$. The amount of turning and the change of size of the small element will be different at different parts of the diagram, so that the flow-net will be of quite different shape in order still to match the differently shaped boundaries in the transformed plane. From the circumstance that small elements in the original and transformed planes conform in shape, such a transformation is referred to as conformal.

As an example, one may take the transformation from the $\rho$ plane to the $W$ plane in accordance with Equations (16.1) and (16.2). In the $\rho$ plane, the paths followed by equipotentials and streamlines might form the flow-net depicted in Figure (16.1), matching the equipotential and streamline boundaries which define the problem illustrated. In the $W$ plane an equipotential, indicated in Figure (16.2), is simply a line parallel to the stream function axis, while a streamline similarly is simply a line parallel to the potential axis. Thus in the $W$ plane, the flow-net is simply a uniform square mesh formed by orthogonal families of straight lines. It follows therefore from the property of conformal transformations that the flow-net in the $\rho$ plane also is formed by families of orthogonal curves which define a network of meshes which are, when very small, square. This result was derived directly from the solution of Laplace's equation described in Section 14.5 and Note 34.

An equation in one plane, say the $\rho$ plane, may be taken through a succession of conformal transformations, each presenting the boundaries of the problem in a different form. The original boundaries may be quite complicated, as for example those which describe the flow of water from the arrival at the water table to the outfall at each of a set of long equidistant parallel drains at uniform depth, as illustrated in Figures (14.1) and (15.5). Nevertheless, at each successive transformation the boundaries may take successively simpler forms, until in the end the problem may be seen as, for example, the flow between a point source and a point sink in an

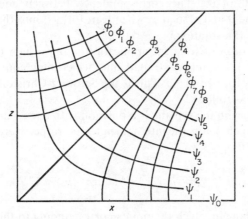

Figure 16.1 A hypothetical flow-net in one corner of a rectangular conductor. The $\rho$ plane where $\rho = x + iz$.

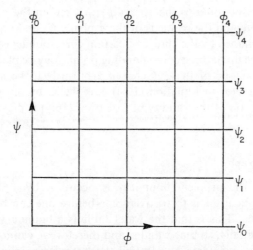

Figure 16.2 The conformal transformation of Figure (16.1) into the $W$ plane where $W = \phi + i\psi$.

infinite plane, the solution of which is well known. Such a set of transformations was, in fact, developed by Gustafsson (1946). The solution may then be transformed back to the $\rho$ plane to give the solution of the problem as originally posed.

Such direct methods of solution are not common, and it is often necessary to transform both the $\rho$ plane and the $W$ plane successively until common ground is found in the form of a plane in which the boundaries of both functions are simple, say one of the axes. Such a plane bounded by an axis is called a half plane. The correspondence between equivalent points derived from the $\rho$ and $W$ planes when transformed into the common half plane provides the solution.

A transformation of frequent application is one which transforms any polygon in one plane into a straight line, which may conveniently be an axis, in the transformed plane. This is the Schwarz-Christoffel transformation, expressed as follows.

If the boundary in the $\rho$ plane is a polygon with apexes at the points $\rho_1$, $\rho_2, \ldots \rho_n$, at which points the internal angles are respectively $\alpha_1, \alpha_2, \ldots \alpha_n$, the transformation,

$$\rho = C_1 \int (\sigma - \mu_1)^{(\alpha_1/\pi - 1)} (\sigma - \mu_2)^{(\alpha_2/\pi - 1)} \ldots (\sigma - \mu_n)^{(\alpha_n/\pi - 1)} \mathrm{d}\sigma + C_2 \qquad (16.4)$$

results in a $\sigma$ plane in which the points corresponding to the apexes of the polygon lie on the real axis. If the $\sigma$ plane is defined by the function,

$$\sigma = \mu + iv$$

the points corresponding to the apexes $\rho_1$, $\rho_2$, etc. in the $\rho$ plane are the points $\mu_1$, $\mu_2$, etc., on the real axis of the $\sigma$ plane. $C_1$ and $C_2$ are constants which are determined by the choice of certain correspondences between the two planes, since up to three corresponding points may be placed arbitrarily on the $\mu$ axis. If one of these is chosen arbitrarily to be at infinity, the corresponding factor in Equation (16.4) is replaced by unity.

A demonstration of the validity of Equation (16.4) is presented in Note 38.


## 16.2 The hodograph

If a part of the problem boundary is occupied by a water table, a fundamental difficulty has to be overcome before one can begin to transform the $\rho$ plane. This is that the water table is a boundary whose shape is not known initially, so that $x$ and $z$, and therefore $\rho$, cannot be specified. This difficulty is overcome by plotting the boundaries in a diagram the axes of which are the horizontal and vertical components of the velocity of flow at each point, that is to say, $v_x$ and $v_z$ respectively, instead of the positions $x$ and $z$ themselves. Such a diagram is called a hodograph. Then as shown in Note 39 the water table in the hodograph plane describes a circle with a known centre and radius, although the fraction of the complete circle which it occupies is usually initially unknown. Even this degree of uncertainty may be removed by specifying a portion of the circle and proceeding to a solution which, when obtained, will indicate the modification of the boundary conditions which the arbitrary specification has imposed. That is to say, the arbitrary choice of the portion of the hodograph circle of the water table forms a part of the specification of the particular problem, and this part is revealed in the final solution.

The boundaries of a problem as they appear in the hodograph plane may be listed as follows. A streamline which makes an angle $\theta$ with the horizontal has the relationship between $v_x$ and $v_z$ of the form

$$v_x/v_z = \tan \theta$$

In the hodograph diagram this is a straight line of slope $\theta$, the same as the slope of the streamline in the $\rho$ plane, but passing through the origin. If a streamline boundary is a straight line of slope $\theta$, its hodograph is a parallel straight line through the origin.

If a boundary is an equipotential with slope $\theta$, then at all points the streamlines intersect it at right angles and therefore have slope $\pi/2+\theta$. The components of flow velocity therefore satisfy the equation,

$$v_x/v_y = \tan(\pi/2+\theta)$$

and the hodograph is a straight line passing through the origin with slope $\pi/2+\theta$, i.e. it is perpendicular to the corresponding boundary in the space diagram.

A water table is not necessarily a streamline. In general there will be a flux across it which, in the circumstances to be discussed here, will be recognizable as a vertical flux $N$, positive if upward, in the unsaturated zone above the water table. Thus $N$ may be the flow necessary to maintain evaporation at the surface of the land, or if negative may be downward percolation due to surface precipitation. The equation of the hodograph of such a water table is, as shown in Note 39,

$$v_x^2 + \{v_z+(K-N)/2\}^2 = \{(K+N)/2\}^2 \tag{16.5}$$

This indicates that the hodograph is a circle with its centre on the axis of $v_z$ at a distance $(K-N)/2$ below the origin, and with radius $(K+N)/2$. As shown in Figure (16.3), this circle cuts the axis of $v_z$ at $+N$ and $-K$. In this diagram a line is shown drawn from the intersection at the point $(0, -K)$ to the point $(v_x, v_z)$ on the circle which corresponds to the point on the water table where the slope is $\theta$. This line itself makes the angle $\theta$ with the axis of $v_z$.

As also is shown in Note 39, Equation (16.5) represents equally the hodograph of the upper boundary of the capillary fringe if this surface, rather than the water table, is the boundary between the upper unsaturated soil and the lower zone of groundwater.

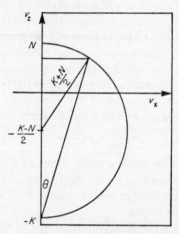

Figure 16.3 In the hodograph plane, $v_x + iv_z$, the water table in soil of conductivity $K$ with vertical flow $N$ upward from it, is represented by the semicircle shown.

If a surface of seepage is plane and slopes at an angle $\theta$, it is represented in the hodograph plane by a line perpendicular to the plane, passing through the point $(0, -K)$ on the axis of $v_z$, as is proved in Note 39.

In order to relate the hodograph plane to the $\rho$ plane, one introduces the concept of a complex velocity of flow $\omega$ which is defined by relating it to the complex potential gradient by its compliance with Darcy's law. Thus,

$$\omega = u + iv = -K \, dW/d\rho \qquad (16.6)$$

From Equation (N34.3) of Note 34, $dW/d\rho$ is simply $\partial W/\partial x$, so that by the use of Equation (16.1), one may write from Equation (16.6),

$$u + iv = -K(\partial\phi/\partial x + i \, \partial\psi/\partial x) \qquad (16.7)$$

Again quoting Note 34, one has from Equation (N34.10),

$$\partial\psi/\partial x = \partial\phi/\partial z$$

so that Equation (16.7) becomes

$$u + iv = -K(\partial\phi/\partial x - i \, \partial\phi/\partial z) \qquad (16.8)$$

Now, since
$$v_x = -K \, \partial\phi/\partial x$$
and
$$v_z = -K \, \partial\phi/\partial z$$

it follows from equating the real and imaginary parts of Equation (16.8) that
$$u = v_x \tag{16.9}$$
and
$$v = -v_z \tag{16.10}$$

Hence the $\omega$ plane is simply the hodograph plane with the sign of $v_z$ reversed, or, in other words, it is the mirror image of the hodograph in the axis of $v_x$. Thus with this reversal, the properties of the hodograph described above can be applied to the $\omega$ plane, with one small reservation. The chord which joins the point $-K$ on the $v_z$ axis of the hodograph to a point on the semicircle which represents the water table makes the angle $\theta$ with the $v_z$ axis as already noted. Reflection of the hodograph in the $v_x$ axis reverses the sign of this angle, so that in the $\omega$ plane the chord makes the angle $-\theta$ with the axis of $v$.

### 16.3 The stationary water table in drained land; van Deemter's analysis

The method of successive conformal transformations, utilizing the hodograph plane, will be illustrated here by solving the problem of constructing the flow-net of the drainage situation shown in Figures (14.1) and (15.5), namely the drainage of locally precipitated rainfall by a system of long parallel drains of uniform separation and depth in land of uniform hydraulic conductivity. In particular, the shape and location of the water table are sought. The case is idealized and generalized by supposing that there is both a known flux at the water table and a known rate of flow at great depth. The former may either be upward as by evaporation at the surface, or downward as by the percolation of rainfall. The latter also may be either upward as in the case of artesian water, or downward as by drainage to a deep aquifer. The flux at the water table may be either greater in magnitude or less than the flux at the lower boundary. The six combinations of these various circumstances provide boundary conditions which are shown sketched in Figure (16.4). The analysis of each case follows the same course, and the particular flow-net emerges from the general solution when the appropriate magnitudes and signs of the boundary fluxes are introduced.

The development of this analysis has owed much to many contributors at various stages, of whom Wedernikov (1936, 1937, 1939), Gustafsson (1946) and Engelund (1951) may be specially mentioned. The analysis to be followed here is that of van Deemter (1950) with slight modifications due to Childs (1959), to take into account the effects of a capillary fringe.

The cross-section of the problem in the $\rho$ plane is shown in Figure (16.5) in which, to minimize sign confusion, both the velocity of flow $N$ as it issues into the upper unsaturated soil and the velocity $M$ of the deep groundwater are taken to be positive, that is to say, upward. $M$ is also taken to be of greater magnitude than $N$, so that the problem represents one of the drainage of artesian foreign water, the efflux at the drain $D$

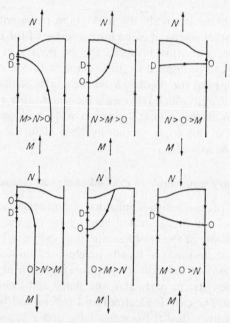

Figure 16.4 The situation resulting from a combination of vertical flux $N$ above the water table and vertical flux $M$ at great depth may be divided into these 6 cases, according to the signs and magnitudes of $N$ and $M$. $D$ represents the drain; the dividing streamline intersects one or other of the bounding planes at a stagnation point $O$, where the flow is zero.

accommodating the excess flow of groundwater above what can be evaporated at the surface. As in Figure (15.5), the complete drainage flownet is produced by a reflection of the half section shown in Figure (16.5) in one of the vertical boundaries, followed by a repetition of the complete cross-section so formed to infinity in both directions of $x$.

The analysis requires a highly artificial restriction, namely that the drain should have a negligibly small radius of cross-section and consequently that the hydraulic potential here should have an infinitely large negative value in order to accommodate the flow of water into the drain through a vanishingly small area of drain wall. Further, the artesian source is taken to be at an infinitely great depth. In the event, neither of these restrictions will be found to invalidate the application of the solution to practical field situations.

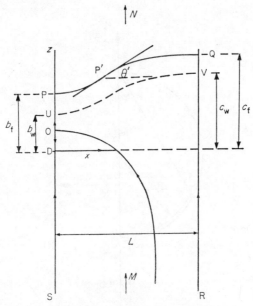

Figure 16.5 The boundaries of the drainage situation shown in the first case of Figure (16.4). The $\rho$ plane.

The diagram shows a capillary fringe boundary $PQ$ at which the potential is given by

$$\phi = H_f + z \tag{16.11}$$

where $z$ is the height relative to $D$, which is taken as the origin of coordinates, and $H_f$ is the negative pressure head at which the soil begins to desaturate, that is to say, the air entry value. The water table is indicated by $UV$ and here, because the hydrostatic pressure is zero,

$$\phi = z$$

The section is shown as tending at $RS$ to infinity in the negative vertical direction. Since the artesian flow $M$ exceeds the ultimate surface evaporation rate $N$, the former is divided into two zones by a streamline which meets $DP$ at $O$. On one side of this the groundwater proceeds to the water table and in due course to the evaporating surface; and on the other side the balance proceeds to the drain. At the dividing point $O$, the velocity of flow is zero; between $O$ and $P$, the velocity is upward and therefore positive; and between $O$ and $D$ it is negative, increasing in magnitude to infinity at $D$. Below $D$ there is a sudden reversal to upward flow with infinitely great magnitude at $D$ itself, decreasing to $M$ at $S$.

The $\omega$ plane of the system is shown in Figure (16.6). The boundaries

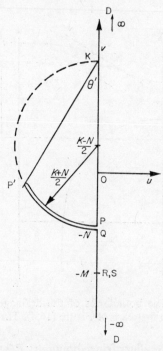

Figure 16.6 The boundaries of Figure (16.5) conformally transformed into the $\omega$ plane, where $\omega = u + iv$ $= v_x - iv_z$. $PP'Q$ is a portion of the hodograph circle shown in Figure (16.3).

corresponding to those in the $\rho$ plane of Figure (16.5) are developed by the hodograph rules. At all boundaries in Figure (16.5), except the capillary fringe, the flow is vertical, so that the corresponding boundary in the $\omega$ plane is some part or other of the axis of $v$. The origin of the $\omega$ plane is clearly the point $O$ which represents the stagnation point where the velocity vanishes. Between $O$ and $P$ the velocity of flow, $v_z$, increases from zero to $N$ so that $v$, which is $-v_z$, decreases from zero to $-N$ as shown. The fringe boundary $PQ$ produces the semicircular boundary with centre at

$-(K-N)/2$ and with intercepts of $N$ and $-K$ on the $v_z$ axis, so that in the $\omega$ plane the boundary is a portion of a circle with centre at $(K-N)/2$ on the $v$ axis and with intercepts at $K$ and $-N$ on that axis. The radius of the circle is $(K+N)/2$. As the slope of the fringe boundary increases to $\theta'$ at the steepest point and then decreases to zero again at $Q$, so does the $\omega$ plane tracer point move along the circle from $P$ to $P'$ and back to $Q$ which is indistinguishably close to $P$. At $P'$ the line drawn from $P'$ to the intercept $K$ on the $v$ axis makes the angle $-\theta'$ where, as shown in Note 39, Equation (N39.6),

$$\tan \theta' = -u/(K-v) \tag{16.12}$$

From $Q$ to $R$ the flow velocity increases from $N$ to $M$, so that $v$ decreases from $-N$ to $-M$ as shown, and as the flow remains at this value across the section $RS$, so does $v$ remain at the same value. Hence $RS$ reduces to a single point in the $\omega$ plane. It is referred to on all subsequent transformations simply as $R$. Along $SD$ the flow velocity increases to infinity and then sharply reverses at $D$ to an infinitely large negative value, and this is shown in the $\omega$ plane by $v$ tending to minus infinity and reappearing at plus infinity. Finally the boundary is completed by closing on the stagnation point $O$ at the origin.

The boundary in the $\omega$ plane already has much of the ultimately required form, since it is for the greater part a straight line coinciding with one of the axes. The point $PQ$ at which it deviates into the circular portion of the boundary may be brought to the origin by adding $iN$ to $\omega$, and at the same time it is convenient to bring the variables to dimensionless form by dividing by $(K+N)$, the diameter of the hodograph circle. The boundary may now be turned into the real axis by multiplying by $i$, so that the real parts become imaginary and vice versa, and the point $PQ$ may be removed to infinity by inversion, which brings the point $D$ from infinity to the origin. This succession of transformations amounts to the single transformation to a $\zeta$ plane where

$$\zeta = 1/\{i(\omega+iN)/(K+N)\}$$
$$= (K+N)/(i\omega-N) \tag{16.13}$$

The boundaries in the $\zeta$ plane are readily drawn from Equation (16.13), except for the water table. From the specification of $\omega$ in Equation (16.6) one has, from Equation (16.13),

$$\zeta = (K+N)/(iu-v-N) \tag{16.14}$$

where $u$ vanishes over the whole of the boundary in the $\rho$ plane except the water table. Thus, with this exception, $\zeta$ is real and the boundary is the real axis. It is shown in Figure (16.7), in which the corresponding points are shown below the axis and the values of $\zeta$ at these points are indicated above. All these values are obtained from Equation (16.14) by substituting the appropriate values of $v$ (i.e. $-v_z$), namely zero at $O$, $-N$ at $P$ and $Q$, $-M$ at $R$ and $S$, and infinity at $D$.

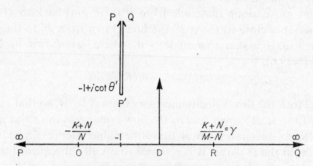

Figure 16.7 The boundaries of Figures (16.5) and (16.6)
transformed into the $\zeta$ plane.

The portion of the boundary $PP'Q$ where $u$ does not vanish is drawn after expanding Equation (16.14) to take the form,

$$\zeta = -\frac{(K+N)(v+N)}{(v+N)^2+u^2} - \frac{iu(K+N)}{(v+N)^2+u^2} \qquad (16.15)$$

where $u$ and $v$ satisfy the conditions expressed in Equations (N39.6) and (N39.7) of Note 39, together with Equations (16.9) and (16.10). These are

$$-u = (K-v)\tan\theta$$

$$-u = (N+v)/\tan\theta$$

These equations permit the expression of $u$ and $v$ in terms of the parameter $\theta$ thus:

$$u = -(K+N)\sin\theta\cos\theta$$

$$v = K\sin^2\theta - N\cos^2\theta$$

Substitution of these values in Equation (16.15) gives the required expression for the fringe boundary $PP'Q$ in the $\zeta$ plane, namely,

$$\zeta = -1 + i\cot\theta \qquad (16.16)$$

The whole boundary is shown in Figure (16.7). That part of it which represents the fringe boundary is, in accordance with Equation (16.16), a straight line parallel with the imaginary axis and passing through the point $-1$ on the real axis. It appears from infinity at $P$, makes its nearest approach to the $u$ axis at $P'$, and then bends back and returns on itself to infinity again at $Q$. At $P'$ the value of $\zeta$ is

$$\zeta = -1 + i \cot \theta'$$

Summarizing the values of $\zeta$ at the various labelled points on the boundary, as shown in Figure (16.7), one has, from Equation (16.12),

$$\zeta_D = 0 \tag{16.17}$$

$$\zeta_0 = -(K+N)/N \tag{16.18}$$

$$\zeta_P = -\infty \tag{16.19}$$

$$\zeta_Q = +\infty \tag{16.20}$$

$$\zeta_R = (K+N)/(M-N) = \gamma \tag{16.21}$$

It is convenient to represent the ratio expressed in Equation (16.21) by the symbol $\gamma$ because it recurs frequently in the analysis.

The geometrical figure shown in the $\zeta$ plane of Figure (16.7) is one in which the three sides are straight lines. Hence it is a triangle notwithstanding that two of the apexes are at infinity. It is therefore a special case of a polygon which can be transformed to a half plane by the use of the Schwarz-Christoffel transformation. Since the internal angle at $P'$ has the magnitude $2\pi$ and the sum of all the internal angles of a triangle amounts only to $\pi$, it follows that each of the internal angles at the infinite apexes $P$ and $Q$ has the magnitude $-\pi/2$. In transforming $\zeta$ to a half plane $\sigma$ in which the boundary is the real axis, one is allowed three arbitrary allocations of apexes on the real axis, and there are only three. It is convenient to choose $Q$ to be at infinity in the $\sigma$ plane, and it is therefore to be ignored in the Schwarz-Christoffel transformation, Equation (16.4). $P$ is placed at the origin with vanishing $\sigma$, and $P'$ is allotted the value $-1$. Thus to summarize, using the notation of Equation (16.4),

$$\mu_1 = \mu_P = 0; \alpha_1 = -\pi/2$$

$$\mu_2 = \mu_{P'} = -1; \alpha_2 = 2\pi$$

$$\mu_3 = \mu_Q = \infty; (\sigma - \mu_3)^{(\alpha_3/\pi - 1)} = 1$$

With these allocations of value, the Schwarz-Christoffel transformation of Equation (16.4) is shown in Note 40(a) to be

$$\zeta = \frac{1}{2}\left(\frac{\sigma-1}{\sigma^{\frac{1}{2}}}\right)\cot \theta' - 1 \tag{16.22}$$

The resulting boundaries in the $\sigma$ plane are shown in The Figure (16.8).

values of $\sigma$ at the points $D$, $R$ and $O$ are written respectively $\lambda^2$, $\mu^2$ and $v^2$ where, as also shown in Note 40(a),

$$\lambda - 1/\lambda = 2 \tan \theta' \qquad (16.23)$$

or

$$\lambda = \tan \theta' + (1 + \tan^2 \theta')^{\frac{1}{2}} \qquad (16.23a)$$

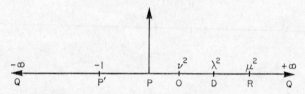

Figure 16.8 The boundaries of Figures (16.5) to (16.7) transformed into the real axis of the $\sigma$ half plane.

The form of Equation (16.23a) is the more suitable for numerical calculation of particular examples. Similarly,

$$\mu - \frac{1}{\mu} = 2(1+\gamma) \tan \theta' = 2\left(\frac{K+M}{M-N}\right) \tan \theta' \qquad (16.24)$$

or

$$\mu = (1+\gamma) \tan \theta' + \{1 + (1+\gamma)^2 \tan^2\theta'\}^{\frac{1}{2}} \qquad (16.24a)$$

and

$$v - 1/v = -2(K/N) \tan \theta' \qquad (16.25)$$

or

$$v = -(K/N) \tan \theta' + \left\{1 + \left(\frac{K}{N}\right)^2 \tan^2\theta'\right\}^{\frac{1}{2}} \qquad (16.25a)$$

The relative positions of $D$, $R$ and $O$ on the real axis of the $\sigma$ half plane may be examined by considering the relative values of $\lambda$, $\mu$ and $v$. This becomes simple when it is remembered that any quantity of the form $\alpha - 1/\alpha$ is positive when $\alpha$ exceeds unity, and increases as $\alpha$ increases; and is negative when $\alpha$ is less than unity and decreases to negative infinity as $\alpha$ decreases to zero.

Now from Equation (16.21), $\gamma$ is positive when $M$ exceeds $N$, that is for all drainage cases. For positive $\theta'$ one has, from Equation (16.23),

$$\lambda > 1$$

Also, from a comparison between Equations (16.24) and (16.23), since $(1+\gamma) \tan \theta'$ exceeds $\tan \theta'$,

$$\mu > \lambda$$

From Equation (16.25), if $N$ is positive, corresponding to evaporation, $v - 1/v$ is negative, so that

$$N > 0$$

$$v < 1$$

Since for drainage it has been shown that $\lambda$ exceeds unity,

$$v < \lambda$$

Equation (16.25) shows that as $N$ passes from small positive values through zero to small negative values, $v - 1/v$ passes from negative infinity to positive infinity. Here $v$ itself passes from small positive values through zero to positive infinity discontinuously. Thus with positive $M$, corresponding to upward artesian flow, the stagnation point lies between $D$ and $P$ in the $\sigma$ plane, i.e. it lies on the boundary $DP$ in the $\rho$ plane, if $N$ is positive; but it lies between $R$ and $Q$ on the $\sigma$ plane, i.e. on the boundary $RQ$ in the $\rho$ plane, if $N$ is negative. This confirms the sketches of Figure (16.4).

If $M$ is vanishingly small, but still in excess of $N$, so that $N$ is necessarily negative and downward, then, from Equations (16.24) and (16.25),

$$v = \mu$$

so that in the $\sigma$ plane $O$ coincides with $R$ and in the $\rho$ plane the stagnation point is at infinite depth.

When $M$ as well as $N$ is negative, i.e. downward, one may write

$$M = -p$$

$$N = -q$$

where $p$ and $q$ are positive.

Then Equations (16.24) and (16.25) become

$$\mu - \frac{1}{\mu} = 2\left(\frac{K-p}{q-p}\right)\tan\theta'$$

and

$$v - \frac{1}{v} = 2\frac{K}{q}\tan\theta'$$

It has been seen in Sections 12.7(b) and 13.1, that $K$ is the upper limit of the rate of infiltration so that $q$ is necessarily less than $K$. Usually it is much less. Hence $\dfrac{K-p}{q-p}$ exceeds $\dfrac{K}{q}$, so that

$$\mu > v$$

Also $K/q$ exceeds unity, so that from Equations (16.23) and (16.25),

$$v > \lambda$$

Hence in the $\sigma$ plane $O$ lies between $D$ and $R$, i.e. it lies on the boundary $DR$ in the $\rho$ plane, again confirming the sketch of this case in Figure (16.4).

In what follows it will be found convenient to express $\mu$ as a fraction of $\lambda$, in the form,

$$\mu/\lambda = 1+\beta \qquad (16.26)$$

The value $\mu$ is not to be confused with the apex points, $\mu_1$, $\mu_2$, and $\mu_3$, mentioned above.

From Equations (16.23), (16.24) and (16.26) it is readily shown that as $\theta'$ increases from zero to $\pi/2$, corresponding to a point in the $\omega$ plane traversing the complete semicircle of the capillary fringe boundary from $P$ to the intercept $K$ on the $v$ axis, so does $\tan \theta'$ increase from zero to infinity and $\beta$ from zero to $\gamma$. Each different limit of $\theta'$ within this range corresponds to a different $\omega$ plane boundary and to a different particular case. The nature of the differences will emerge in due course.

The complex potential must now be transformed to coincide with the $\sigma$ plane. The $W$ plane itself is shown in Figure (16.9), which is drawn from

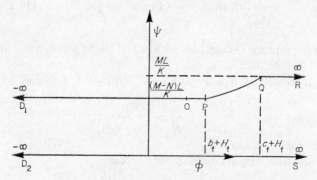

Figure 16.9 Boundaries of Figure (16.5) plotted in the $W$ plane where $W = \phi+i\psi$.

information conveyed in Figure (16.5). The potential at $RS$ is infinitely great since this boundary is infinitely remote, and if the bounding streamline $SD$ is taken as the zero streamline and the bounding streamline $RQ$ as the maximum, the total rate of flow included between them is evidently $ML$, where $L$ is the width of the semi-section. Following the rule demonstrated in Section 14.6, Equations (14.13), (14.22), (14.30) and (14.43), the maximum stream function is labelled $(ML)/K$. In similar fashion it is demonstrable that the total flow escaping from the capillary fringe boundary is $NL$, so that the flow contained between the zero streamline and the streamline which passes through the stagnation point $O$ is $(M-N)L$. This latter streamline is therefore labelled with the stream function $(M-N)L/K$. It bifurcates to form the two branches $OP$ and $OD$. The potential at $P$ is $b_f + H_f$ and that at $Q$ is $c_f + H_f$ where $b_f$ and $c_f$ are the heights of the fringe boundary at the respective points and $H_f$ is the negative pressure which characterizes the fringe boundary.

The departure of this boundary from linearity between $P$ and $Q$ may readily be rectified when it is remembered that as $x$ increases from $O$ to $L$ along the water table, the stream function increases linearly from $(M-N)L/K$ to $ML/K$ and it is this increase which causes the kink. Hence the increase of stream function may be neutralized by subtracting from the complex potential the quantity $iN\rho/K$. The result is the $\Omega$ plane where

$$\Omega = W - iN\rho/K = \phi + Nz/K + i(\psi - Nx/K) \qquad (16.27)$$

The boundary in the plane resulting from this transformation is shown in Figure (16.10). It is necessary to distinguish between $D_1$ and $D_2$ in this

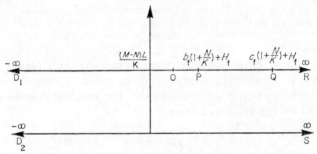

Figure 16.10 Conformal transformation of Figure (16.9) to the $\Omega$ plane to straighten the boundary $D_1 R$.

diagram since although they are one and the same point, namely the point sink, in the $\rho$ plane, $D_2$ is the termination of the zero streamline while $D_1$ is the termination of the streamline whose stream function is $L(M-N)/K$. Similarly, it is required to distinguish here between the infinitely distant points $R$ on the maximum streamline and $S$ on the zero streamline.

The boundary of Figure (16.10) is to be regarded as a polygon with only two parallel sides, whose two internal angles at $D$ and $R$ at infinity are each zero. Hence the Schwarz-Christoffel transformation may again be applied, and this time the point $D$ is chosen to be at infinity on the real axis of the half plane $\eta$, resulting from the transformation, while $R$ is placed at the origin. An application of Equation (16.4) then leads, as shown in Note 40(b), to the equation,

$$\eta = e^{-\pi K\Omega/\{(M-N)L\}} \qquad (16.28)$$

The values of $\eta$ at the points $P$ and $Q$, which do not feature as apex points of the boundary in the $\Omega$ plane for the purpose of the Schwarz-Christoffel transformation, are shown in Note 40(b) to be,

$$\eta_Q = -e^{-\pi KH_f/\{(M-N)L\}-\pi c_f\gamma/L} \tag{16.29}$$

$$\eta_P = -e^{-\pi KH_f/\{(M-N)L\}-\pi b_f\gamma/L} \tag{16.30}$$

These points are labelled accordingly in Figure (16.11), the $\eta$ half plane.

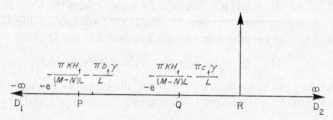

Figure 16.11 Conformal transformation of the boundaries in Figure (16.10) to the real axis of the $\eta$ half plane.

Both $\omega$ and $W$ have now been transformed into half planes $\sigma$ and $\eta$ respectively, in which the boundaries of the problem form the real axis, but corresponding points do not coincide. The required coincidence may be achieved by the transformation,

$$\eta = -\left(\frac{\sigma - \mu^2}{\sigma - \lambda^2}\right)e^{-\pi c_f\gamma/L - \pi KH_f/\{(M-N)L\}} \tag{16.31}$$

This transformation is derived in Note 40(c).

Thus the object has in part been attained, for Equation (16.29) relates the potential $W$ to the velocity $\omega$ via the intermediate functions $\sigma$ and $\eta$. Before proceeding to relate $W$ to $\rho$, one may usefully derive one of the ultimately desired relationships directly from Equation (16.31). The point $P$ has not featured in matching the $\eta$ plane to the $\sigma$ plane and may therefore be used as an independent check. At this point $\sigma$ vanishes, hence from Equation (16.30) and (16.31), it follows that

$$-e^{-\pi b_f\gamma/L - \pi KH_f/\{L(M-N)\}} = -(\mu/\lambda)^2 e^{-\pi c_f\gamma/L - \pi KH_f/\{L(M-N)\}}$$

or, using Equation (16.26),

$$(1 + \beta)^2 = e^{(c_f - b_f)\pi\gamma/L}$$

This in turn is expressed most simply in the form,

$$(\pi/L)(c_f - b_f) = (2/\gamma)\ln(1+\beta) \tag{16.32}$$

Thus when the parameters of the problem, namely $K$, $M$, $N$ and $L$, are postulated, and a value of $\theta'$ has been specified, $\gamma$ is known from Equation (16.21) and $\beta$ from Equations (16.24), (16.23) and (16.26). Equation (16.32) then gives the difference between the maximum and minimum height of the upper boundary of the groundwater, over the medial plane and the drain respectively.

A final transformation, made solely for the convenience of subsequent computation, is

$$t = (\sigma)^{\frac{1}{2}}/\lambda \qquad (16.33)$$

This turns the negative part $PQ$ of the real axis of $\sigma$, which corresponds to the groundwater boundary, through a right angle to form the imaginary axis of the $t$ plane, as shown in Figure (16.12).

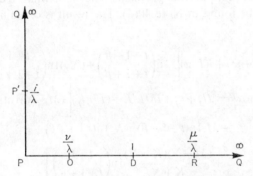

Figure 16.12 Conformal transformation of Figure (16.8) to the $t$ quadrant.

The succession of transformations, Equations (16.13), (16.22) and (16.33), is equivalent to the single transformation,

$$(K+N)/(i\omega - N) = (t^2\lambda^2 - 1)(\cot\theta')/(2\lambda t) - 1$$

which reduces to

$$\omega = -i\left(N + \frac{K+N}{(\lambda^2 t^2 - 1)(\cot\theta')/2\lambda t - 1}\right) \qquad (16.34)$$

On the other hand, the succession of transformations, Equations (16.28), (16.31) and (16.33), is equivalent to the single transformation,

$$e^{-\pi K\Omega/\left\{(M-N)L\right\}} = -\left(\frac{\lambda^2 t^2 - \mu^2}{\lambda^2 t^2 - \lambda^2}\right)e^{-\pi c_f\gamma/L - \pi KH_f/\left\{(M-N)L\right\}}$$

which reduces to

$$\Omega = H_f + c_f(K+N)/K - \left(\frac{(M-N)L}{\pi K}\right)\ln\left(\frac{\mu^2/\lambda^2 - t^2}{t^2 - 1}\right) \qquad (16.35)$$

One now uses the transformation, Equation (16.18), in conjunction with Equation (16.27), namely,

$$\omega = -K(d\Omega/d\rho + iN/K) \tag{16.36}$$

This operates on Equation (16.34) to express the right hand side of that equation in terms of $\rho$ and $W$, via $\Omega$ instead of $\omega$, which was only introduced in order to take advantage of the properties of the hodograph plane. Thus from Equations (16.34) and (16.36), one has

$$i(1+N/K)d\rho = -d\Omega + (d\Omega/dt)\{(\lambda^2 t^2 - 1)(\cot\theta')/2\lambda t\}dt \tag{16.37}$$

The integration of Equation (16.37) with the help of Equation (16.35) provides the solution of the problem. The somewhat tedious detailed procedure will be found in Note 40(d). The result is contained in the two equations,

$$\rho = x + iz = L + ic_f + i(L/\pi)\left[\; \ln\left(\frac{t-1-\beta}{t+1+\beta}\right) + (2/\gamma)\ln\left(\frac{t+1}{t+1+\beta}\right)\right] \tag{16.38}$$

$$W = \phi + i\psi = H_f + c_f + iML/K + (L/\pi K)\Bigg[\;(M-N)\ln(t-1) +$$

$$- M\,\ln(t-1-\beta) + N\,\ln(t+1+\beta) +$$

$$+ \frac{(M-N)(K-N)}{K+N}\ln\left(\frac{t+1}{t+1+\beta}\right)\Bigg] \tag{16.39}$$

where, to recapitulate Equation (16.21) and (16.26),

$$\gamma = (N+K)/(M-N)$$

$$1+\beta = \mu/\lambda$$

$$= \frac{(1+\gamma)\tan\theta' + [1+(1+\gamma)^2\tan^2\theta']^{\frac{1}{2}}}{\tan\theta' + (1+\tan^2\theta')^{\frac{1}{2}}}$$

Thus with selected values of the parameters $K$, $L$, $M$, $N$, $H_f$ and $\theta'$, one enters Equation (16.38) with a chosen value of $t$ within the quadrant of Figure (16.12), and by equating the real and imaginary parts one arrives at the coordinates $x$ and $z$ of the point in the $\rho$ plane to which $t$ refers. Entry into Equation (16.39) with this same value of $t$, reveals the potential and stream functions $\phi$ and $\psi$ at this point. From a repetition with a sufficient number of values of $t$ referring to a sufficient number of points the whole flow-net may be revealed.

Since $t$ is in general complex, the computations require the evaluation of logarithms of complex numbers, which are subject to the rule,

$$\ln(A+iB) = \ln(P\;e^{i\theta}) = \ln P + i\theta$$

where

$$P^2 = A^2 + B^2$$

$$\tan \theta = B/A$$

Also,

$$\ln(-A+iB) = \ln\{(-1)(A-iB)\}$$
$$= \ln(-1) + \ln(P\,e^{-i\theta})$$

Using Equation (14.3) one thus has

$$\ln(-A+iB) = i\pi + \ln P - i\theta$$

Without going further into this kind of detail one may derive some of the most important relationships of the flow-net by confining the chosen values of $t$ to the real axis of the $t$ plane of Figure (16.12). Thus the boundary $DP$ between the drain and the free surface $PQ$ as shown in the $\rho$ plane corresponds to real values of $t$ between zero and unity in the $t$ plane; while the boundary $QR$ in the $\rho$ plane corresponds to real values of $t$ in excess of $1+\beta$ in the $t$ plane. Substituting these values of $t$ in Equations (16.38) and (16.39) one may arrive, as shown in Note 40(e), at the following results.

Firstly,

$$\pi c_f/L = \ln(1+2/\beta) + (2/\gamma)\ln(1+\beta/2) \qquad (16.40)$$

Thus the height of the ground water fringe boundary is discovered at the medial plane. Hence although this quantity occurs in the equations for $\rho$ and $W$, Equations (16.38) and (16.39), and in those for special cases derived therefrom, it is known in terms of the imposed parameters and therefore does not prevent the calculation of $\rho$ and $W$ by appeal to those equations.

Next,

$$\pi b_f/L = \ln(1+2/\beta) + (2/\gamma)\ln\left(\frac{1+\beta/2}{1+\beta}\right) \qquad (16.41)$$

Then the heights $z_{DP}$ on the boundary $DP$ at which the potential is $\phi_{DP}$ and the hydrostatic pressure head is $H_{DP}$ are given by the set of equations,

$$\pi z_{DP}/L = \pi c_f/L + \ln\left(\frac{\beta+1-t}{\beta+1+t}\right) + (2/\gamma)\ln\left(\frac{1+t}{\beta+1+t}\right) \qquad (16.42)$$

$$\pi\phi_{DP}/L = (\pi/L)(H_f+c_f) + (1/K)\Bigg[ (M-N)\ln(1-t) - M\,\ln(\beta+1-t) +$$
$$+ N\,\ln(\beta+1+t) + \frac{(M-N)(K-N)}{K+N}\ln\left(\frac{1+t}{\beta+1+t}\right)\Bigg] \qquad (16.43)$$

N

$$\pi H_{DP}/L = \pi H_f/L + \frac{M-N}{K}\ln\left(\frac{1-t}{1+t}\right) + \frac{M+K}{K}\ln\left(\frac{\beta+1+t}{\beta+1-t}\right) \quad (16.44)$$

Similarly, along the boundary $QR$,

$$\pi z_{QR}/L = \pi c_f/L + \ln\left(\frac{t-1-\beta}{t+1+\beta}\right) + (2/\gamma)\ln\left(\frac{t+1}{t+1+\beta}\right) \quad (16.45)$$

$$\pi\phi_{QR}/L = (\pi/L)\,(H_f+c_f) + (1/K)\Bigg[(M-N)\ln(t-1) - M\,\ln(t-1-\beta) +$$

$$+N\,\ln(t+1+\beta) + \frac{(M-N)(K-N)}{K+N}\ln\left(\frac{t+1}{t+1+\beta}\right)\Bigg] \quad (16.46)$$

$$\pi H_{QR}/L = \pi H_f/L + \frac{M-N}{K}\ln\left(\frac{t-1}{t+1}\right) + \frac{M+K}{K}\ln\left(\frac{t+1+\beta}{t-1-\beta}\right) \quad (16.47)$$

Further, on the boundary $DS$ below the drain the value of $t$, as indicated in the $t$ plane drawn in Figure (16.12), is real and lies between 1 and $1+\beta$, and within this range Equations (16.38) and (16.39) give

$$\pi z_{DS}/L = \pi c_f/L + \ln\left(\frac{1+\beta-t}{1+\beta+t}\right) + (2/\gamma)\ln\left(\frac{t+1}{t+1+\beta}\right) \quad (16.48)$$

$$\pi\phi_{DS}/L = (\pi/L)(H_f+c_f) + (1/K)\Bigg[(M-N)\ln(t-1) - M\,\ln(1+\beta-t) +$$

$$+N\,\ln(t+1+\beta) + \frac{(M-N)(K-N)}{K+N}\ln\left(\frac{1+t}{\beta+1+t}\right)\Bigg] \quad (16.49)$$

$$\pi H_{DS}/L = \pi H_f/L + \frac{M-N}{K}\ln\left(\frac{t-1}{t+1}\right) + \frac{M+K}{K}\ln\left(\frac{\beta+1+t}{\beta+1-t}\right) \quad (16.50)$$

For purposes of illustration of the manipulation of these equations to provide information in particular cases, one may choose at will a set of parameters $M/K$, $N/K$, and therefore $\gamma$. Next one may specify a particular value of $\theta'$ within the permissible range 0 to $\pi/2$, thus determining the value of $\beta$ in accordance with Equations (16.23), (16.24), and (16.26). Then by choosing successive values of $t$, one may calculate pairs of values of $z$ and $H$ along the boundaries by using either Equations (16.42), (16.45) or (16.48) in conjunction with Equation (16.40), together with either Equations (16.44), (16.47) or (16.50) respectively. Thus the distribution of pressure along the boundary may be plotted. In Figure (16.13) the results of such a procedure are sketched for the boundary $DP$ above the drain and $DS$ below it, for a number of decreasing values of $\theta'$, using Equations (16.42) and (16.44) for the former and Equations (16.48) and (16.50) for the latter. The first feature that emerges is that the pressure increases from the negative value $H_f$ at the fringe boundary, reaches a maximum and then decreases to assume eventually the infinitely great negative value at the $z$ axis. Below the drain it increases steadily from the infinitely great negative value to become eventually positive. Next, if $\theta'$ is sufficiently small, less than the value labelled $\theta'_{opt}$ in Figure (16.13), the pressure becomes positive before reaching the peak value above the drain, and in general

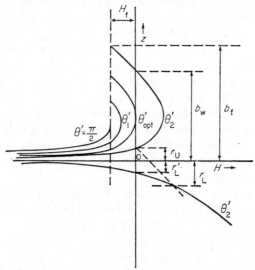

Figure 16.13 The distribution of pressure $H$ up the plane through the drain, *SDP* of Figure (16.5). The different curves are for different imposed values of $\theta'$.

there are two heights at which the pressure is just zero. The upper of these is evidently the location of the water table, which may be written $b_w$ to distinguish it from the height $b_f$ of the fringe boundary at this plane, and these heights are so marked in the diagram. The lower point at zero pressure is equally evidently a surface of seepage, that is to say, it is the location $r_U$ of the roof of a drain into which the ground water is seeping at zero pressure. The region of soil at steadily increasing negative pressure between this point and the axis has no real existence excepting in the highly artificial circumstance of the removal of water by a drain at which high negative pressures are maintained by pumping in such a way as to exclude air; in that case the drain wall would merely be removed to a value of $z$ corresponding to that high negative pressure. Similarly, the intercept of the curve on the negative $z$ axis marks the location of the floor $r'_L$ of a drain which is everywhere a surface of seepage, which implies that no depth of standing water is allowed to build up in it. If, however, the drain is just full of water so that the perimeter is an equipotential, then the pressure at the floor is equal to the depth of the floor below the roof, and the position of the floor $r_L$ is then the value of $z$ at which this pressure occurs. These alternative positions of the drain floor are shown in the diagram. In either case, the diagram indicates the drain size which imposes the particular pressure distribution for the chosen parameters. Thus the particular value of $\theta'$ which results in the calculation of this particular pressure distribution is the parameter by which the drain size is introduced into the theory.

Next one may notice that as $\theta'$ increases, the pressure at the peak above the drain decreases, the heights $b_f$ and $b_w$ decrease, and the drain roof rises. At an angle $\theta'_{\mathrm{opt}}$, the pressure at the peak just vanishes, and for this curve the water table and the drain roof come together at a single point. Here $b_w$ and $r_U$ have the same value. For still larger values of $\theta'$ there is no region of positive pressure, and therefore no point which can represent the location of a surface of seepage. These curves therefore do not correspond to any possible practicable case, since it is impossible to identify the location of a drain which is necessary to remove the ground water at the rate specified in the boundary conditions.

The distribution of pressure down the boundary $QR$ may be found in the same way, by appeal to Equations (16.45) and (16.47), and the height of the water table here, namely $c_w$, emerges as the value of $z_{QR}$ at which the pressure is just zero. As $\theta'$ increases to $\theta'_{\mathrm{opt}}$, the heights $c_f$ and $c_w$ are found to decrease just as do the heights $b_f$ and $b_w$, but to a less marked degree, until the maximum permissible angle $\theta'_{\mathrm{opt}}$ is reached. It is for this reason that the nomenclature $\theta'_{\mathrm{opt}}$ is used, since smaller angles produce higher ground water levels, while larger angles do not correspond to practical situations.

In order to present curves showing the dependence of the height of the upper boundary of the groundwater on the parameters, $K$, $M$ and $N$, it is convenient first to take the case of a capillary fringe of negligible thickness, that is to say, with vanishingly small $H_f$. It will be seen from Figure (16.13) that in this case the angle $\theta'$ can assume the maximum value of $\pi/2$ and still there will be a possible drain perimeter. This perimeter will also coincide with the water table and, in this case, the capillary fringe boundary. Thus the heights $c$ and $b$, which need not be differentiated into $c_f$ and $c_w$, and $b_f$ and $b_w$, will be the optimum values. When $\theta'$ has this maximum value, $\beta$ becomes identical with $\gamma$, as is shown in the discussion following Equation (16.26), which is to say that the ratio $\gamma/\beta$ assumes the value unity. In this optimum case, Equations (16.40) and (16.41) become respectively

$$\pi c_{\mathrm{opt}}/L = \ln(1+2/\gamma) + (2/\gamma)\ln(1+\gamma/2) \qquad (16.51)$$

$$\pi b_{\text{opt}}/L = \ln(1+2/\gamma)+(2/\gamma)\ln\left(\frac{1+\gamma/2}{1+\gamma}\right) \qquad (16.52)$$

Following van Deemter, one may then plot curves of $c_{\text{opt}}/L$ and $b_{\text{opt}}/L$ as a function of $\gamma$, and then on the same diagram plot further curves of $c/L$ and $b/L$ versus $\gamma$ for various different values of the ratio $\gamma/\beta$, using the more general equations, Equations (16.40) and (16.41). Van Deemter's own curves of this kind are shown in Figure (16.14). These curves are for decreasing values of $\theta'$ and of drain radius, but the corresponding drain radii are not indicated directly.

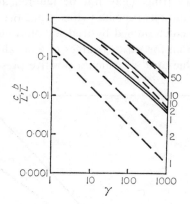

Figure 16.14 Curves showing $\dfrac{c}{L}$ and $\dfrac{b}{L}$ as functions of $\gamma$, after van Deemter. Full lines represent $\dfrac{c}{L}$, broken lines represent $\dfrac{b}{L}$. The figures against the curves indicate the values of $\dfrac{\gamma}{\beta}$.

Alternatively, one may adopt the procedure of Engelund (1951) and present curves for more narrowly restricted parameters than the combination $\gamma$. Thus if one considers drainage of groundwater with local precipitation alone at the rate $q$, then $M$ has the value zero and $N$ has the value $-q$, so that

$$\gamma = K/q - 1$$

Curves of $c$ and $b$ may then be plotted as a function of the perspicuous ratio $q/K$ for various values of either $\beta/\gamma$ or $\theta'$ as may be preferred. Figures (16.15) and (16.16), after Engelund, show respectively $c_{\text{opt}}/L$ and $b_{\text{opt}}/L$ as functions of $q/K$, $H_f$ again being taken to be zero, and the ratios $c/c_{opt}$, $b/b_{opt}$ and $r_U/_{opt}r_U$ as functions of $\theta'$ for various values of $q/K$. The ratio $b_{opt}/L$ in Figure (16.15) is of course the same curve as $_{opt}r_U/L$. The quantity $r_U$ is not truly the drain radius since the distance $r_L$ of the drain floor below the axis is not in general equal to $r_U$, but in fact the difference is usually small. Thus the ratio of $c_{opt}/L$ may be obtained for a given ratio $q/K$, for which ratio one may also read off the value of $_{opt}r_U$. The actual drain radius $r_U$ provides a ratio $r_U/_{opt}r_U$ from which the value of $\theta'$ may be read off and the appropriate correction factor $c/c_{opt}$ determined. The required value of $c/L$ then emerges.

When the capillary fringe may not be ignored and $H_f$ may not be neglected, the computations become tedius, since the value of $\theta'_{opt}$ is less than $\pi/2$ and must be determined by trial, by plotting the pressure distributions above the drain for various values of $\theta'$ as already described, and by choosing that value of $\theta'$ at which a positive pressure appears at one

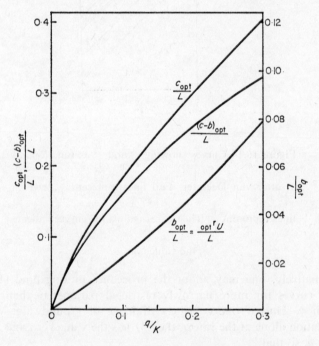

Figure 16.15 Curves of $\dfrac{c_{opt}}{L}$ and $\dfrac{b_{opt}}{L}$ or $\dfrac{_{opt}r_U}{L}$ as a

function of $\dfrac{q}{K}$, after Engelund. There is no artesian flow.

point only. Childs (1960b) programmed the calculations for a digital computer so that computing time was no objection and the true position of the drain floor could also be included in the results. The modifications in the more important features of the results due to the existence of capillary fringes of different thicknesses are presented in Figures (16.17) and (16.18). The calculations are presented only for the optimum values of $\theta'$, which are of course, as shown in Figure (16.13), different for each fringe thickness. The differentiating subscript *opt* may therefore be omitted here.

The height $c_f$ of the fringe boundary at the medial plane is presented in Figure (16.17) in the form of $-(c_f-c_0)/H_f$ where $c_0$ is the height which would have been observed in the absence of a capillary fringe, i.e. with negligibly small $H_f$. Similarly $c_w$ is presented in the form $(c_w-c_0)/H_f$, since $c_0$ is the height of the water table as well as of the capillary fringe when the latter is negligibly thin. It will be observed that the ratio $(c_f-c_w)/H_f$ is the vertical distance between the curves of Figure (16.17). and that this distance is nowhere greatly different from unity. Thus the thickness of the capillary fringe is approximately equal to the magnitude of $H_f$ over the range shown.

The proportion of the capillary fringe which is accounted for by a depression of the water table below $c_0$ is the greatest for thin fringes, when the total effect is in any case negligible, and decreases rapidly as the fringe thickens, until at the maximum of the range of $H_f$ shown, eighty per cent of the fringe thickness lies above the position which the water table would

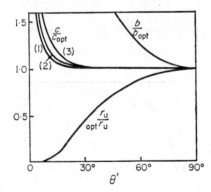

Figure 16.16 Curves of $\dfrac{c}{c_{opt}}$, $\dfrac{b}{b_{opt}}$ and $\dfrac{r_U}{opt\,r_U}$ as functions

of $\theta'$. Curve (1) is for $\dfrac{q}{K}$; $= 0.001$; curve (2) for $\dfrac{q}{K} = 0.01$;

and curve (3) for $\dfrac{q}{K} = 0.1$.

have occupied in the absence of a fringe. Hence it is a reasonable working approximation to neglect the fringe, to use the simpler formulas resulting in Figures (16.14), (16.15) and (16.16), and to add to the water table height so emerging a supplement of about eighty per cent of the magnitude of the air entry pressure $H_f$, in order to arrive at the height of the fringe boundary.

Figure (16.18) shows the effect of a capillary fringe of increasing thickness

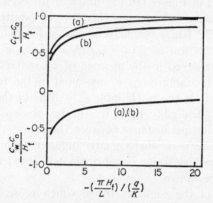

Figure 16.17 The position of the capillary fringe boundaries, relative to the position of the water table in the absence of a capillary fringe, as a function of the capillary fringe thickness. Curve (a) is for $\dfrac{q}{K} = 0\cdot1$; and curve (b) for $\dfrac{q}{K} = 0\cdot01$ and $0\cdot001$.

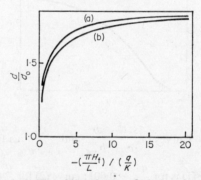

Figure 16.18 The dependence of the optimum drain diameter, expressed as a ratio of the optimum diameter in the absence of a capillary fringe, on the capillary fringe thickness.

on the optimum drain diameter. It will be noted from Figure (16.16) that the most important of the items of the solution, namely $c_f$, is not greatly affected by a change of the drain size until quite small sizes are reached. A reduction of $r$ to as little as one-fifth of $r_0$ produces an increase of $c_f$ of no more than ten per cent. The effect of the capillary fringe on the optimum drain size is relatively small, so that repercussions on the value of $c_f$ on this account are of secondary importance.

Finally, it may be noted that over a range of $q/K$ up to $0.3$ the curve of $c_0$ or of $c_f$ as a function of $q/K$ is approximately linear, with a slope which depends on the sub-range within these limits. Thus between a zero value of $q/K$ and $0.001$, one may write very approximately,

$$\left.\begin{aligned} c_0/L &= 4(q/K) \\ c_f/L &= 4(q/K) - 0.8\ H_f/L \\ &0 < q/K < 0.001 \end{aligned}\right\} \qquad (16.53)$$

and similarly,

$$\left.\begin{aligned} c_0/L &= 3(q/K) \\ c_f/L &= 3(q/K) - 0.8\ H_f/L \\ &0 < q/K < 0.01 \end{aligned}\right\} \qquad (16.54)$$

and

$$\left.\begin{aligned} c_0/L &= 2(q/K) \\ c_f/L &= 2(q/K) - 0.8\ H_f/L \\ &0 < q/K < 0.10 \end{aligned}\right\} \qquad (16.55)$$

These are often useful working approximations.

When $q/K$ exceeds about $0.5$, the height of the fringe boundary begins to increase more rapidly and becomes infinitely great when $q$ becomes equal to $K$. This is yet another reflection of the fact, noted in Sections 12.7(b) and 13.1, that the rate of infiltration from a flooded surface, which may be regarded as a surface water table, settles to a steady rate equal to the saturated conductivity, and this rate represents the maximum possible rate of infiltration in the absence of a deeply-ponded surface. In such circumstances, the only immediate relief that a drainage system can afford is the removal of surface water. When it is remembered that the saturated conductivity of some clay soils can be as low as a millimetre or two a day, it will be seen that this limitation of function of drainage can be of major significance, since in humid climates the average precipitation over a drainage stress period of two or three weeks may well exceed this rate. Such soils are recognized as presenting, in the main, problems of surface drainage to which the application of groundwater theory has little relevance.

The analysis of irrigation problems follows the same course and results in the same equations, but in such cases $N$ exceeds $M$ and $\gamma$ is therefore negative. In these circumstances, a consideration of Equations (16.23) to

N*

(16.25) leads to the establishment of a range of possible values of $\beta$ which are quite different from those appropriate to drainage problems. The results are

$$0 < \beta < \infty$$

$$\text{for } \gamma < -1$$

and

$$0 < \beta < -\gamma/(1+\gamma)$$

$$\text{for } -1 \leqslant \gamma < 0$$

with

$$\theta' < 0$$

The pressure of water at the irrigation pipe is necessarily substantial.

The analysis is an interesting academic exercise, but it will not be discussed further here, since irrigation by raising the level of the ground water through the application of water at high pressure to buried pipelines is not a commonly practicable procedure.

## 16.4 Van Deemter's analysis: drainage by ditches

Drainage by equally spaced ditches is amenable to analysis by the method of successive transformations of the hodograph and potential planes, in a manner similar to that described in the preceding section. The resulting integrals are not, however, of an elementary kind, and for details the reader may be referred to the original paper by van Deemter (1950). The results only are presented here.

The solution is provided for the case of drainage to a ditch of width $2d$, with a separation of $2(d+L)$ between the axes of neighbouring ditches. One assumes the incidence of rainfall at a rate $q$ which corresponds to the flux $-N$ of Section 16.3. It remains necessary to postulate an infinitely great depth of soil, but in this case there is an absence of artesian water or of deep drainage, i.e. $M$ is given the value zero. The section of the ditch system in a vertical plane perpendicular to the ditch axis is shown in Figure (16.19). It must also be assumed that the water level in the ditch does not rise above the ditch floor. Rainfall which is directly incident on the ditch does not, of course, contribute to the problem, since it is run away directly without penetrating the soil.

The solution is shown in Figure (16.20) in the form of curves which relate $c/L$ to $q/K$ for various ratios $d/L$. As before, $c$ is the height of the water table above the sink level, in this case the floor of the ditch, at the highest point, namely midway between neighbouring drains. No account is here taken of the presence of a capillary fringe, but it is unlikely that the results of the

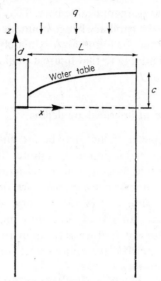

Figure 16.19 A cross-section in the $\rho$ plane of the groundwater between parallel ditches. The soil is infinitely deep and the ditch water is level with the ditch bottom.

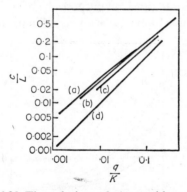

Figure 16.20 The solution of the problem shown in Figure (16.19). Curve (a) is for $\dfrac{d}{L} = 0$; curve (b) for $\dfrac{d}{L} = 0{\cdot}002$; curve (c) for $\dfrac{d}{L} = 0{\cdot}1$; and curve (d) for $\dfrac{d}{L} = \infty$; curve (a) is barely distinguishable from the curve for tile drains.

preceding section should be inapplicable here. That is to say, if the air entry pressure, which defines the capillary fringe boundary, is $H_f$, then the height of the fringe boundary may be found approximately by calculating $c$ in the absence of a fringe and adding to that figure a thickness equal to eighty per cent of the magnitude of $H_f$.

### 16.5 Impermeable strata at intermediate depths

When any impermeable layer that may be present in the soil profile is so deep that it may be regarded as infinitely remote, the solutions of Sections 16.3 and 16.4 are applicable. When on the other hand such a layer is so near the surface that the drains penetrate to or lie on it, then the Dupuit-Forchheimer approximation is very satisfactory. At intermediate depths there is neither an exact analytical procedure nor a simple acceptable approximation, and one must either rely upon the reiterative numerical or analogue methods described in Sections 15.1 and 15.2 or try to devise further approximate formulas.

First one may enquire in what range a depth must lie for it to be regarded as intermediate. Solutions of drainage problems involving lines of equidistant parallel drains, using analogue methods, have indicated that if an impermeable layer is deeper below the drains than about one-sixth of the distance separating neighbouring drains, that is to say if the depth $p$ exceeds about $0\cdot3\,L$, it may for all practical purposes be regarded as infinitely deep. This is shown in Figure (16.21), to be discussed in detail later. An examina-

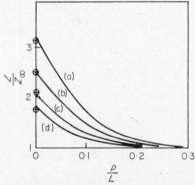

Figure 16.21 The dependence of the water table height, $Z$, on the depth of the impermeable bed $p$. $Z_\infty$ is the water table height with an infinitely deep bed. Curve (a) is for $q/K = 0\cdot01$; curve (b) for $q/K = 0\cdot02$; curve (c) for $q/K = 0\cdot05$; and curve (d) for $q/K = 0\cdot1$. The circles on the axis of $Z/Z_\infty$ are the ratios calculated from Equations (15.16) and (16.51).

tion of the flow-nets shows that this is because no more than about ten per cent of the surface precipitation, namely the amount falling on the middle ten per cent of the catchment, ever flows in a path any part of which penetrates to a greater depth than $0\cdot3\ L$, even though such greater depths are available, so that an impediment at or below that depth can have little effect. Hence an intermediate depth in the present context is one which is greater than the depth of the drains, but less than one-sixth of the drain separation. It is convenient to express $p$ in terms of the fraction $p/L$, so that an intermediate depth is one for which $p/L$ lies between 0 and $0\cdot3$, all measurements being made from the drain level.

The magnitude of the effect of varying the bed depth may be indicated, as in Table 3, by comparing the water table heights in the limiting cases of infinite and zero bed depth, at various intensities of precipitation. For the former, one may use Equation (16.51) of the hodograph analysis, ignoring the capillary fringe, and for the latter Equation (15.16) which results from the Dupuit-Forchheimer treatment. This table shows that the difference is relatively the greatest for the least intense rates of precipitation, as would be expected, since there is the greatest difference between the thickness of the groundwater zone available to flow in the limiting cases when the height of the water is least. In this table $c$ indicates the water table height at the medial plane between drains in the absence of a capillary fringe and for the optimum drain size, that is to say it refers to $c_{opt}$ of Equation (16.51).

*Table 3*

| $q/K$ | $\gamma$ | $c/L$ hodograph | $Z_m/L$ Dupuit-Forchheimer | $Z_m/c$ |
|---|---|---|---|---|
| $0\cdot0001$ | 9999 | $0\cdot00062$ | $0\cdot010$ | $16\cdot1$ |
| $0\cdot001$ | 999 | $0\cdot0045$ | $0\cdot032$ | $7\cdot0$ |
| $0\cdot01$ | 99 | $0\cdot032$ | $0\cdot100$ | $3\cdot2$ |
| $0\cdot02$ | 49 | $0\cdot057$ | $0\cdot140$ | $2\cdot5$ |
| $0\cdot05$ | 19 | $0\cdot11$ | $0\cdot220$ | $2\cdot0$ |
| $0\cdot10$ | 9 | $0\cdot18$ | $0\cdot320$ | $1\cdot75$ |

Figure (16.21), after Collis-George and Youngs (1958), shows the influence of varying the bed depth on the height of the water table over a tenfold range of rate of precipitation, as observed in a series of experiments with electric and hydraulic analogues. In this diagram, the water table height $Z$ refers to the medial plane between parallel drains, that is to say to $Z_m$. It is plotted as the ratio $Z/Z_\infty$, where $Z_\infty$ is the water table height when the bed is effectively at infinitely great depth. It will be noted that, as mentioned above, the curves all approach the value unity asymptotically and depart only negligibly from that value for ratios of $p/L$ in excess of $0\cdot3$.

Curves for values of $q/K$ other than those for which results are presented may be interpolated, and the interpolation is rendered simple from the fact that the curves are all of similar shape and may be reduced to a single empirical expression. Marked on the axis of $Z/Z_\infty$ are values of this ratio calculated from the hodograph theory for infinite bed depth and the Dupuit-Forchheimer approximation for zero bed depth, i.e. zero $p/L$. It will be seen that these values conform reasonably well with the observed ratios for the appropriate values of $q/K$. Secondly, it will be noted that the curves are all approximately of the same shape when referred to the asymptote $Z/Z_\infty = 1$. That is to say, for all of them the ratio $(Z/Z_\infty - 1)/\{(Z/Z_\infty)_{p=0} - 1\}$ is almost the same at the same value of $p/L$. Hence they may all be reduced to the same curve by plotting this dimensionless ratio as a function of $p/L$, and in this reduction the calculated value of $(Z/Z_\infty)_{p=0}$ may be used by virtue of the first-mentioned feature. In fact, this reduced curve is found empirically to conform well enough to the equation,

$$\log\frac{Z/Z_\infty - 1}{(Z/Z_\infty)_{p=0} - 1} = -5\cdot8p/L \qquad (16.56)$$

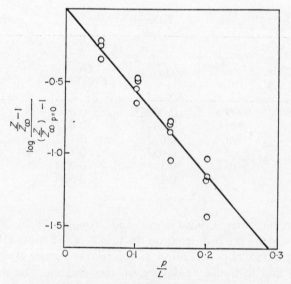

Figure 16.22 The shapes of the curves of Figure (16.21) expressed in a semi-logarithmic diagram.

This is demonstrated in Figure (16.22), which is a plot of points taken off Figure (16.21) scattered around the straight line drawn from Equation (16.56).

Thus for a given value of $q/K$ the ratio $(Z/L)_\infty$ may be evaluated from hodograph theory, for it is the value of $c/L$ of Figure (16.15), or of Equations (16.40) or (16.41), or of one or other of the approximations of Equations (16.53) to (16.55), as may be appropriate or preferred. The ratio $(Z/L)_{p=0}$ is determined from the Dupuit-Forchheimer formula, Equation (15.16). The ratio $(Z/L)_{p=0}/(Z/L)_\infty$ is the ratio $(Z/Z_\infty)_{p=0}$ which occurs in Equation (16.56). The ratio $Z/Z_\infty$ for any other value of $p/L$ is therefore obtainable from Equation (16.56), and finally the ratio $Z/L$ emerges since $(Z/L)_\infty$ is known.

For approximate work in the absence of detailed references, it is not too difficult to retain in the memory the Dupuit-Forchheimer formula, Equation (15.16), Equation (16.54) as a mean approximation to the hodograph results, and Equation (16.56).

An early alternative approach to the problem, due to Hooghoudt (1940), relies upon an assumption that the ratio $Z/L$ is so small that the water table may be regarded as a flat surface. This approach seeks to build the flow pattern from an appropriate array of horizontal cylindrical sinks treated as pumped wells. If the rate of flow of water emerging per unit length of such a tube, sometimes called the strength, is $2Q$, then according to Equation (14.11) the potential at a distance $r$ from the axis of a single tube is $\phi$, given by

$$\phi = \{Q/(\pi K)\}(\ln r + C) \tag{16.57}$$

where $K$ is the hydraulic conductivity of the soil and $C$ is an arbitrary constant which enables one to assign zero potential to any desired point to suit one's convenience. An array of equidistant tubular drains in a horizontal plane, buried in a limitless soil, is shown in Figure (16.23), and

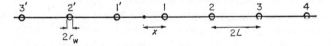

Figure 16.23 An array of parallel drains in a plane in an infinite soil.

is regarded as removing water from a distant source. The potential at any given point due to the array of tubes of equal strength may be determined as the sum of the potentials due to the individual tubes, each contribution being calculated by Equation (16.57) with the appropriate values of $r$. In particular, the potential distribution may be calculated in the plane of the drains. Thus at a point on this plane distance $x$ from the axis of a specified drain, which may be indicated by the subscript 1, the potential $\phi_x$ is given by

$$\phi_x = \{Q/(\pi K)\}\{\ln x + C_1 + \ln(2L+x) + C_2 + \ln(4L+x) + C_3 + \ldots$$
$$+ \ln(2L-x) + C'_1 + \ln(4L-x) + C'_2 + \ldots\}$$

where $C_1$, $C_2$, etc., refer to the first and second drains to the right of the point considered, and $C'_1$, $C'_2$, etc., refer to drains to the left of that point. As a particular case, the potential $\phi_w$ at the surface of the specified drain, of radius $r_w$, is, if $r_w$ is very much less than $2L$ as is usually the case,

$$\phi_w = \{Q/(\pi K)\}\{\ln r_w + C_1 + \ln(2L) + C_2 + \ln(4L) + C_3 + \ldots$$
$$+ \ln(2L) + C'_1 + \ln(4L) + C'_2 + \ln(6L) + C'_3 + \ldots\}$$

Thus the potential at $x$, referred to the potential at the drain surface as datum, is

$$\phi_x - \phi_w = \{Q/(\pi K)\}\{\ln(x/r_w) + \ln(1+x/2L) + \ln(1+x/4L) + \ldots$$
$$+ \ln(1-x/2L) + \ln(1-x/4L) + \ldots\}$$
$$= \{Q/(\pi K)\}\{\ln(x/r_w) + \ln[1-(x/2L)^2] + \ln[1-(x/4L)^2] +$$
$$+ \ln[1-(x/6L)^2] + \ldots\} \tag{16.58}$$

From the symmetry of the situation the plane of the drains is a plane of streamlines, since there is no straying of flow from one side of the plane to the other. If the problem were one of the flow of electricity in a conductor with electrodes corresponding to the drains, the conducting sheet could be cut along the plane of the "drains" without any effect whatever on the flow-net, since the cut would be along a line where there is no tendency of current to cross from one side to the other. The hydraulic case is different because the distribution of potential along the now exposed plane of the drains requires that the pressure be higher at points more remote from the drain, since the potential is higher, but the height is the same. Hence the exposed surface would be a surface of seepage, and the leakage of water would upset the flow-net. This can be prevented by cutting along the line of zero pressure, instead of along the plane of the drains, but now the cut is not along a streamline and this would upset the flow-net. However, provided that the water table is almost indistinguishable from the plane of the drains, the difference of the shape of the boundary introduces only negligibly small errors in the use of Equation (16.58) to describe the potential distribution along this boundary, and in this case the potential distribution is the same thing as the distribution of the height, since the pressure is everywhere zero in the water table. In particular, the height $Z_m$

of the water table at its highest point, midway between drains, is obtained from Equation (16.58) by giving the variable $x$ the particular value $L$, with the result,

$$Z_m - Z_0 = \phi_L - \phi_w = \frac{Q}{\pi K}\left(\ln\frac{L}{r_w} + \sum_{n=1}^{n=\infty}\ln(1-(1/2n)^2)\right)$$

$$= \frac{Q}{\pi K}\left(\ln\frac{L}{r_w} - 0.454\right) \tag{16.59}$$

$Z_0$ is the height of the roof of the drain, which is supposed to be just filled with water without back-pressure, and the datum from which $Z$ is measured is quite arbitrary.

This then would be the equation for the height of the water table due to the flow of artesian water from vertically below the plane of the drains, with each drain removing water at the rate $Q$ per unit length. The other half of the $2Q$ originally postulated is, of course, due to the flow from the upper part of the complete section imagined in the development, which is now removed.

If now a uniform rate of flow $q$ vertically downward is superimposed on this flow-net, of such a magnitude as to neutralize exactly the rate of artesian flow at depth, the flow-net would be modified everywhere without changing the potential distribution on the water table, since this is supposed to be essentially flat, and would therefore coincide with an equipotential of the superimposed flow-net. But the resultant flow-net would now correspond to the drainage of rainfall arriving at such a rate $q$ as would produce the drain flow $Q$ per unit length with an impermeable substratum at a remote depth. The equation of the potential distribution on the water table, Equation (16.58), and the equation for the water table height, Equation (16.59), still hold, with $Q$ equal to $2Lq$.

As a check, the water table height predicted by Equation (16.59) may now be compared with that obtained from the rigorous hodograph analysis, with a suitably small value of $q/K$, say $0.001$. Taking the case of rainfall without artesian water, and in the absence of a capillary fringe, one may use Equation (16.51) to show that

$$c/L = 0.0046$$

From Equation (16.59), on the other hand, one may derive a result by taking a value of $L$ of, say 50 metres, which would give a reasonable value of $Z$ and by substituting the product $2Lq$ for $Q$. A commonly acceptable value of $r_w$ might be $0.05$ metre. Substitution of these values in Equation (16.59) results in

$$\frac{Z_m - Z_0}{L} = 0.0041$$

This fair agreement indicates that up to this stage the logarithmic formula is not unacceptable as an approximation. If the soil is of limited depth $p$, owing to the presence of an impermeable bed, consider the consequences of taking a thickness $2p$ and imposing the same conditions at the lower surface as are known to be imposed at the upper. Then by symmetry the medial plane at the depth $p$ is a streamline, and is therefore equivalent to an impermeable barrier in its effects. The lower bounding surface is said to be the mirror image of the upper surface, being mirrored in the medial plane. As an approximation one may therefore recalculate the potential distribution due, not only to the upper surface drains of strength $2Q$, but also to the lower array. That this is a rather poor approximation is due to the fact that the logarithmic distribution of potential Equation (14.11) on which the analysis is based is valid only for arrays of sinks or sources in an infinite medium, and is not strictly appropriate to the slice of limited thickness $2p$. The error of the approximation is greater for smaller thicknesses $p$.

Again one must be concerned chiefly with the potential distribution at the water table, and this is the sum of the potentials due to the actual drains, as given by Equation (16.58), and the mirror drains. The latter contribution, $(\phi_x - \phi_w)'$, is readily derived by an application of Equation (16.57) similar to that which led to Equation (16.58), and is

$$(\phi_x - \phi_w)' = (Z - Z_0)' = \frac{Q}{2\pi K}\left[ \ln\{1 + (x/2p)^2\} + \right.$$

$$\left. + \sum_{n=1}^{n=\infty} \ln\frac{[(2p)^2 + (2nL+x)^2][(2p)^2 + (2nL-x)^2]}{[(2p)^2 + (2nL)^2]^2}\right] \quad (16.60)$$

The total potential at the point on the water table distant $x$ from the drain is thus the sum of Equations (16.58) and (16.60), where $Q$ has the value $2Lq$. While the datum from which $Z$ is measured is entirely arbitrary, it now seems convenient to allocate it either to the impermeable bed, in conformity with Section 15.3 and Equations (15.15) to (15.17), or to the drain level in conformity with Section 16.3. If the former, then $Z_0$ has approximately the value $p$, as in Equation (15.17).

This treatment proves to be a rather poor approximation of the truth when the depth $p$ is too small a fraction of $L$, while the use of the Dupuit-Forchheimer approximation, Equation (15.15), gives a poor approximation when $p$ is too large a fraction of $L$, since in effect it assumes that the drain penetrates the whole depth of the soil to the impermeable bed.

Hooghoudt's analysis demonstrates that the approximate formula based on the sum of Equations (16.58) and (16.60) is acceptable for values of $x$ not exceeding $p/(2)^{\frac{1}{2}}$, while the Dupuit-Forchheimer approximation,

Equation (15.15), is acceptable for values of $x$ greater than this. Hence a composite solution may be constructed. Up to the distance $p/(2)^{\frac{1}{2}}$ the radial formula is used. At the limit of this range,

$$Z_{0.707p} - Z_0 = A + B \tag{16.61}$$

where $A$ is given by Equation (16.58) and $B$ by Equation (16.60) with $x$ allotted the value $p/\sqrt{2}$. Thus,

$$A = \left(\frac{2Lq}{\pi K}\right)\{\ln(0.707p/r_w) + \sum_{n=1}^{n=\infty} \ln[1 - p^2/(8n^2L^2)]\} \tag{16.62}$$

$$B = \frac{Lq}{\pi K}\left[\ln 1.13 + \sum_{n=1}^{n=\infty} \ln \frac{(2p)^2 + (2nL + 0.707p)^2}{(2p)^2 + (2nL)^2} + \right.$$

$$\left. + \sum_{n=1}^{n=\infty} \ln \frac{(2p)^2 + (2nL - 0.707p)^2}{(2p)^2 + (2nL)^2}\right] \tag{16.63}$$

Beyond this point one may use the Dupuit-Forchheimer formula, Equation (15.14), referred to the displaced origin at $x = p/\sqrt{2}$, namely

$$Z^2 - Z_{0.707p}^2 = \frac{q}{K}\{2(L - 0.707p)(x - 0.707p) - (x - 0.707p)^2\}$$

$$= \frac{q}{K}(2Lx - x^2 - 1.414pL + p^2/2)$$

By assigning to $x$ the value $L$ one arrives at $Z_m$, the maximum water table height at the medial plane, in the form,

$$Z_m{}^2 - Z_{0.707p}^2 = \frac{q}{K}(L - p/\sqrt{2})^2$$

This may be put into the form,

$$Z_m - Z_{0.707p} = \frac{q}{2K\bar{Z}}(L - p/\sqrt{2})^2 \tag{16.64}$$

where $\bar{Z}$ is the mean water table height in the range of $x$ between $0.707p$ and $L$. Thus

$$\bar{Z} = (Z_m + Z_{0.707p})/2$$

Thus the water table height at the medial plane is given as the sum of Equations (16.61) and (16.64) by the equation,

$$Z_m - Z_0 = A + B + \frac{q}{2K\bar{Z}}(L - p/\sqrt{2})^2 \tag{16.65}$$

If the difference $Z_m - Z_0$ is negligibly small compared with the depth $p$ of the impermeable bed, which is itself sufficiently expressed by $Z_0$, the mean water table height $\overline{Z}$ may be replaced in Equation (16.65) by $p$, with the result

$$Z_m - Z_0 = A + B + \frac{q}{2Kp}(L - p/\sqrt{2})^2$$

If this approximation is not reasonable, then $\overline{Z}$ is not known until $Z_m$ is known, so that it appears that one is trapped in a circular argument. However, a series of approximations with trial values of $\overline{Z}$ will permit the calculation of a correct value of $Z_m$ as that which provides consistency between the assumed and emergent values of $\overline{Z}$.

Ernst (1954) has proceeded in this problem in a manner somewhat similar to that of Collis-George and Youngs, in this case solving a range of situations by relaxation instead of by analogue. His range of $q/K$ is more extended than that presented in Figure (16.21), being from 0·14 to 0·001. Curves privately communicated to the author show that the Dupuit-Forchheimer approximation in the form Equation (15.15) is quite good for values of $p/L$ below 0·05 combined with values of $q/K$ above 0·01. In general, a formula similar to Hooghoudt's, described above, but somewhat simpler, proves to be satisfactory. This provides for a simply logarithmic increase in potential in accordance with Equation (16.57) for a distance of 0·6$p$ from the drain, with a Dupuit-Forchheimer increase, Equation (15.14), beyond that distance. Thus from Equation (15.14)

$$Z_m^2 - Z_{0\cdot6p}^2 = \frac{q}{K}(L - 0\cdot6p)^2$$

The assumption of axial symmetry of flow about the drain centre in its near neighbourhood requires that the flow rate $Lq$ is one-quarter of the complete cylindrical flow rate $Q$, so that

$$Z_{0\cdot6p} - Z_0 = \frac{2Lq}{\pi K}\ln\frac{0\cdot6p}{r_w}$$

For additional accuracy $Z_0$ is taken to be, not $p$ simply, but the sum of $p$ and $r_w$, so that the drain is assumed to be a cylindrical channel just filled with water to the roof. These equations together provide the solution, as in the case of Equation (16.65). The drain radius was taken to be approximately that given by the hodograph analysis in the optimum situation. This approximation was generally satisfactory for values of $p/L$ up to 0·2, while above 0·2 the hodograph solution of Section 16.3, appropriate to infinite depth of impermeable bed, was applicable.

The curves obtained by Ernst have been used by Boumans (1954) to prepare nomograms from which the distance $L$ may be read off for any chosen values of $q/K$, $p$, and allowable $Z_m$. These nomograms have been presented by Visser (1954) and reproduced by van Schilfgaarde (1957).

Yet other approximate treatments may be mentioned without being described at length. The original papers may be consulted. The flow-net for a combination of rainfall and artesian water in infinitely deep soil may be determined rigorously by van Deemter's analysis, and it may be seen by reference to Figure (16.4) that there is a stagnation point $O$ at a height $z_0$ at the medial plane between the drains, and that from this point there is a streamline to the drain which divided the artesian water from the rain-fed groundwater. This is confirmed by the analysis in Section 16.3 using Equation (16.45). This streamline is equivalent to an impermeable bed, which is not, however, flat. The height $z_0$ may be calculated using $v/\lambda$ given by Equations (16.23a) and (16.25a), for this ratio is the value $t_0$ for substitution in Equation (16.45). Childs (1960b) has identified $z_0$ approximately with the depth of the impermeable bed for the purpose of discussing the effect of such a bed on the influence of the capillary fringe.

List (1964) has developed this idea further to simulate a reasonably flat impermeable bed by introducing a second and lower drainage system to intercept the artesian water. This drain system is vertically below that which intercepts the rainfall-fed groundwater, and the depth and shape of the simulated bed are controllable through the choice of the depth of the auxiliary drains and the intensity of the flow of artesian water relative to the rainfall. For additional kinds of approximation, the reader may be referred to Kirkham's review (1966).

## 16.6 The drainage of groundwater under a flat water table

When the water table is truly flat and level, it must also be an equipotential, since both the height and the pressure at every point on it are uniform. The only situation in nature where such a circumstance prevails is where the water table has risen to the surface and is there ponded. The infiltration rate is not imposed as a boundary condition, but is distributed over the surface in a manner which is determined by the potential boundary condition, and remains to be discovered as an important item in the solution of the problem. In practice such a circumstance is not common, since by hypothesis the drainage system is ineffective as regards its usual function of containing the groundwater within acceptable limits. However, it sometimes happens that control of the water table is not the immediately important end, but rather, for example, the removal of excess salt during the reclamation of salinized land. Here the speed of leaching is at its maximum when the water table is at its highest, and there is no objection to surface ponding. It remains true, of course, that the leaching is at its most intense near the drain lines and is more or less complete here while far from complete near the medial plane, so that maintenance of a surface flood is wasteful of water if maintained beyond this stage.

The determination of the flow-net in this case may be achieved by the application of the radial logarithmic formula for the potential in the neighbourhood of a single cylindrical drain, in a manner similar to that described in the preceding section. One has to construct a system of mirror images in unbounded space which will result in the known boundary conditions. Thus the flat equipotential water table would result from an extension of the soil cross-section vertically upward to infinity, and installing in this extension a set of drains to form a mirror image of the true drains, the ponded surface water table being the mirror. The mirror "drains" must be of opposite strength to the true drains, so that if the latter are removing water at a rate $Q$ per unit length, then the mirror images must add water at this rate.

The fact that such a system does indeed produce a flat level water table is readily confirmed, for the potential at a point is

$$\phi = (\phi/2\pi K)(\ln r_1 - \ln r_2)$$

where $r_1$ is the distance of the point from the drain axis of strength $Q$ and $r_2$ is the distance from the drain of strength $-Q$. At the mirror surface $r_1$ and $r_2$ are equal, so that $\phi$ vanishes everywhere on this surface, which is therefore the zero equipotential.

The presence of an impermeable floor may be simulated in the unbounded space by introducing a set of image drains, the mirror surface being the plane of the impermeable floor. The "irrigator" drains mirrored in the water table would also, of course, be mirrored in the floor, and drains mirrored in the floor would also be mirrored in the water table, and so *ad infinitum*. The net result is therefore an infinite number of mirrored systems, as is experienced optically by an observer between two parallel mirrors.

No new principles are illustrated in this treatment, which involves in essence the summation of the potential contributions of a doubly infinite array of drains in a regular pattern, each contribution of which obeys the logarithmic law of Equation (14.11). This is an exercise in mathematical manipulation rather than in physical insight and will not be treated further in detail here. For a review of the matter, the reader is referred to Kirkham (1957), who has himself been active in this development.

# NOTES

### Note 37. The conformal transformation

Any path drawn in the $\rho$ plane, as for example, an equipotential or a stream-line, will appear in the transformed $\sigma$ plane as a different but equivalent path. Thus the points $\rho_1$ and $\rho_2$ define a path element $\rho_2 - \rho_1$, for

$$\rho_2 - \rho_1 = (x_2 - x_1) - i(z_2 - z_1)$$

and in the Argand diagram, shown in Figure (N37.1), this expresses the line joining the points $\rho_1$ and $\rho_2$. If the interval is short, it may be indicated in the conventional way by the symbol $\delta\rho$. The corresponding interval of the cor-

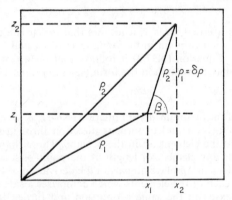

Figure N37.1 An element of boundary $\delta\rho$ in the Argand diagram where $\rho = x + iz$. $\beta$ is the angle with the $x$ axis.

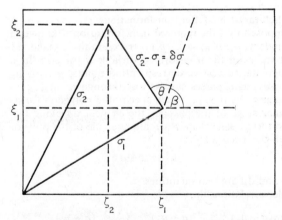

Figure N37.2 The element $\delta\rho$ of Figure (N37.1) con-formally transformed into the element $\delta\sigma$ in the $\sigma$ plane, where $\sigma = \zeta + i\xi$. The transformation turns the element through the angle $\theta$.

responding path in the $\sigma$ plane is $\delta\sigma$ and is shown as the element of different length and different orientation in Figure (N37.2). Since one may write,

$$\delta\sigma = (d\sigma/d\rho)\delta\rho \tag{N37.1}$$

it is evident that the differential coefficient is a multiplier which has the effect of turning the element $\delta\rho$ into the direction of $\delta\sigma$ and of increasing its length until it gives the magnitude $\delta\sigma$. Now from Equation (16.2),

$$\partial\sigma/\partial x = (d\sigma/d\rho)(\partial\rho/\partial x)$$

$$= d\sigma/d\rho \tag{N37.2}$$

From Equation (16.3),

$$\partial\sigma/\partial x = \partial\zeta/\partial x + i\,\partial\xi/\partial x$$

Since both $\zeta$ and $\xi$ are real, as is $x$, it follows that $\partial\sigma/\partial x$ is a complex quantity which varies from point to point in the $\rho$ plane as $\partial\zeta/\partial x$ and $\partial\xi/\partial x$ vary in that plane. Hence from Equation (N37.2) it follows that $d\sigma/d\rho$, which is this same complex number, may be written in the form (see Equations 14.4 and 14.5),

$$d\sigma/d\rho = r\,e^{i\theta} \tag{N37.3}$$

Thus the effect of the factor $d\sigma/d\rho$ in Equation (N37.1) is to increase the magnitude of $\delta\rho$ by the factor $r$ and to turn its direction through the angle $\theta$, thus transforming it into the element $\delta\sigma$ in the $\sigma$ plane. These changes take place to the same degree for any element of length at the point $\rho$, since the orientation and length of the element $\delta\rho$ plays no part in the derivation of Equation (N37.3). Thus for example, each of the elements which comprises a side of a small square at $\rho$ becomes increased in the same proportion and turned through the same angle in being transformed into an element in the $\sigma$ plane. Hence the elementary square in the $\rho$ plane is transformed into another elementary square in the $\sigma$ plane, of different size and reorientation.

### Note 38. The Schwarz-Christoffel transformation

Let it be supposed that the required transformation is known, and that as the polygonal boundary, $\rho_1$, $\rho_2$, $\rho_3$, ... $\rho_n$, is traced in the $\rho$ plane in Figure (N38.1), the calculated values of the transformed variable $\sigma$ traverse the real axis of the $\sigma$ plane through the successive corresponding points, $\mu_1$, $\mu_2$, $\mu_3$, ... $\mu_n$. Or of course, conversely, as $\sigma$ passes successively through $\mu_1$, $\mu_2$, ..., so does $\rho$ pass through the apexes, $\rho_1$, $\rho_2$, ..., which specify the polygon of $n$ sides.

At a particular stage of the passage, the element $d\sigma$ along the real axis of $\sigma$ corresponds to the passage element $d\rho$ along a side of the polygon which makes an angle $\theta$ with the $x$ axis. Thus

$$d\rho = A\,d\mu\,e^{i\theta} \tag{N38.1}$$

where both $A$ and $d\mu$ are real quantities.

Now consider a relationship,

$$d\rho = C_1\,d\sigma(\sigma-\mu_1)^{(\alpha_1/\pi-1)}(\sigma-\mu_2)^{(\alpha_2/\pi-1)}\cdot\cdot(\sigma-\mu_r)^{(\alpha_r/\pi-1)}\cdot\cdot$$

$$\cdot\cdot(\sigma-\mu_n)^{(\alpha_n/\pi-1)} \tag{N38.2}$$

where $\alpha_r$ is the internal angle of the $r$th apex of the polygon.

If one considers the passage of the real axis $\mu$ then the element $d\sigma$ is simply the real element $d\mu$. Take the stage at which $\sigma$ lies in the interval between $\mu_r$ and $\mu_{r+1}$, so that $(\sigma - \mu)$ is real and positive for all apex values of $\mu$ up to and including $\mu_r$, and real and negative for all values of $\mu$ greater than $\mu_r$. Hence all the factors on the right-hand side of Equation (N38.2), excepting $C_1$, which will be considered later, are real and positive up to and including $(\sigma - \mu_r)^{(\alpha_r/r - 1)}$ and

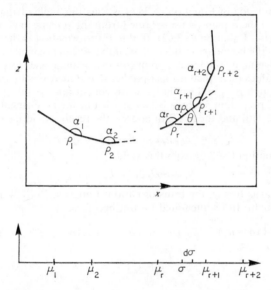

Figure N38.1 The conformal transformation of a polygon in the $\rho$ plane into the real axis of the $\sigma$ half plane.

the rest will generally be complex, as indeed may be $C_1$. A complex factor may, however, be written in the form, using Equation (14.3),

$$(\sigma - \mu)^{(\alpha/\pi - 1)} = (-1)^{(\alpha/\pi - 1)}(\mu - \sigma)^{(\alpha/\pi - 1)}$$

$$= (e^{i\pi})^{(\alpha/\pi - 1)}(\mu - \sigma)^{(\alpha/\pi - 1)}$$

or

$$(\sigma - \mu)^{(\alpha/\pi - 1)} = e^{i(\alpha - \pi)}(\mu - \sigma)^{(\alpha/\pi - 1)} \qquad (N38.3)$$

The second factor on the right-hand side of Equation (N38.3) is simply a real positive quantity. Also $C_1$ may be written $C_1' e^{i\chi}$ where $C_1'$ is real. Equation (N38.2) may therefore be written in the form,

$$d\rho = B \, d\sigma \, e^{i(\chi + \alpha_{r+1} - \pi + \alpha_{r+2} - \pi + \, \cdots \, + \alpha_n - \pi)} \qquad (N38.4)$$

where the factor $B$ is the product of all those factors of Equation (N38.2) which are simply real positive numbers.

An inspection of Figure (N38.1) shows that $(\alpha_{r+1} - \pi)$ is the angle, say $-\beta_{r+1}$, through which the side of the polygon between $\rho_{r+1}$ and $\rho_{r+2}$ must be turned to bring it into line with the side between $\rho_r$ and $\rho_{r+1}$. Thus, from Equation (N38.4),

$$d\rho = B \ d\sigma \ e^{i(\chi - \beta_{r+1} - \beta_{r+2} - \cdots - \beta_n)} \tag{N38.5}$$

A comparison between Equations (N38.1) and (N38.5) shows that the slope $\theta$ is given by

$$\theta_{r+1} = \chi - \beta_{r+1} - \beta_{r+2} - \cdots - \beta_n \tag{N38.6}$$

The right-hand side of Equation (N38.6) is thus the angle of the slope of the element $d\rho$ when the $\sigma$ plane is transformed from the $\rho$ plane in accordance with the transformation Equation (N38.2). If this transformation is the one which is required to turn the boundary in the $\rho$ plane into coincidence with the consecutive sides of the polygon, as the tracing point in the $\sigma$ plane passes through the consecutive apex values of $\mu$, then it must predict that $\theta$ increases by the angle $\beta_{r+1}$ as $\sigma$ passes through $\mu_{r+1}$ and similarly at subsequent apex points.

Now had the preceding analysis been carried out for an element in the stage between the $(r+1)$th and the $(r+2)$th apexes, the result would have been

$$\theta_{r+2} = \chi - \beta_{r+2} - \beta_{r+3} - \cdots - \beta_n \tag{N38.7}$$

Subtracting Equation (N38.6) from this yields,

$$\theta_{r+2} - \theta_{r+1} = \beta_{r+1}$$

as required. That is to say, the transformation, Equation (N38.2), is indeed the one which is sought. In its integrated form it becomes,

$$\rho = C_1 \int (\sigma - \mu_1)^{(\alpha_1/\pi - 1)}(\sigma - \mu_2)^{(\alpha_2/\pi - 1)} \ldots (\sigma - \mu_n)^{(\alpha_n/\pi - 1)} d\sigma +$$

$$+ C_2 \qquad\qquad \left\{ \begin{matrix} \text{(N38.8)} \\ \text{(16.4)} \end{matrix} \right.$$

### Note 39. The hodograph of the water table

Consider a point O on the water table as shown in Figure (N39.1). The slope

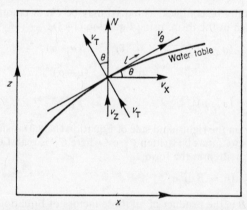

Figure N39.1 The components of the velocity of flow at the water table, just inside and just outside the groundwater.

at this point is $\theta$ and the horizontal and vertical components of the resultant flow velocity of the groundwater are respectively $v_x$ and $v_z$. The rate of flow, $v_\theta$, in a direction tangential to the water table, is thus given by

$$v_\theta = v_x \cos \theta + v_z \sin \theta \qquad \text{(N39.1)}$$

Also, from Darcy's law,

$$v_\theta = -K \, d\phi/dl$$

where $l$ is distance measured in the direction of the tangent. Since the pressure does not change with distance in this direction, being zero at all points on the water table, it follows that

$$v_\theta = -K \, dZ/dl = -K \sin \theta \qquad \text{(N39.2)}$$

where $Z$ is the value of $z$ at the water table. Hence, from Equations (N39.1) and (N39.2),

$$v_x \cos \theta + v_z \sin \theta = -K \sin \theta \qquad \text{(N39.3)}$$

Similarly the flow velocity, $v_T$, normal to the water table, is

$$v_T = -v_x \sin \theta + v_z \cos \theta \qquad \text{(N39.4)}$$

If the resultant flow in the region above the water table is at a velocity $N$ in the vertical direction, then, since $v_T$ must be continuous across the water table,

$$v_T = N \cos \theta \qquad \text{(N39.5)}$$

From Equation (N39.3) it follows that

$$\tan \theta = -v_x/(v_z + K) \qquad \text{(N39.6)}$$

while from Equations (N39.4) and (N39.5),

$$\tan \theta = (v_z - N)/v_x \qquad \text{(N39.7)}$$

when, by elimination of $\tan \theta$,

$$v_x^2 + v_z^2 + v^2(K - N) - NK = 0$$

This may be expressed in the form,

$$v_x^2 + [v_z + (K-N)/2]^2 = [(K+N)/2]^2 \qquad \left\{ \begin{array}{l} \text{(N39.8)} \\ \text{(16.5)} \end{array} \right.$$

This is the equation of a circle with its centre at $[0, -(K-N)/2]$ and of radius $(K+N)/2$. As drawn in Figure (16.3) it is seen to cut the axis of $v_z$ at $N$ and $-K$. A line drawn from the latter point to a point on the circle corresponding to velocity components $v_x$ and $v_z$ makes an angle $\alpha$ with the $v_z$ axis such that

$$\tan \alpha = -v_x/(v_z + K) \qquad \text{(N39.9)}$$

The negative sign arises from the fact that with respect to the axis of $v_z$, $\alpha$ is the result of a clockwise rotation and is a negative angle. A comparison of Equation (N39.9) with Equation (N39.6) shows that $\alpha$ and $\theta$ are identical. Thus as a point moves along the water table in Figure (N39.1), passing through points of increasing slope $\theta$, the corresponding point in the hodograph of Figure (16.3) moves round the circle in such a way that the chord drawn therein takes successive positions making this same angle $\theta$ with the $v_z$ axis.

Had the boundary shown in Figure (N39.1) been not the water table, but the upper boundary of the capillary fringe where the suction is constant and approximately equal to the air entry value, the analysis would have taken precisely the same course since at the only stage at which this boundary condition enters, namely the establishment of Equation (N39.2), one would have had

$$\phi = z_f + H_f$$

where $H_f$ is the constant negative pressure at the fringe boundary, so that

$$d\phi/dl = dz_f/dl = \sin\theta$$

and as before, in Equation (N39.2),

$$v_\theta = -K\sin\theta$$

Hence all the subsequent equations still follow.

A surface of seepage is a surface of constant zero pressure, like a water table, but nothing can usefully be stated about the rate of flow on the emergent side of the surface. Equations (N39.1) and (N39.2) apply here as well as to the water table, so that the consequent Equation (N39.3) is valid. However, in this case the angle $\theta$ is specified by the slope of the surface, and if this is constant, it follows from Equation (N39.3) that

$$v_z = -v_x\cot\theta - K$$

or

$$v_z = v_x\tan(\theta + \pi/2) - K$$

This is the equation of a straight line of slope $\theta + \pi/2$, that is to say, it is perpendicular to the surface in the $\rho$ plane which slopes at the angle $\theta$. Its intercept on the axis of $v_z$ is at $-K$, the same point at which the hodograph of the water table intercepts this axis.

### Note 40. Details of van Deemter's analysis
*(a) The transformation from the $\zeta$ to the $\sigma$ plane*

The Schwarz-Christoffel transformation, Equation (16.4), with the allotted correspondences preceding Equation (16.22), takes the form,

$$\zeta = C_1\int\sigma^{-3/2}(\sigma+1)d\sigma + C_2$$

$$= 2C_1(\sigma-1)/\sigma^{\frac{1}{2}} + C_2 \tag{N40.1}$$

It is readily verified by substitution that this equation gives the correct values of $\zeta$ at $P$ and $Q$ whatever may be the values allotted to $C_1$ and $C_2$. These constants may be evaluated by appeal to the correspondence at $P'$ where $\sigma$ is $-1$ and $\zeta$, from Equation (16.16), is $-1+i\cot\theta'$. Thus substitution of these values in Equation (N40.1) gives

$$-1+i\cot\theta' = 4iC_1 + C_2$$

Equating the real and imaginary parts separately, one arrives at

$$C_1 = \frac{\cot\theta'}{4}$$

$$C_2 = -1$$

Thus the final form of Equation (N40.1) is

$$\zeta = \frac{(\sigma - 1)\cot\theta'}{2\sigma^{\frac{1}{2}}} - 1 \qquad \left\{ \begin{array}{l} \text{(N40.2)} \\ \text{(16.22)} \end{array} \right.$$

At the point $D$ the value of $\zeta$ is zero, so that from Equation (N40.2) the value of $\sigma$ is $\lambda^2$ where

$$0 = \frac{\lambda^2 - 1}{2\lambda}\cot\theta' - 1$$

or

$$\lambda - 1/\lambda = 2\tan\theta' \qquad \left\{ \begin{array}{l} \text{(N40.3)} \\ \text{(16.23)} \end{array} \right.$$

The value of $\sigma$ is written $\lambda^2$ rather than $\lambda$ because at a later stage the variable is to be transformed to $\sigma^{\frac{1}{2}}$. Equation (N40.2) may be treated as a quadratic equation with the solution,

$$\lambda = \tan\theta' + (1 + \tan^2\theta')^{\frac{1}{2}} \qquad \left\{ \begin{array}{l} \text{(N40.4)} \\ \text{(16.23a)} \end{array} \right.$$

Similarly, at the point $R$, the value of $\sigma$ is indicated by $\mu^2$ while $\zeta$ has the value $\gamma$ as defined in Equation (16.21). Substitution of these values in Equation (N40.2) now leads to

$$\mu - 1/\mu = 2(1 + \gamma)\tan\theta' \qquad \left\{ \begin{array}{l} \text{(N40.5)} \\ \text{(16.24)} \end{array} \right.$$

This has the solution,

$$\mu = (1 + \gamma)\tan\theta' + [1 + (1 + \gamma)^2 \tan^2\theta']^{\frac{1}{2}} \qquad \left\{ \begin{array}{l} \text{(N40.6)} \\ \text{(16.24a)} \end{array} \right.$$

Lastly, at 0 the values of $\sigma$ and $\zeta$ are respectively $\nu^2$ and $-(K+N)/N$, and substitution in Equation (N40.2) this time leads to

$$\nu - 1/\nu = -2(K/N)\tan\theta' \qquad \left\{ \begin{array}{l} \text{(N40.7)} \\ \text{(16.25)} \end{array} \right.$$

with the solution

$$\nu = -(K/N)\tan\theta' + [1 + (K/N)^2 \tan^2\theta']^{\frac{1}{2}} \qquad \left\{ \begin{array}{l} \text{(N40.8)} \\ \text{(16.25a)} \end{array} \right.$$

*(b) The transformation from the $\Omega$ to the $\eta$ plane*
The correspondences at $D$ and $R$ respectively are

$$\eta = \pm\infty; \ \alpha_1 = 0; \ \Omega = -\infty$$
$$\eta = 0; \ \alpha_2 = 0; \ \Omega = \infty$$

In this case the Schwarz-Christoffel transformation, Equation (16.4), with the factor due to the apex $D$ equated to unity, because this point is at infinity in the $\eta$ plane, is

$$\Omega = C_1 \int \eta^{-1}\, d\eta + C_2$$

$$= C_1 \ln\eta + C_2$$

This may be rewritten in the form,

$$\eta = e^{(\Omega - C_2)/C_1} \qquad \text{(N40.9)}$$

At $D_2$ the values of $\Omega$ and $\eta$ are respectively $-\infty$ and $+\infty$. Substitution in Equation (N40.9) yields

$$\infty = e^{-\infty/C_1}$$

from which it is evident that $C_1$ must be negative, say $-1/C$. Thus Equation (N40.9) takes the form,

$$\eta = e^{-C(\Omega - C_2)} \qquad \text{(N40.10)}$$

where $C$ is positive.

At $D_1$ the value of $\eta$ becomes $-\infty$ while $\Omega$ takes the value $-\infty + i(M-N)L/K$. Thus at this point, Equation (N40.10) becomes

$$-\infty = e^{\infty} e^{-iC(M-N)L/K}$$

$$= e^{\infty}\left(\cos\frac{C(M-N)L}{K} - i\sin\frac{C(M-N)L}{K}\right)$$

This equation may be satisfied with

$$\frac{C(M-N)L}{K} = \pi \qquad \text{(N40.11)}$$

With this value of $C$ and with $C_2$ given the value zero, since its specification is arbitrary, Equation (N40.10) becomes

$$\eta = e^{-\frac{\pi K \Omega}{(M-N)L}} \qquad \begin{cases} \text{(N40.12)} \\ \text{(16.28)} \end{cases}$$

At $Q$ the value of $\Omega$ from Equation (16.27) is, as marked in Figure (16.10),

$$\Omega_Q = c_f(1 + N/K) + H_f + iL(M-N)/K$$

Direct substitution of this value of $\Omega$ in Equation (N40.12) gives

$$\eta_Q = e^{-i\pi} e^{-\frac{\pi KH}{L(M-N)} - \frac{\pi c_f(K+N)}{L(M-N)}}$$

$$= -e^{-\frac{\pi KH_f}{L(M-N)} - \frac{\pi c_f \gamma}{L}} \qquad \begin{cases} \text{(N40.13)} \\ \text{(16.29)} \end{cases}$$

Similarly the value of $\Omega$ at $P$, when substituted in Equation (N40.12), leads to

$$\eta_P = -e^{-\frac{\pi KH_f}{L(M-N)} - \frac{\pi b_f \gamma}{L}} \qquad \begin{cases} \text{(N40.14)} \\ \text{(16.30)} \end{cases}$$

(c) *The transformation from the $\sigma$ to the $\eta$ plane*

The point $R$ in the $\sigma$ plane is at $\mu^2$ on the real axis, but is at the origin in the $\eta$ plane. Thus subtracting $\mu^2$ from $\sigma$ brings the two planes into coincidence with respect to $R$. The point $D$ is at $\lambda^2$ in the $\sigma$ plane and at infinity in the $\eta$ plane. This correspondence may be effected by subtracting $\lambda^2$ from $\sigma$ to bring it to the origin, and then inverting the result. Thus the transformation,

$$\chi = \frac{\sigma - \mu^2}{\sigma - \lambda^2} \qquad \text{(N40.15)}$$

produces a $\chi$ half plane with $R$ and $D$ at the same points on the real axis as in the $\eta$ plane. However, at $Q$, where $\sigma$ is infinitely great, $\chi$ has the value unity whereas $\eta$ has the value given in Equation (N40.13). Hence the further transformation,

$$\eta = -\chi \, e^{-\frac{\pi K H_f}{L(M-N)} - \frac{\pi c_f \gamma}{L}} \qquad \text{(N40.16)}$$

is required. The net result of Equations (N40.15) and (N40.16) is

$$\eta = -\frac{(\sigma - \mu^2)}{(\sigma - \lambda^2)} e^{-\frac{\pi K H_f}{L(M-N)} - \frac{\pi c_f \gamma}{L}} \qquad \begin{cases} \text{(N40.17)} \\ \text{(16.31)} \end{cases}$$

### (d) The integration of the differential equation

The equation to be integrated, Equation (16.37), is

$$i(1 + N/K)d\rho = -d\Omega + \left(\frac{d\Omega}{dt}\right)\left(\frac{(\lambda^2 t^2 - 1)\cot \theta'}{2\lambda t}\right)dt \qquad \begin{cases} \text{(N40.18)} \\ \text{(16.37)} \end{cases}$$

$d\Omega/dt$ may be determined by differentiating Equation (16.35) with respect to $t$. The result is

$$\frac{d\Omega}{dt} = \left(\frac{L(M-N)}{\pi K}\right)\left(\frac{2t}{\mu^2/\lambda^2 - t^2}\right)\left(\frac{\mu^2/\lambda^2 - 1}{t^2 - 1}\right) \qquad \text{(N40.19)}$$

Also, from Equation (16.23),

$$\cot \theta' = 2\lambda/(\lambda^2 - 1) \qquad \text{(N40.20)}$$

Substituting Equations (N40.20) and (N40.19) in Equation (N40.18), one arrives at the equation,

$$i(1 + N/K)d\rho = -\frac{2L(M-N)(\mu^2/\lambda^2 - 1)}{\pi K(\lambda^2 - 1)}F(t)dt - d\Omega \qquad \text{(N40.21)}$$

where

$$F(t) = \frac{\lambda^2 t^2 - 1}{(t^2 - \mu^2/\lambda^2)(t^2 - 1)}$$

The function $F(t)$ may be factorized, and then takes the form,

$$F(t) = \left(\frac{1}{2(1 - \mu^2/\lambda^2)}\right)\left\{\left(\frac{1 - \mu^2}{\mu/\lambda}\right)\left(\frac{1}{t - \mu/\lambda} - \frac{1}{t + \mu/\lambda}\right) + (\lambda^2 - 1)\left(\frac{1}{t - 1} - \frac{1}{t + 1}\right)\right\}$$

With this form of $F(t)$ transferred to Equation (N40.21), the equation becomes

$$i(1 + N/K)d\rho = -d\Omega + \left(\frac{L(M-N)}{\pi K}\right)\left(\frac{1}{t - 1} - \frac{1}{t + 1}\right)dt + \left(\frac{L(M-N)}{\pi K}\right) \times$$

$$\times\left(\frac{1/\mu-\mu}{\lambda-1/\lambda}\right)\left(\frac{1}{t-\mu/\lambda}-\frac{1}{t+\mu/\lambda}\right)dt \tag{N40.22}$$

From Equations (16.23) and (16.24), and further from Equation (16.21),

$$\frac{1/\mu-\mu}{\lambda-1/\lambda} = -(1+\gamma)$$

$$= -(K+M)/(M-N)$$

This substitution in Equation (N40.22) results in

$$i(1+N/K)d\rho = -d\Omega+\left(\frac{L(M-N)}{\pi K}\right)\left(\frac{1}{t-1}-\frac{1}{t+1}\right)dt +$$

$$-\left(\frac{L(M+K)}{\pi K}\right)\left(\frac{1}{t-\mu/\lambda}-\frac{1}{t+\mu/\lambda}\right)dt$$

This equation may now be integrated without further ado, with the result,

$$i(1+N/K)\rho = -\Omega+\left(\frac{L(M-N)}{\pi K}\right)\ln\left(\frac{t-1}{t+1}\right)-\left(\frac{L(M+K)}{\pi K}\right)\ln\left(\frac{t-\mu/\lambda}{t+\mu/\lambda}\right)+A$$

where $A$ is a constant of integration.

The expression for $\Omega$ itself may now be substituted from Equation (16.35), the final form of Equation (N40.22) becoming

$$i(1+N/K)\rho = -H_f-c_f(1+N/K)+\frac{L(M-N)}{\pi K}\left[\ln\frac{(\mu^2/\lambda^2-t^2)}{(t^2-1)}+\right.$$

$$\left.+\ln\left(\frac{t-1}{t+1}\right)\right]+\frac{L(M+K)}{\pi K}\ln\left(\frac{t+\mu/\lambda}{t-\mu/\lambda}\right)+A \tag{N40.23}$$

From the known condition that at the point $Q$ the value of $t$ is infinitely great as may be seen in Figure (16.12), while in the $\rho$ plane its position is at $L+ic_f$, it follows from the substitution of these values in Equation (N40.23) that

$$A = iL(1+N/K)+H_f-\frac{L(M-N)}{\pi K}\ln(-1)$$

With this substitution, and after dividing through by $i(1+N/K)$, Equation (N40.23) takes the form,

$$\rho = L+ic_f-\frac{iL(M-N)}{\pi(N+K)}\left[\ln\frac{(t^2-\mu^2/\lambda^2)}{(t^2-1)}+\ln\left(\frac{t-1}{t+1}\right)\right]+$$

$$-\frac{iL(M+K)}{\pi(N+K)}\ln\left(\frac{t+\mu/\lambda}{t-\mu/\lambda}\right) \tag{N40.24}$$

To recapitulate Equation (16.21),

$$(K+N)/(M-N) = \gamma$$

so that

$$(M+K)/(N+K) = 1+1/\gamma$$

Also $\beta$ is defined in Equation (16.26) as

$$\mu/\lambda = 1 + \beta$$

With these substitutions, Equation (N40.24) takes the form, after some algebraic manipulations,

$$\rho = x + iz$$

$$= L + ic_f + \frac{iL}{\pi}\left[\ln\left(\frac{t-1-\beta}{t+1+\beta}\right) + \left(\frac{2}{\gamma}\right)\ln\left(\frac{t+1}{t+1+\beta}\right)\right] \qquad \begin{cases}(N40.25)\\(16.38)\end{cases}$$

An expression for $W$ may be derived from the transformation Equation (16.27) together with Equation (16.35) for $\Omega$. Thus

$$W = \Omega + iN\rho/K$$

$$= H_f + c_f(1 + N/K) - \frac{L(M-N)}{\pi K}\ln - \frac{(t+1+\beta)(t-1-\beta)}{(t+1)(t-1)} +$$

$$+ \frac{iN}{K}\left[L + ic_f + \frac{iL}{\pi}\left(\ln\frac{t-1-\beta}{t+1+\beta} + \frac{2}{\gamma}\ln\frac{t+1}{t+1+\beta}\right)\right]$$

Algebraic rearrangement results in

$$W = \phi + i\psi$$

$$= H_f + c_f + \frac{iML}{K} + \frac{L}{\pi K}\Bigg[(M-N)\ln(t-1) - M\ln(t-1-\beta) +$$

$$+ N\ln(t+1+\beta) + \frac{(M-N)(K-N)}{(K+N)}\ln\left(\frac{t+1}{t+1+\beta}\right)\Bigg] \qquad \begin{cases}(N40.26)\\(16.39)\end{cases}$$

*(e) Conditions on the boundaries in the $\rho$ plane*

Reference to the $t$ plane of Figure (16.12) shows that along the boundary $QR$ the value of $t$ is real, positive, and greater than $1 + \beta$. Hence all the logarithms in Equations (N40.25) and (N40.26) are logarithms of real positive numbers and may be found from tables. Thus, equating the imaginary parts gives

$$\pi z/L = \pi c_f/L + \ln\left(\frac{t-1-\beta}{t+1+\beta}\right) + \left(\frac{2}{\gamma}\right)\ln\left(\frac{t+1}{t+1+\beta}\right) \qquad \begin{cases}(N40.27)\\(16.45)\end{cases}$$

Equating the real parts, one has simply

$$x = L$$

in confirmation of what is known from the boundaries in the $\rho$ plane. Similarly, Equation (N40.26) yields, from the real parts,

$$\pi\phi/L = \frac{\pi}{L}(H_f + c_f) + \frac{1}{K}\Bigg[(M-N)\ln(t-1) - M\ln(t-1-\beta) +$$

$$+ N\ln(t+1+\beta) + \frac{(M-N)(K-N)}{K+N}\ln\left(\frac{t+1}{t+1+\beta}\right)\Bigg] \qquad \begin{cases}(N40.28)\\(16.46)\end{cases}$$

o

Since $\phi$ is the sum of $H$ and $z$, $H$ may readily be derived by subtracting $z$ from $\phi$ using Equations (N40.28) and (N40.27). The result is

$$\pi H/L = \pi H_f/L + \frac{M-N}{K}\ln\left(\frac{t-1}{t+1}\right) + \frac{M+K}{K}\ln\left(\frac{t+1+\beta}{t-1-\beta}\right) \qquad \begin{cases}(\text{N40.29})\\(16.47)\end{cases}$$

Along the boundary $DP$ the value of $t$ is real and positive, but lies between zero and unity. Hence $t-1$ and $t-1-\beta$ are both negative numbers, and one must write,

$$\ln\left(\frac{t-1-\beta}{t+1+\beta}\right) = i\pi + \ln\left(\frac{1+\beta-t}{1+\beta+t}\right)$$

$$\ln(t-1) = i\pi + \ln(1-t)$$

and

$$\ln(t-1-\beta) = i\pi + \ln(1+\beta-t)$$

in order that all the logarithms in Equations (N40.25) and (N40.26) may be expressed in real terms. Then the following results accrue in a manner similar to the preceding.

$$\pi z/L = \pi c_f/L + \ln\left(\frac{\beta+1-t}{\beta+1+t}\right) + \left(\frac{2}{\gamma}\right)\ln\left(\frac{1+t}{\beta+1+t}\right) \qquad \begin{cases}(\text{N40.30})\\(16.42)\end{cases}$$

$$\pi\phi/L = \frac{\pi}{L}(H_f+c_f) + \frac{1}{K}\bigg[(M-N)\ln(1-t) - M\ln(\beta+1-t) +$$

$$+ N\ln(\beta+1+t) + \frac{(M-N)(K-N)}{K+N}\ln\left(\frac{1+t}{\beta+1+t}\right) \qquad \begin{cases}(\text{N40.31})\\(16.43)\end{cases}$$

$$\pi H/L = \pi H_f/L + \frac{M-N}{K}\ln\left(\frac{1-t}{1+t}\right) + \frac{M+K}{K}\ln\left(\frac{\beta+1+t}{\beta+1-t}\right) \qquad \begin{cases}(\text{N40.32})\\(16.44)\end{cases}$$

It is known that at $D$ on this boundary, namely the drain axis, $z$ vanishes while $t$ assumes the value unity. Substitution of these values in Equation (N40.30) leads to the important result

$$\frac{\pi c_f}{L} + \ln\frac{\beta}{\beta+2} + \left(\frac{2}{\gamma}\right)\ln\frac{2}{\beta+2} = 0$$

or

$$\pi c_f/L = \ln(1+2/\beta) + (2/\gamma)\ln(1+\beta/2) \qquad \begin{cases}(\text{N40.33})\\(16.40)\end{cases}$$

Combining this result with Equation (16.23), one arrives at the complementary expression for $b_f$, namely

$$\pi b_f/L = \ln(1+2/\beta) + (2/\gamma)\ln\left(\frac{1+\beta/2}{1+\beta}\right) \qquad \begin{cases}(\text{N40.34})\\(16.41)\end{cases}$$

Finally along the boundary $DS$ the value of $t$ is real and positive, and lies between 1 and $1+\beta$. Thus only the logarithm of $t-1-\beta$ needs to be rationalised to the form $i\pi + \ln(1+\beta-t)$. The result this time leads to Equations (16.48) to (16.50) of the text.

# CHAPTER 17

# Some three-dimensional drainage flow-nets

## 17.1 The Dupuit-Forchheimer approximation in three dimensions

WHEN the drainage system is any other than a set of parallel equidistant drainlines of uniform depth and separation, with the separation small in comparison with the length of individual lines, it is no longer true that the streamlines are confined to planes perpendicular to the direction of the drains. The problem is in general one of the variation of the potential and stream functions in all three Cartesian dimensions. No exact analysis, comparable to the hodograph analysis in two dimensions, is known, but progress may be made by pursuing the Dupuit-Forchheimer approximation and extending the results to cases of substantial bed depths, for which this approximation is not appropriate, by appeal to relationships which have been discovered in the case of two-dimensional problems.

As in the case of two dimensions the Dupuit-Forchheimer assumption is that flow is limited to the horizontal direction, and it follows that there must be an impermeable bed to which the drains penetrate, and that the equipotential surfaces are vertical and are labelled with the height at which they intersect the water table. The horizontal direction of flow will differ at different points and will have components in the directions of the two horizontal Cartesian axes, $x$ and $y$. In accordance with Darcy's law, these component velocities $v_x$ and $v_y$ will be proportional to the components of the slope of the water table, $\partial Z/\partial x$ and $\partial Z/\partial y$, since on the Dupuit-Forchheimer assumption these are the components of the potential gradient.

If one considers a vertical prism of soil of height $Z$ between the impermeable bed and the water table, with rectangular section corresponding to sides $\delta x$ and $\delta y$, then, just as in the two-dimensional case, the rate of flow of liquid in the $x$ direction across the vertical face of sectional area $Z\,\delta y$ at $x$ is

$$Q_x = -KZ\,\delta y(\partial Z/\partial x)$$

while at the parallel section at $x + \delta x$, the rate of flow is

$$Q_{x+\delta x} = -K\,\delta y\{Z(\partial Z/\partial x) + \{\partial[Z(\partial Z/\partial x)]/\partial x\}\delta x\}$$

The first of these two equations represents the rate of flow into the prism, and the second represents the rate of flow out, so that the rate of water storage, $(dS/dt)_x$, on account of the varying rate of flow in the $x$ direction is, by difference,

$$(dS/dt)_x = K\,\delta y\{\partial[Z(\partial Z/\partial x)]/\partial x\}\delta x \qquad (17.1a)$$

and similarly, for variation of flow in the $y$ direction,

$$(dS/dt)_y = K\,\delta x\{\partial[Z(\partial Z/\partial y)]/\partial y\}\delta y \qquad (17.1b)$$

Further the storage in the profile, $(dS/dt)_q$, due to rainfall on the upper surface of the prism at the rate $q$ is

$$(dS/dt)_q = q\,\delta x\,\partial y \qquad (17.1c)$$

Hence the net storage rate from all causes is the sum of these three contributions, namely

$$dS/dt = (\delta x \delta y)[q + K\{\partial[Z(\partial Z/\partial x)]/\partial x\} + \\ + K\{\partial[Z(\partial Z/\partial y)]/\partial y\}] \qquad (17.2)$$

In the particular case where a steady state has been attained, the rate of storage is zero, so that Equation (17.2) takes the form, after substitution of $\frac{1}{2}\partial Z^2/\partial x$ for $Z(\partial Z/\partial x)$ and similarly for the $y$ direction,

$$\partial^2 Z^2/\partial x^2 + \partial^2 Z^2/\partial y^2 = -2q/K \qquad (17.3)$$

This is the basic equation which must be solved in order that the water table height $Z$ may be known at any specified point $(x,y)$.

## 17.2 The circular peripheral drain

Where a body of drained soil is contained within a circular drain which penetrates to the impermeable bed, as shown in Figure (17.1), the symmetry of the situation permits a very simple solution. It is evident that the flow is at all points radial, directed from the axis of symmetry at the centre of the circular periphery. Hence the equipotential surfaces are coaxial cylinders and the flow of liquid across them is uniformly distributed. The total rate of flow across such an equipotential surface outward towards the peripheral drain is equal to the rate of precipitation on the catchment area of the upper surface within this cylindrical boundary. Hence an application of Darcy's law to the rate of flow across a cylinder of radius $r$ yields the equation,

$$\pi r^2 q = 2\pi r Z K(dZ/dr) \qquad (17.4)$$

where the rainfall rate is $q$ and the height of the water table at $r$ is $Z$. If one takes the case of a peripheral ditch distant $R$ from the centre with water standing at the height $Z_0$, which is also the height of the water table at this boundary, the equation is readily integrated in the form,

$$(q/K)\int_r^R r\,\mathrm{d}r = -2\int_Z^{Z_0} Z\,\mathrm{d}Z$$

or

$$2(Z^2 - Z_0^2)/(R^2 - r^2) = q/K \qquad (17.5)$$

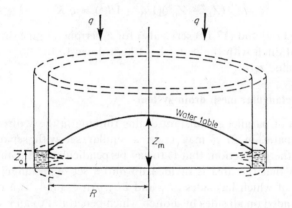

Figure 17.1 Three-dimensional Dupuit-Forchheimer diagram. The cross-section of the groundwater draining to a circular peripheral ditch.

At the centre of the drained circular plot where $r$ vanishes, the water table achieves its greatest height $Z_m$ given by

$$2(Z_m^2 - Z_0^2)/R^2 = q/K \qquad (17.6)$$

When the water level in the ditch is reduced to the level of the impermeable bed, the solution is equally applicable to the situation of drainage by a peripheral pipe drain laid on the bed. With vanishing $Z_0$ the solution is

$$2Z^2/(R^2 - r^2) = q/K \qquad (17.7)$$

$$2Z_m^2/R^2 = q/K \qquad (17.8)$$

## 17.3 The elliptical drain

When the peripheral drain, seen in plan view, follows an elliptical path, the solution may be obtained by extending the preceding result for a circular path intuitively (Childs and Youngs, 1961). If the ellipse has major and minor axes, respectively $2a$ and $2b$, then, as shown in Note 41, the solution is

$$(Z^2 - Z_0^2)(1/a^2 + 1/b^2)/(1 - x^2/a^2 - y^2/b^2) = q/K \qquad (17.9)$$

The origin of coordinates is the centre of the elliptical drain path, $x$ is measured in the direction of the major axis and $y$ is in the direction of the minor axis.

Again the maximum height of the water table, $Z_m$, is given by allowing both $x$ and $y$ to vanish, that is to say at the centre of the elliptical plot. The result is

$$(Z_m^2 - Z_0^2)(1/a^2 + 1/b^2) = q/K \qquad (17.10)$$

Equations (17.9) and (17.10) serve also for a peripheral pipe drain and for a peripheral ditch with the water level kept down to the floor when $Z_0$ is given the value zero.

## 17.4 The rectangular mesh drain system

A system of parallel equidistant drains running in the $x$ direction, with uniform separation $2L_2$, may cross a similar set with separation $2L_1$ running in the $y$ direction, that is to say perpendicular to the first set. The net effect is that the area is divided up into a set of rectangular drained plots, each of which has sides $2L_1$ and $2L_2$. An example is a rectangular field surrounded on all sides by a ditch which penetrates to an impermeable bed, in which ditch the water stands to a depth $Z_0$. A plan view of such a drainage system is shown in Figure (17.2).

A solution of Equation (17.3), subject to such boundary conditions, has been presented by Carslaw and Jaeger (1959). It is

$$Z^2 - Z_0^2 = (q/K)(L_1^2 - x^2) - 32(q/K)(L_1^2/\pi^3) \sum_{n=0}^{n=\infty} \left[ (-1)^n \cosh\{(2n+1)\pi x/2L_1\} \right.$$

$$\left. \cosh\{(2n+1)\pi y/2L_2\}/[(2n+1)^3 \cosh\{(2n+1)\pi L_2/2L_1\}] \right] \qquad (17.11)$$

The origin of coordinates is taken to be at the centre of the rectangle. At this centre where $x$ and $y$ vanish the water table has its maximum height, $Z_m$, given by

$$(Z_m^2 - Z_0^2)/L_1^2 = (q/K)\left[ 1 - (32/\pi^3)\sum_0^\infty (-1)^n/\{(2n+1)^3 \cosh(2n+1)\pi L_2/2L_1\} \right]$$

The series converges rapidly, and if $L_1$ is always taken to be the lesser of the two separations it will be found that no more than the first term is required. The solution then takes the form

$$(Z_m^2 = Z_0^2)/L_1^2 = (q/K)[1 = 32/\{\pi^3 \cosh(\pi L_2/2L_1)\}] \qquad 17.12$$

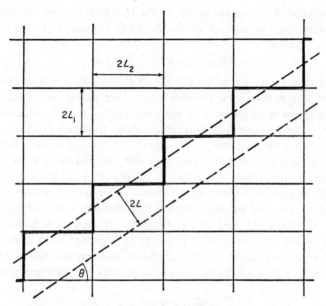

Figure 17.2 A drainage system of two sets of parallel
drains of different separations forming a rectangular
mesh system. The system may also be regarded as a set
of parallel step-shaped drainlines, one of which is shown
heavily lined.

As an example, when the mesh is square with sides of length $2L$ the solution
gives the numerical value

$$(Z_m{}^2 - Z_0{}^2)/L^2 = 0 \cdot 59 \, q/K \tag{17.13}$$

As the ratio $L_2/L_1$ increases to large values, the series tends to zero, with
the result, as expected, that the formula reduces to that which is appropriate to a single set of parallel drains of separation $L_1$, namely Equation
(15.15).

A very simple expression results from approximating the rectangular
mesh to an elliptical peripheral drain. An ellipse inscribed in the rectangle
follows the rectangular path fairly closely over much of its length. Such
an ellipse has major and minor axes of $2L_2$ and $2L_1$ respectively. Substitution of these values in Equations (17.9) and (17.10) respectively gives

$$(Z^2 - Z_0{}^2)(1/L_1{}^2 + 1/L_2{}^2)/(1 - x^2/L_1{}^2 - y^2/L_2{}^2) = q/K \tag{17.14}$$

and

$$(Z_m{}^2 - Z_0{}^2)(1/L_1{}^2 + 1/L_2{}^2) = q/K \tag{17.15}$$

This may be compared with the exact expression by considering the particular case of the square mesh of side $2L$ which may be compared with the circular drained plot of radius $L$. Thus Equation (17.15) gives the approximate value,

$$(Z_m^2 - Z_0^2)/L^2 = 0 \cdot 5 \, q/K$$

for comparison with Equation (17.13). The error is less than twenty per cent, and this is the worst possible case, since as the eccentricity of the ellipse increases with increase of the ratio $L_2/L_1$, both Equations (17.15) and (17.12) tend to the same formula, namely Equation (15.15).

Yet another approximation is useful, although naïve, because it provides a basis for extending the three-dimensional analysis to take into account an impermeable bed which is deeper than the floor of the drain ditch. The system of crossing drains shown in Figure (17.2) may equally be regarded as a set of parallel step-shaped drainlines, with "treads" and "risers" of length $2L_1$ and $2L_2$ respectively. If each step-shaped drain is replaced by the straight drain which may be regarded as its average path, then the result is a set of parallel drains of uniform separation $2L$. If the mean "angle of rise" of the "staircase" is $\theta$, then from the geometry of the figure,

$$L/L_2 = \cos \theta$$

$$L/L_1 = \sin \theta$$

whence

$$1/L^2 = 1/L_1^2 + 1/L_2^2 \tag{17.16}$$

Substitution of this separation in the formula, Equation (15.15), for parallel drains, gives the result,

$$(Z_m^2 - Z_0^2)(1/L_1^2 + 1/L_2^2) = q/K$$

and this repeats the formula, Equation (17.15), obtained from the elliptical approximation. It has already been shown that this approximation is not unacceptably poor.

The concept of substitution of straight drains following the average path of the step-shaped drains does not depend on the prevalence of Dupuit-Forchheimer conditions. It so happens that when such conditions prevail, it is possible to test the conclusions against those derived from an alternative method. Hence it seems likely that such a substitution may be equally useful when the impermeable bed is at great depth and the results of the hodograph analysis are applicable. Thus in the results of Section 16.3, if one substitutes for $c$ the value of $L$ resulting from Equation (17.16), one would have the corresponding result for the rectangular mesh drain. In particular, the specially comparable formula, Equation (16.51), for the drainage of locally incident rainfall without artesian complication becomes, with $\gamma$ equal to $K/q-1$,

$$\pi Z_m (1/L_1{}^2 + 1/L_2{}^2)^{\frac{1}{2}} = \ln\left(\frac{K/q+1}{K/q-1}\right) + \frac{2}{(K/q-1)} \ln \frac{(K/q+1)}{2} \qquad (17.17)$$

In neither of these applications does the depth of the impermeable bed enter into the analysis, since in the one case it is infinitely large compared with the separation of the drains, and in the other it is vanishingly small. Hence a finite change of the drain separation between the limits $L_1$ and $L_2$ does not affect the ratio of bed depth to drain separation, which, as shown in Section 16.5, is the ratio in which this important parameter has to be expressed. A comparison by Childs and Youngs (1961) between observed and theoretically predicted water tables is presented in Figures (17.3) and (17.4). To a system of parallel drains of separation $L_1$ were added other perpendicular systems of varying separations $L_2$. Figure (17.3) shows the variation of the water table height with intensity of rainfall when

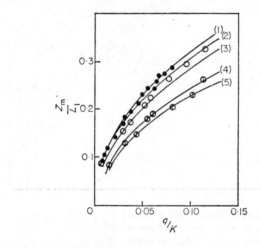

● observed points for $L_1/L_2 = 0$
○ observed points for $L_1/L_2 = 0.5$
⊕ observed points for $L_1/L_2 = 1.0$
curve (1), theoretical curve from Equation (17.15),
    $L_1/L_2 = 0$
curve (2), from Equation (17.12), $L_1/L_2 = 0.5$
curve (3), from Equation (17.15), $L_1/L_2 = 0.5$
curve (4), from Equation (17.12), $L_1/L_2 = 1.0$
curve (5), from Equation (17.15), $L_1/L_2 = 0.5$

Figure 17.3 The water table height over a rectangular mesh drainage system as a function of $q/K$. The water table height is expressed as a fraction of the constant separation, $L_1$, of the more closely spaced drains.

O*

the drains are laid on the impermeable bed for three ratios of $L_1/L_2$, together with the theoretical curves given by Equation (17.12) and the very approximate Equation (17.15). Figure (17.4) shows the ratio of the water table height expressed as a ratio of the height over the parallel system $L_1$ by itself, at various ratios of $L_1/L_2$. The calculated curves are for Equation (17.12) which is valid for the Dupuit-Forchheimer conditions alone, for Equation (17.15) which is valid also for very deep impermeable beds, and, with the help of Equation (16.56) together with Equation (17.16), for some intermediate bed depths.

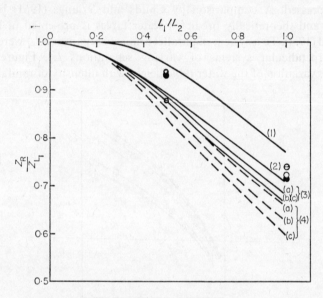

● average of experimental points, $p/L_1 = 0$
○ average of experimental points $p/L_1 = 0.3$
⊖ average of experimental points, $p/L_1 = 0.15$
curve (1), from Equation (17.12), $p/L_1 = 0$
curve (2), from Equation (17.15), $p/L_1 = 0$ or $p/L_1 > 0.3$
group (3), from Equations (17.16) and (16.56) with
   $q/K = 0.1$
group (4), from Equations (17.16) and (16.56) with
   $q/K = 0.01$
curve (a), $p/L_1 = 0.02$; curve (b), $p/L_1 = 0.05$; curve
   (c), $p/L_1 = 0.1$

Figure 17.4 The ratio $Z_R/Z_{L_1}$ as a function of the ratio $L_1/L_2$. $Z_{L_1}$ is the water table over the parallel drain system of separation $L_1$, and $Z_R$ is the water table when to this system is added the orthogonal system with separation $L_2$. $p$ is the depth of the impermeable bed.

## 17.5 Drainage by an array of pumped wells

It sometimes happens, as for example in Sind and Punjab, that a relatively permeable stratum is overlaid by a few feet of soil of rather low hydraulic conductivity. It may then be advantageous to drain the groundwater in the lower permeable stratum by pumping it from a number of widely spaced tube wells of considerable depth, instead of by intercepting it by surface drains after it has risen into the upper layer. If the tube wells are sited in a regular array, the system may be analysed quite simply on the basis of the Dupuit-Forchheimer approximation.

In Figure (17.5) are shown two different regular arrays of wells, one in

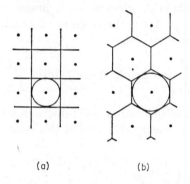

(a)          (b)

(a) square mesh system
(b) equilateral triangular mesh
Figure 17.5 Patterns of wells in regular array showing catchment areas.

which the sites lie at the intersections of lines which divide the area into a set of equilateral triangles; and one in which the wells are at the intersections of a square mesh grid. In the former case, the area is divided into a set of hexagonal catchments, each of which is drained by one of the wells, and in the latter the individual catchments are square.

The rainfall rate $q$, together with extraction from each well at the rate $Q$, must eventually set up a steady state provided that $Q$ equals the total rate of precipitation over a single catchment served by one well. When the steady state is achieved it is evident by symmetry that the boundaries between catchments are lines of zero flow and of level water table, and that the rate of flow across a vertical equipotential surface between the well and the boundary must just equal the rate of precipitation on the surface area between the boundary and the intersection of the equipotential with the surface.

As in Section 17.4, the boundary may be replaced without much error by the inscribed circle, and the equipotential surfaces are a set of vertical circular cylinders coaxial with the tube well. If the separation of neighbouring wells of the array is $2L$, then the radius of the circle inscribed in either the square or hexagonal element is of length $L$. Taking the well as the origin of coordinates, the direction of the flow is opposite to that of increasing $r$ and is therefore algebraically negative. The equation of flow across a cylindrical equipotential surface of radius $r$ is therefore, in accordance with Darcy's law,

$$\Delta Q = -q(A - \pi r^2) = -2\pi r Z K (dZ/dr)$$

where $A$ is the total area of the element of catchment, square or hexagonal as the case may be, and $\Delta Q$ is the fraction of $Q$ caught by the surface between $r$ and the boundary. Thus

$$(q/K) \int_{r_w}^{r} [(A/2\pi)(dr/r) - (r/2)dr] = \int_{Z_w}^{Z} Z \, dZ$$

where $r_w$ is the radius of the well in which the water level is $Z_w$, assumed to be identical with the water table level at this point. Integration results in the equation,

$$\frac{q}{K}\left(\frac{A}{\pi}\ln\frac{r}{r_w} - \frac{r^2 - r_w^2}{2}\right) = Z^2 - Z_w^2 \qquad (17.18)$$

The maximum water table height is at the boundary where, in the light of the approximation, one may equate $r$ to $L$. Thus,

$$\frac{q}{K}\left(\frac{A}{\pi}\ln\frac{L}{r_w} - \frac{L^2 - r_w^2}{2}\right) = Z_m^2 - Z_w^2 \qquad (17.19)$$

It is generally the case that the separation of the neighbouring wells is of the order of kilometres, while the well radius is of the order of a metre, so that in Equation (17.19) the term $r_w$ may be neglected in comparison with $L$, although not, of course, in the ratio $L/r_w$. The factor $A$ is $3 \cdot 5 L^2$ in the case of hexagonal catchments and $4L^2$ for square catchments, while the area of the inscribed circle to which the catchment may be approximated is $\pi L^2$. Hence for $A$, one may write $BL^2$ where $B$ is either $\pi$, $3 \cdot 5$, or $4$, according to the geometry of the well array or the degree of approximation. Then Equation (17.19) takes the form,

$$\frac{q}{K} = \frac{(Z_m^2 - Z_w^2)/L^2}{(B/\pi)\ln(L/r_w) - \frac{1}{2}} \qquad (17.20)$$

# NOTES

**Note 41. The solution of the Dupuit-Forchheimer problem for an elliptical boundary**

Let $X$ and $Y$ be the values of $x$ and $y$ at a point on the elliptical path of the drain, so that

$$X^2/a^2 + Y^2/b^2 = 1 \qquad \text{(N41.1)}$$

The solution must satisfy Equation (17.3) subject to the condition that the water table height $Z$ has the value $Z_0$ at the boundary given by Equation (N41.1).

Following the general treatment of Section 15.3 one considers an elliptical surface within and coaxial with the elliptical periphery of the drained plot and with its major and minor axes in the same directions as those of the periphery. If its equation is postulated to be

$$x^2/a^2 + y^2/b^2 = 1/n^2 \qquad \text{(N41.2)}$$

then a straight radial line drawn from the common centre must cut this ellipse at a point $(x,y)$ which satisfies both Equation (N41.2) and the equation to the straight line, namely

$$y = mx$$

where $m$ is a constant.

The solution of these simultaneous equations is

$$x^2(1/a^2 + m^2/b^2) = (y^2/m^2)(1/a^2 + m^2 b^2) = 1/n^2$$

Similarly the peripheral ellipse is cut at point $(X,Y)$ where

$$X^2(1/a^2 + m^2/b^2) = (Y^2/m^2)(1/a^2 + m^2/b^2) = 1$$

Thus, from these two sets of equations,

$$x^2/X^2 = y^2/Y^2 = 1/n^2$$

from which it follows that

$$(x^2 + y^2)/(X^2 + Y^2) = 1/n^2 \qquad \text{(N41.3)}$$

From a comparison of Equations (N41.2) and (N41.3), one may write

$$(x^2 + y^2)/(X^2 + Y^2) = x^2/a^2 + y^2/b^2 \qquad \text{(N41.4)}$$

Equation (14.5), the solution for the circular drained plot, may be written in the form,

$$[2(Z^2 - Z_0^2)/R^2]/(1 - r^2/R^2) = q/K \qquad \text{(N41.5)}$$

In the elliptical case, the ratio $(x^2 + y^2)/(X^2 + Y^2)$ plays the role of $r^2/R^2$, while some unspecified constant $A$ may be substituted for $R^2/2$. Making these substitutions in Equation (N41.5) and using Equation (N41.4), one may be guided to a tentative trial solution of the form,

$$[(Z^2 - Z_0^2)/A]/(1 - x^2/a^2 - y^2/b^2) = q/K \qquad \text{(N41.6)}$$

If this is indeed a solution, then $A$ may be found by differentiation and substitution in the basic differential equation of the problem, Equation (17.3).

Successive differentiation of Equation (N41.6) with respect to $x$ and $y$ separately yields

$$\partial^2 Z^2/\partial x^2 = -2(q/K)(A/a^2)$$

$$\partial^2 Z^2/\partial y^2 = -2(q/K)(A/b^2)$$

whence substitution in Equation (17.3) gives

$$A = 1/(1/a^2 + 1/b^2)$$

With this value of $A$ the trial solution, Equation (N41.6), becomes

$$(Z^2 - Z_0^2)(1/a^2 + 1/b^2)/(1 - x^2/a^2 - y^2/b^2) = q/K \qquad \left\{ \begin{matrix} \text{(N41.7)} \\ \text{(17.9)} \end{matrix} \right.$$

This solution satisfies the differential equation, and furthermore at the boundary where $x$ and $y$ take the values $X$ and $Y$ which satisfy Equation (N41.1), it correctly provides the boundary condition that $Z$ has the value $Z_0$. Hence Equation (N41.7) is indeed the solution of the problem.

# CHAPTER 18

# The non-steady flow of groundwater

## 18.1 A general statement of the problem

THE studies described in Chapters 14 to 17 were brought within the scope of available mathematical methods by the simplifying limitation to cases where the rate of flow into the system equals the rate of flow out, so that a stationary state of the flow-net prevails. Such stationary states must be rare in nature and fortuitous where they occur at all. The relationships deduced between the boundary conditions and the configuration of the flow-net, and in particular the height of the water table and of the capillary fringe, may usefully be interpreted as relationships between the average flow-net over a period of time and the average boundary conditions when the latter are known to be varying, as for example, when rainfall is intermittent. Such an interpretation is at this stage, however, intuitive and needs to be substantiated.

In this chapter it is proposed to elucidate, as far as is at present possible, the manner in which the height of the water table or of the capillary fringe responds to temporal changes of the rate of rainfall in cases of the drainage of locally precipitated groundwater. In the process, the solutions for steady states will be used insofar as they may be shown to apply to momentary stages of the varying flow-net.

It is evident that the rate of inflow into the flow-net must be known as a function of time before any progress can be attempted. At the surface momentarily occupied by the water table or capillary fringe boundary an additional flux contribution is introduced by reason of the fact that this boundary is moving and consequently there is release or storage of water in the zone above the groundwater. Thus an upward movement of the boundary introduces an upward flow out of the groundwater zone while a downward movement introduces a downward flux, additional to whatever may be contributed by rainfall.

Again, as the water table rises or falls there is a change of the potential distribution at this boundary and therefore a change of potential distribution throughout the flow-net. In general, this change demands a change of the distribution of the pressure component of the potential, since only at the water table itself and at the drain boundaries is the pressure controlled. The speed at which a pulse of pressure change is transmitted through the water is known to be very great, since the passage of sound is of just the nature of the passage of a pressure pulse. The speed of sound in water is known to be about 1500 metres per second, which is infinitely fast compared with any likely speed of movement of the water table, so that the pressure changes are able to occur at all points instantaneously as may be required to satisfy Laplace's equation in accordance with the instantaneously known boundary conditions. Hence each stage of the flow-net may be studied as a momentary state which satisfies Laplace's equation with known momentary boundary conditions of potential or of flux, provided that the flux may be satisfactorily inferred from the rate of movement of the water table.

As to this last point, it has been shown in Section 12.11 that the specific yield $Y$ is defined by the equation,

$$\Delta V = Y \Delta Z$$

where $\Delta V$ is the volume of water per unit area of the water table which is released when the water table falls by the distance $\Delta Z$. Conversely, $\Delta V$ is the volume which must move upward across the initial position of the water table when the latter rises by the distance $\Delta Z$. Thus, having regard to the sign convention that both the velocity and the water table height are positive when measured upward, one may write

$$v_z = \mathrm{d}V/\mathrm{d}t = Y\,\mathrm{d}Z/\mathrm{d}t \tag{18.1}$$

where $v_z$ is the contribution to the flux at the momentary water table position which is due to water table movement.

It was emphasized in Section 12.11 that while $Y$ might satisfactorily be regarded as a constant, characteristic of the soil, during more or less steady water table movements of long duration, it may vary between wide limits when water tables which are near the surface flucutate rapidly and irregularly. However, the magnitude of the errors that may be expected to ensue when the motion of the water table is analysed on an uncritical assumption of the constancy of $Y$ has not yet been carefully examined. In studies of the non-steady state it has been customary to proceed on the basis that $Y$ is constant, and this course will be followed here.

## 18.2 The non-steady water table for locally precipitated rain

Apart from the non-steady stages of the pumping of water from wells, which is a subject treated at length by Todd (1959), the most thoroughly studied situation in the field of the non-steady flow of groundwater has been the drainage of locally precipitated water by a system of parallel equidistant drains at uniform depth, such as featured in Chapters 15 and 16 and is illustrated in Figures (14.1) and (15.5). The analytical solutions of the steady state, whether by transformation of the hodograph or by the Dupuit-Forchheimer approximation, were obtainable only because the flux across the water table, which in those cases was due to rainfall alone, could reasonably be assumed to be uniformly distributed over that surface. Progress could have been made in the case of the Dupuit-Forchheimer method had the distribution been not uniform but a known analytical function of the distance from the drain. Had the distribution been expressible only numerically, as for example, by a graph of rainfall intensity versus distance, a solution could have been obtained only by numerical methods of integration or by analogue.

In the non-steady state, the water table movement is not known initially and it is the object of the exercise to determine this feature. Since this movement determines one of the components of the flux at the water table, this flux also is incapable of initial specification, and even when known is hardly likely to be expressible in terms of a simple algebraic formula. Hence it is characteristic of solutions in the non-steady state that they should proceed by numerical methods or by the use of analogues.

The boundary conditions must now include a specified initial water table stage. As an illustration of procedure, one may consider the solution by the method of electric analogues of the problem of the course of the fall of a water table from a steady state appropriate to a steady rainfall rate $q$, when this rainfall suddenly ceases (Childs, 1947). The initial and some subsequent stages are shown in Figure (18.1). At each stage the flux at the water table is due solely to the rate of fall of the water table since the rainfall rate is zero. If $\delta t$ is the time taken to fall from stage 1 to stage 2, then the mean rate of fall during the stage interval is $\delta Z/\delta t$, where $\delta Z$ is not necessarily the same at all distances from the drain. If the interval is not excessively long and the stage difference not too large, this mean rate of fall may be taken as the actual rate of fall when the water table is passing through the stage which is drawn midway between 1 and 2. Thus this mean stage is governed by the conditions that the hydrostatic pressure is zero and the flux is proportional to the difference between the heights of the water tables at stages 1 and 2, since the time interval is the same at all points. An analogue, as described in Section 15.2, may be constructed to scale, and the water table boundary adjusted by trial and error until these two conditions are fulfilled simultaneously, namely that the current fed in at the water table boundary is proportional to the distance by which the stage falls below the preceding stage, and the voltage at the water table boundary is eve proportional to the height above the drain electrode. When this stage is discovered, it then serves as the initial stage for the determination of the next stage in a like manner. Since the time interval between stages is not specified, but remains to be determined, the height of the mid-point of the water table may be quite arbitrarily specified at each stage, although it is natural to adopt equal intervals between stages.

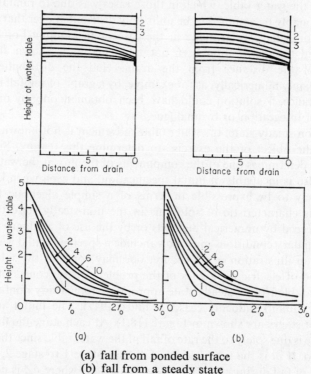

(a) fall from ponded surface
(b) fall from a steady state

Figure 18.1 Stages in the descent of a transient water table over parallel drains. The upper diagrams show the shapes of the water table stages; the lower diagrams show the descent as a function of time. In the lower diagrams the figures against the curves indicate the distance from the drain.

At the conclusion of this procedure, one has a succession of water table stages, but the time scale is lacking. The time intervals between stages are determined as follows.

The distribution of the flux at the mean water table position between two stages is known, since it is the same as the distribution of the electric currents which simulate the flux in the analogue. Hence the water table may be divided into an integral number of elements each of which transmits the same amount of water per unit time. Thus the streamlines dividing equal increments of stream function begin at known points on the water table. The equipotentials separated by equal increments of hydraulic potential are derived directly from the measured potential distribution in the analogue, so that the complete flow-net may be drawn in the manner described in Sections 15.1 and 15.2. From this one may measure the average ratio of width to length, $W/l$, of the mesh of the net; the interval of hydraulic potential between the equipotentials, which is the vertical interval of height, $\delta z$, between the intersection of the equipotentials with the water table; and the height $\delta Z$ and width $\delta A$ of the elements of soil volume bounded by the water table stages and neighbouring stream lines.

The volume of water in the element $\delta A\ \delta Z$ is, by definition of the specific yield $Y$, given by

$$\delta V = Y\ \delta A\ \delta Z$$

and therefore the mean flow across the area $\delta A$ during the interval $\delta t$ between the stages is

$$\delta V/\delta t = Y\ \delta A\ \delta Z/\delta t \tag{18.2}$$

Because the streamlines have been drawn at equal intervals of stream function, the flow $\delta V/\delta t$ is the same for all elements of the water table and consequently the product $\delta A\ \delta Z$ is the same for all. It is therefore most accurately calculated from the experiment, which is inevitably subject to error, as the mean of all the elements.

An application of Darcy's law to a mesh of the flow-net gives

$$\delta V/\delta t = K(W/l)\delta z$$

whence, from Equation (18.2),

$$\delta t = Y\ \delta A\ \delta Z(l/W)/(K\ \delta z) \tag{18.3}$$

The value of $K$ may, if desired, be expressed in terms of the rainfall which was the cause of the initial steady state. Thus from a specification of the hydraulic conductivity and the specific yield of the aquifer, both of which are independent of the stage of the water table, one may assign an absolute time interval between each of the successive stages.

In the initial state due to steady rainfall at the rate $q$, the ratio $q/K$ is expressed, in accordance with Section 15.1, in terms of the then prevailing flow-net by the equation,

$$q/K = (W'/l')(\delta z'/\delta A') \tag{18.4}$$

where $W'/l'$ is the ratio of the width to the length of the meshes of the flow-net in the initial stage, and $\delta A'$ is the horizontal intercept of the stream-lines and $\delta z'$ is the vertical intercept of the equipotentials at the water table. Hence from Equations (18.3) and (18.4), one may express the stage intervals in terms of the stage flow-nets in the form,

$$\delta t = (Y\,\delta Z/q)(\delta z'/\delta z)(\delta A/\delta A')(l/W)/(l'/W')$$

Some results of some studies of this kind are shown in Figures (18.1a) and (18.1b), the latter being for an initial steady state due to constant uniform rainfall and the former for an initially flooded surface. It will be noted that for much of the descent the water table tends to fall as a whole without appreciable change of shape, so that the displacement flux is uniform and the stage shape is similar to that which is appropriate to the steady state with uniformly distributed rainfall. Only at the extremes of the period of descent are there substantial departures from this shape. Toward the end of the period the shape must tend to the ultimate flat water table, while if the initial stage is one of a flat water table at a flooded surface, the early stages will be influenced by this shape. Hence it is not surprising that Collis-George and Youngs (1958) found that, except at these limiting stages, the curve of water table height versus average surface flux during the non-steady stages, with flux calculated by dividing the drain efflux rate by the catchment area, also fitted the curve for the steady states.

## 18.3 The relationship between average and steady states

If the rainfall rate which determines the configuration of the water table varies with time, then the water table will fluctuate in sympathy. Since at each stage position the momentary flux is related to the water table height, it is reasonable to explore the possibility of expressing the water table height as a function of the incident or imposed rainfall rate which is itself a known function of time. Certain approximations are necessary.

From what was said at the end of the preceding section, it seems likely that no great error will be caused by supposing that even in transient stages of the moving water table, the rate of rise or fall is approximately the same at all points at a given moment and that, in consequence, the displacement flux due to this movement is uniformly distributed. Thus formulas which have been derived on the assumption of such a uniformity of distribution of the flux at the water table position may be employed. Also $dZ/dt$ may be equated to $dZ_m/dt$, where $Z$ is the water table height at a given point and $Z_m$ is its particular value at the highest point, midway between neighbouring drains of a parallel system.

The equation relating $Z_m$ to the flux at the momentary position of the water table depends upon the depth to the impermeable bed, as discussed in Section 16.5, but at one extreme, for great bed depths, it may be expressed very approximately by Equations (16.53) to (16.55), i.e.

$$Z_m/L = B(q/K) \tag{18.5}$$

with $B$ lying between 2 and 4, according to the range of $q/K$. In conformity with the usage of Chapters 15 and 16, $L$ is half the distance between neighbouring drains. The quantity $q$ is the net flux, in this case comprising the algebraic sum of the incident rainfall and the displacement fluxes at the given moment.

At the opposite extreme where the impermeable bed coincides with the level of the drain system, the Dupuit-Forchheimer equation is appropriate, namely Equation (15.16):

$$Z_m/L = (q/K)^{\frac{1}{2}} \tag{18.6}$$

One must approximate this to a linear form,

$$Z_m/L = A + B(q/K) \tag{18.7}$$

where $B$ is the slope of Equation (18.6) over the range of $q/K$ for which the curve may be taken as approximately linear. Figure (18.2), which illustrates Equation (18.6), shows that the linear form, Equation (18.7), fits within about five per cent between the limits of 20% and 100% of the range of $q/K$. For a range of $q/K$ up to 0·1, the value of $B$ is about 2; for $q/K$ within the range 0·01, $B$ is about 6·5; while for $q/K$ not exceeding 0·001, $B$ is about 20. Thus for much of the range, the magnitude of $B$ of Equation (18.7) is of the same order as that of $B$ of Equation (18.5). The former equation will be taken as the general form, with $A$ decreasing as the bed depth increases until it vanishes for great depths.

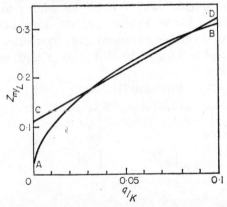

Figure 18.2 The dependence of water table height on intensity of rainfall with drains on the impermeable floor. Over much of the curve $AB$ the straight line $CD$ is an acceptable approximation.

At a given moment when the water table is rising with a speed $dZ/dt$, the flux due to the water table displacement is $Y(dZ/dt)$ and is upward, opposing the rainfall flux $q_r$. The net downward flux $q$ is therefore

$$q = q_r - Y(dZ/dt)$$

and substitution in the equation for the water table height, Equation (18.7), gives, if one drops the subscripts since one may understand $dZ/dt$ to be the same as $dZ_m/dt$,

$$\frac{Z}{L} = A + \frac{B(q_r - Y\,dZ/dt)}{K} \tag{18.8}$$

This is the basic equation, and the solutions will take different forms according to the form of variation of $q_r$.

Equation (18.8) may be rewritten in the form,

$$-Y\frac{dZ}{dt} = \frac{K}{B}\left(\frac{Z}{L} - A\right) - q_r \tag{18.9}$$

Integration between two limits of $t$ gives

$$\int_{t_0}^{t_1} Y\,dZ + \frac{K}{B}\int_{t_0}^{t_1}\left(\frac{Z}{L} - A\right)dt - \int_{t_0}^{t_1} q_r\,dt = 0 \tag{18.10}$$

Over a sufficient length of time the variations of $q_r$ will result in a water table which sometimes rises and sometimes falls, so that $dZ$ is sometimes positive and sometimes negative, while $dt$ is always positive. Hence the first of these integrals may be neglected in comparison with the others. Provided that the specific yield $Y$ when the water table is rising is not systematically greater (or less) than when the water table is falling, this neglect of the first integral is warranted quite irrespective of whether $Y$ is constant or not, and nothing in the analysis of $Y$ leads one to expect such systematic variation.

Hence one may write Equation (18.10),

$$\frac{K}{B}\frac{\displaystyle\int_{t_0}^{t_1}(Z/L - A)dt}{\displaystyle\int_{t_0}^{t_1} dt} - \frac{\displaystyle\int_{t_0}^{t_1} q_r\,dt}{\displaystyle\int_{t_0}^{t_1} dt} = 0$$

That is to say,

$$\left(\frac{K}{B}\right)\left(\frac{\bar{Z}}{L} - A\right) - \bar{q}_r = 0$$

$$\bar{Z}/L = A + B(\bar{q}_r/K) \tag{18.11}$$

where $\bar{Z}$ is the average water table height and $q_r$ the average rainfall rate during the period. Thus the equation which relates the average water table height to the average rainfall rate, Equation (18.11), is the same as that which relates the steady-state values, Equation (18.7). This justifies the intuitive interpretation, referred to in Section 18.1, of steady-state formulas to apply equally to average states.

## 18.4 The aperiodic variation of rainfall

The cases to be studied here are those where a steady state of the water table due to a constant rainfall rate $q_0$ gives way suddenly to a changing water table level when the rainfall rate changes instantaneously to $q_u$, at which value it thereafter remains. The appropriate substitution for $q_r$ in the basic equation, Equation (18.9), yields,

$$Y\frac{dZ}{dt}+\left(\frac{K}{B}\right)\left(\frac{Z}{L}-A\right)-q_u = 0 \qquad (18.12)$$

where $q_u$ is constant. As shown in Note 42, the solution of this equation is

$$Z-Z_u = (Z_0-Z_u)\,e^{-Kt/YBL} \qquad (18.13)$$

where $Z_0$ is the initial height at the moment that the rainfall is changed to the rate $q_u$, and $Z_u$ is the ultimate height at the steady state with rainfall rate $q_u$. This ultimate height is given by

$$\frac{Z_u}{L} = A+q_u\frac{B}{K} \qquad (18.14)$$

This exponential approach to the final state is approximately what is commonly observed, and Isherwood (1959) obtained such a relationship when he computed the successive stages of the water table by the relaxation method based on the principles described in Section 15.1. Where marked departure from an exponential relationship is found, usually at the extremes of the time range, the discrepancy is in general in the direction that would be expected if the cause is the variability of the specific yield, as described in Section 12.11.

It will be observed from Equation (18.14) that for a given value of $q_u/K$ the ultimate water table height is controlled by a choice of the drain spacing $2L$. A low value of $L$ produces a low water table. Equation (18.13) shows that the rate of approach to the level so controlled is also affected, a close spacing producing a rapid rate of approach through its effect on the exponential term. A close spacing is thus doubly beneficial.

**18.5 The simple harmonic fluctuation of rainfall**

While natural rainfall is not in detail of a regularly fluctuating character, such as may be expressed as a simple harmonic function, seasonal average rainfall may approach such a form and controlled seasonal applications of irrigation water may also often be so expressed. In these cases, one may write,

$$q_r = \bar{q} + q_0 \sin(\omega t) \tag{18.15}$$

where $\bar{q}$ is the mean rate, $q_0$ is the amplitude of variation, and the time period, $T$, of a complete cycle of variation is $2\pi/\omega$. The basic equation of the non-steady state, Equation (18.8), now becomes

$$Y\frac{dZ}{dt} + \frac{K}{B}\left(\frac{Z}{L} - A\right) - \bar{q} - q_0 \sin(\omega t) = 0 \tag{18.16}$$

In Note 43 it is shown that this equation has the solution,

$$Z = \bar{Z} + Z_0 \sin(\omega t + \theta) \tag{18.17}$$

where

$$\frac{\bar{Z}}{L} = A + B\frac{\bar{q}}{K} \tag{18.18}$$

$$\frac{Z_0}{L} = \frac{Bq_0/K}{(1 + Y^2\omega^2B^2L^2/K^2)^{\frac{1}{2}}} \tag{18.19}$$

and

$$\tan\theta = -\frac{Y\omega BL}{K} \tag{18.20}$$

For a given soil and intensity of application of water, one may vary the frequency of application, $\omega$, and the separation of the drains. At very low frequency when $\omega L$ is so small that the term $Y^2\omega^2B^2L^2/K^2$ may be neglected in comparison with unity, then Equation (18.20) shows that $\tan\theta$, and therefore $\theta$, tends to zero so that the water oscillation comes into phase with the fluctuation of the application. At the same time it will be noted from Equation (18.19) that

$$Z_0/L = B(q_0/K) \tag{18.21}$$

and hence from Equations (18.17), (18.18) and (18.21)

$$Z/L = A + B(\bar{q} + q_0 \sin \omega t)/K$$

Further, using Equation (18.15),

$$Z/L = A + B(q_r/K)$$

That is to say, the water table is at all times at just the height that is required to satisfy the steady-state equation with the momentarily prevailing rainfall rate.

At the opposite extreme, at very high frequency of alternation, when $\omega L$ is very large, $\tan \theta$ tends to minus infinity and $\theta$ itself approaches $-\pi/2$ or $-90°$, so that the water table oscillation lags behind the rainfall fluctuation by a quarter of a cycle.

From Equation (18.19) the amplitude $Z_0$ of oscillation is seen to be greatly reduced as compared with Equation (18.21), and is zero when $Y^2\omega^2B^2L^2/K^2$ is infinitely great. The water table oscillations are then said to be heavily damped. Then from Equations (18.17) and (18.18),

$$Z/L = \bar{Z}/L = A + B(\bar{q}/K)$$

so that the water table remains stationary at the level appropriate to a constant rate of application of magnitude equal to the average of the rates actually applied.

The magnitude of $\omega L$ which may be regarded as roughly dividing the damped from the undamped water table oscillations is that at which

$$Y^2\omega^2B^2L^2/K = 1$$

In a typical case, the value of $Y$ might be about $0\cdot1$, $B$ is, as has been seen in Section 18.3, likely to be about $5\cdot0$, and for the sake of an example one may take $K$ to be about $1\cdot0$ metre per day. With these substitutions, one has

$$\omega^2 L^2 = 2$$

or

$$T = 2^{\frac{1}{2}}\pi L \tag{18.22}$$

As an example of the application of this analysis, one may consider the design of a drainage system for irrigation with a mean excess of application over consumptive use of about $1\cdot0$ mm per day to control salinization. With the above quoted hydraulic conductivity of $1\cdot0$ metre per day, this corresponds to a value of $q/K$ of about $0\cdot001$. If there is no impermeable bed within a depth of fifty metres or so, one may apply Equation (16.53), namely

$$4\,q/K = 0\cdot004 = \bar{Z}/L \tag{18.23}$$

If the impermeable bed is at drain depth, the Dupuit-Forchheimer result, Equation (15.16), gives

$$q/K = 0\cdot001 = \bar{Z}^2/L^2 \tag{18.24}$$

It can be taken as a reasonable requirement that the water table might be allowed to rise to a height of not more than half a metre above drains that are a metre or more deep. Thus for a deep impermeable bed, Equation

(18.23) gives

$$L = \overline{Z}/0\cdot004 = 0\cdot5/0\cdot004$$

or

$$L = 125 \text{ metre}$$

On the other hand, for a bed at drain level, Equation (18.24) gives

$$L = \overline{Z}/(0\cdot001)^{\frac{1}{2}} = 0\cdot5/0\cdot032$$

or

$$L = 16 \text{ metre}$$

Substituting these values in Equation (18.22) one discovers that for a deep impermeable layer, the critical time period $T$ is about 560 days or more than a year and a half, while for the shallow impermeable bed it is about 70 days or just over two months. Thus in the former case an annual cycle of irrigation or of rainfall would be at a higher frequency than the critical, the sympathetic water table oscillation would be highly damped, and certainly a monthly or fortnightly cycle would be heavily damped. In the case of the shallow bed, the annual cycle would be at a lower frequency than the critical and would be scarcely damped, but a monthly or fortnightly irrigation cycle would be at relatively high frequency and would be sensibly damped.

### 18.6 Drain flow characteristics; flood control

At each stage of the transient water table, the flow-net is a momentary steady state and the rate of flow is equal to the rate of emission of water from the drains as well as to the flux across the momentary position of the water table. The drain flow rate is thus governed by the water table height in just the same way that the water table height is related to the water table flux. Thus the behaviour of the water table described in the preceding sections may be interpreted also as the behaviour of the drain efflux rate.

An instantaneous increase of rainfall rate is not at once reflected in increased drain outflow, since the water table does not at once rise to the new steady-state position, but, as shown in Section 18.4, approaches the new state exponentially. The drain outflow thus also approaches the changed rate exponentially, the first flush of the increased rainfall being stored as groundwater under the rising water table. Similarly, when rainfall ceases the drain efflux rate does not immediately fall to zero, but is maintained by drawing upon stored groundwater, both water table height and drain efflux rate falling together exponentially to zero. Thus the groundwater acts as a reservoir which buffers the drainage rate from the rapid fluctuations of rainfall.

Quantitatively, it is sufficient here to examine the hypothetical case of simple harmonic fluctuation. When, as shown in Section 18.5, the drainage system is intensive, with the separation very small, then the term $Y^2\omega^2B^2L^2/K^2$ tends to be negligible except for quite high frequency fluctuations, and the water table tends to reflect the rainfall rate closely. Thus the stream flow in the drained catchment is subject to fluctuations as intense as those of the rainfall, and there is little control of flood peaks. When, however, drainage is no more intense than is necessary, so that $L$ is as large as is consistent with the required degree of water table control, the term $Y^2\omega^2B^2L^2/K^2$ may be sufficiently large to damp the water table oscillations, thus stabilizing also the drain efflux rate and buffering the streamflow in the drained catchment against the rainfall fluctuations.

If the drainage system is insufficiently intense to prevent the water table from rising to the surface, then the considerations of the reservoir effect of the groundwater are inoperative, for the excess water is removed as surface run-off and suffers only the slight impediment to flow exerted by the uneven land surface. Thus both excessive and deficient drainage systems may be expected to exert a degree of flood control which is less than optimal. Hence drainage which prevents surface flooding, and no more, may well result in improved flood control, while drainage which is required to lower a water table for agricultural reasons, in circumstances where the said water table did not in any case reach the surface, may exacerbate floods marginally. There is little experimental evidence on this point. Conway and Millar (1960) present curves of stream flow in peat catchments with different degrees of drainage. Although they describe the drained catchments as much more liable to flash floods than the undrained, in fact all are characterized by very sharp drainage and the difference is marginal. Insufficient evidence is presented to enable one to judge whether this is because the drainage of the drained catchments is insufficient or excessive. In a survey of the subject, Heikurainen (1964) stresses the variability of experience and opinion on the effects of the drainage of peat on the flood characteristics of the catchment.

## NOTES

**Note 42. The aperiodic movement of the water table**
The equation of movement of the water table, Equation (18.12), is

$$Y\frac{dZ}{dt}+\left(\frac{K}{B}\right)\left(\frac{Z}{L}-A\right)-q_u = 0 \tag{N42.1}$$

Introduce the temporary variable,

$$\frac{\chi}{L} = \frac{Z}{L} - A - q_u \frac{B}{K} \tag{N42.2}$$

so that

$$\frac{dZ}{dt} = \frac{d\chi}{dt}$$

With these substitutions Equation (N42.1) becomes,

$$Y \frac{d\chi}{dt} + \frac{K\chi}{BL} = 0 \tag{N42.3}$$

the solution of which is

$$\chi = \alpha \, e^{\beta t} \tag{N42.4}$$

where $\alpha$ and $\beta$ are constants which remain to be determined.

From Equation (N42.4),

$$\frac{d\chi}{dt} = \alpha\beta \, e^{\beta t}$$

Substituting these values of $\chi$ and $d\chi/dt$ in the differential equation, Equation (N42.3), one has,

$$Y\alpha\beta \, e^{\beta t} + \frac{K\alpha}{BL} \, e^{\beta t} = 0$$

whence,

$$\beta = -\frac{K}{YBL} \tag{N42.5}$$

Substituting this value of $\beta$ in Equation (N42.4) and restoring the original variable $Z_u$ by using Equation (N42.2), one arrives at

$$\frac{Z}{L} - A - q_u \frac{B}{K} = \frac{\alpha}{L} e^{-Kt/YBL} \tag{N42.6}$$

After an infinitely long time interval, the exponential factor vanishes so that $Z$ settles down to the constant value $Z_u$ given by

$$\frac{Z_u}{L} = A + q_u \frac{B}{K} \qquad \begin{cases} \text{(N42.7)} \\ \text{(18.14)} \end{cases}$$

This confirms what is already known, since $Z_u$ is the ultimate steady-state height of the water table for the steady rainfall rate $q_u$ and is consequently given by Equation (18.7).

Thus from Equation (N42.6),

$$\frac{Z}{L} - \frac{Z_u}{L} = \frac{\alpha}{L} e^{-Kt/YBL} \tag{N42.8}$$

The initially observed value of $Z$, when $t$ is arbitrarily assigned the value zero, is $Z_0$, so that the above equation becomes

$$Z_0 - Z_u = \alpha \tag{N42.9}$$

Hence the final form of the solution, substituting $\alpha$ in Equation (N42.8) from Equation (N42.9), is

$$Z-Z_u = (Z_0-Z_u)e^{-Kt/YBL} \qquad \left\{\begin{array}{l}(N42.10)\\(18.13)\end{array}\right.$$

### Note 43. The oscillating water table

The differential equation of movement of the water table, Equation (18.16), is

$$Y\frac{dZ}{dt}+\frac{K}{B}\left(\frac{Z}{L}-A\right)-\bar{q}-q_0\sin(\omega t) = 0 \qquad \left\{\begin{array}{l}(N43.1)\\(18.16)\end{array}\right.$$

Intuitively one expects the water table to settle down to an oscillation in sympathy with the harmonic application of water, with the same frequency but not necessarily in phase. The phase difference $\theta$, the mean water table height $\bar{Z}$, and the amplitude of oscillation $Z_0$ remain to be found. The equation which describes such a motion is

$$Z = \bar{Z}+Z_0\sin(\omega t+\theta) \qquad \left\{\begin{array}{l}(N43.2)\\(18.17)\end{array}\right.$$

in accordance with which,

$$\frac{dZ}{dt} = \omega Z_0\cos(\omega t+\theta) \qquad (N43.3)$$

Substitution of these values of $Z$ and $dZ/dt$ in Equation (N43.1) followed by the expansion of the circular functions, produces the form,

$$Y\omega Z_0(\cos\omega t\cos\theta-\sin\omega t\sin\theta)+$$

$$+\frac{K}{B}\left(\frac{\bar{Z}}{L}+\frac{Z_0}{L}(\sin\omega t\cos\theta+\cos\omega t\sin\theta-A\right)+$$

$$-\bar{q}+q_0\sin\omega t = 0 \qquad (N43.4)$$

This equation must be satisfied at all times, no matter what the values of $\sin\omega t$ and $\cos\omega t$, so that the constant terms and the coefficients of $\sin\omega t$ and of $\cos\omega t$ must separately equal zero. This condition gives rise to the three equations,

$$Y\omega Z_0\cos\theta+\frac{KZ_0}{BL}\sin\theta = 0 \qquad (N43.5)$$

$$-Y\omega Z_0\sin\theta+\frac{KZ_0}{BL}\cos\theta-q_0 = 0 \qquad (N43.6)$$

$$\frac{K}{B}\left(\frac{\bar{Z}}{L}-A\right)-q = 0 \qquad (N43.7)$$

It follows at once from Equation (N43.5) that

$$\tan \theta = -\frac{Y \omega BL}{K} \qquad \begin{cases} \text{(N43.8)} \\ \text{(18.20)} \end{cases}$$

One may deduce from Equation (N43.8) that

$$\sin \theta = -\frac{Y \omega BL}{(K^2 + Y^2 \omega^2 B^2 L^2)^{\frac{1}{2}}}$$

$$\cos \theta = \frac{K}{(K^2 + Y^2 \omega^2 B^2 L^2)^{\frac{1}{2}}}$$

Insertion of these values for $\sin \theta$ and $\cos \theta$ in Equation (N43.6) results in a value for $Z_0$, namely,

$$\frac{Z_0}{L} = \frac{B q_0 / K}{(1 + Y^2 \omega^2 B^2 L^2 / K^2)^{\frac{1}{2}}} \qquad \text{(N43.9)}$$

Equation (N43.7) leads directly to

$$\frac{\bar{Z}}{L} = A + B \frac{\bar{q}}{K} \qquad \begin{cases} \text{(N43.10)} \\ \text{(18.18)} \end{cases}$$

# CHAPTER 19

# Methods of measuring hydraulic conductivity

## 19.1 A general survey of the problem

Methods of measurement of the hydraulic conductivity of soil materials may be divided into those which are performed in the laboratory and those which are carried out in the field. The former are generally capable of refinement and accuracy to a degree not possible in the field, since procedures are under close control. The degree of complication which is permissible in field experiments is severely restricted, and in fact anything more elaborate than the boring of tube wells and the observation of perturbed groundwater, or the response of land to surface infiltration, can seldom be entertained.

The choice of laboratory or field method is not, however, merely a choice between different levels of accuracy, but rather a matching of the method to the object of the enquiry. If the purpose is to determine the hydraulic conductivity of soils for the purpose of designing field engineering works or for elucidating observed field phenomena, the measurement must in general be performed in the field by a suitable field method. The reason for this is that the extraction of samples for later examination in the laboratory, even when carried out by implements which purport to retain the field structure, can hardly be performed without some alteration of the structure, and the hydraulic conductivity is so profoundly affected even by apparently minor changes of structure that the results of the laboratory measurement can never be applied with confidence in the field. In this connection, it may be stressed that even such apparently structureless material as coarse sand may pack appreciably differently beneath an overburden of surface material as compared with the closest packing in the absence of compressive stress.

A second reason is that Darcy's law is applicable to statistically averaged materials, and the measured conductivity is that of the sample. If the sample is not so large that it represents fairly the soil from which it was extracted, then the results are not applicable to the field. If the structural units are large, the required size of the sample may be formidably large.

433

The alternative procedure, of measuring a formidably large number of smaller samples, may well provide an acceptably repeatable average value of conductivity, but that average does not necessarily represent accurately the value which is effective in the field, as may readily be shown by reference to some hypothetical structures. Thus in Figure (19.1) is shown a regular

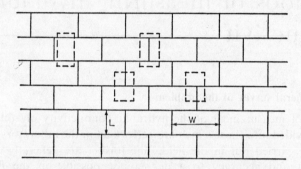

Figure 19.1 A hypothetical structure showing that a sample of the size shown in the dotted outlines cannot truly represent the whole material. However placed, the sample will have a zero conductivity.

structure of vertical and horizontal fissures, with separation $W$ and $L$ respectively. The horizontal fissures are continuous and the vertical are displaced by the distance $W/2$ in successive layers, so that the fissures in alternative layers lie vertically above one another. It is now obvious that a sample of width less than $W/2$ and length exceeding $L$ can never provide a continuous vertical flow path and the measured conductivity will be quite repeatedly zero, while the field value, or that of a satisfactorily large sample, will be appreciable.

On the other hand, the measurement of the conductivity may be required merely as a tool in the examination of some other property. For example, one way of examining the stability of aggregation of soil material is to select crumbs within a quite arbitrary size range and to subject them to a chosen disruptive process. The degree of disruption may be estimated in a number of ways, one of which is to trace the reduction of hydraulic conductivity as the breakdown of the aggregates modifies the pore space. The measured quantity is a property of the sample and not a field state, and the measurement demands laboratory manipulations.

In this chapter a number of methods will be described, some appropriate to the laboratory and some to the field, while another division will be made between the measurement of saturated and of unsaturated soils.

## 19.2 Laboratory permeameters

The definition of hydraulic conductivity implicit in the statement of Darcy's law leads directly to a form of apparatus for measuring this quantity. It is in fact of the kind used by Darcy in his original demonstration. It is shown in Figure (19.2).

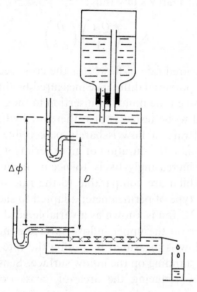

Figure 19.2 The constant head permeameter. In this example the head is maintained by a Mariotte bottle.

In this form the potential difference between the two ends of the vertical column of material is maintained constant by ensuring the constancy of the level of the water surface into which the lower end projects and the level of the water which stands on the upper surface. This latter is attained either by an overflow weir and a rate of supply in excess of the rate of transmission, or by a Mariotte bottle.

The sample must be supported on a base which must retain the grains of the material but allow the passage of the water. It must be permeable, but its hydraulic conductivity cannot be the same as that of the sample except fortuitously. Sometimes a similar retainer rests on the upper surface. Thus the column is, in effect, one of non-uniform conductivity, and the overall potential difference between the ends, measured by the difference of water level in the exit and entrance reservoirs, can provide only an average potential gradient down the composite column and not the true gradient within the sample. Hence manometers are inserted near the ends of the

P

sample column, and the difference of level of the water in the open limbs measures the difference of potential or head between the points at which the manometers are inserted. Thus the true potential gradient is revealed.

The rate of flow is recorded as the rate of collection of water at the overflow of the lower reservoir, and the hydraulic conductivity is calculated by a direct application of Darcy's law thus:

$$K = \left(\frac{Q/t}{A}\right)\left(\frac{D}{\Delta\phi}\right)$$

where $Q/t$ is the measured rate of flow, $A$ is the cross-section of the sample in the tube, $\Delta\phi$ is the potential difference measured by the manometers, and $D$ is the distance between the points of insertion in the column.

When the material to be tested is of rather low hydraulic conductivity, so that the observed rate of flow is barely perceptible with conveniently small potential gradients, the duration of an experiment becomes tediously long and errors are increasingly likely to occur, including errors due to evaporation losses which are comparable to the rate of collection of the transmitted water. A type of permeameter adapted to such circumstances is shown in Figure (19.3), and is known as a variable head permeameter.

The depth of water on the upper inflow surface is not maintained, but instead the rate of flow is measured by noting the rate of decrease of the volume of this water standing on the inflow surface. Sensitivity of measurement is increased by reducing the area of cross-section of the input chamber at a distance above the input surface, so that the upper surface

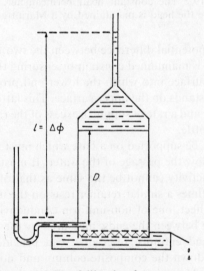

Figure 19.3 The falling head permeameter.

of the standing water forms a meniscus in a tube of suitably narrow bore, as shown. In consequence, the passage of a small amount of water through the porous column of relatively large cross-section is rendered visible as a relatively large movement of the meniscus. Thus a calibration of the narrow bore tube enables one to estimate the rate of flow with enhanced accuracy. Since, however, the height of this meniscus measures the potential difference between the ends of the sample column, both the potential gradient and the rate of flow vary continuously. Hence, measurements must be taken instantaneously or else a formula must be developed to enable the hydraulic conductivity to be determined from measurements of finite volumes of water transmitted during finite time intervals.

Instantaneous values of the potential gradient and of the rate of flow may be deduced by plotting both the volume of water transmitted and the height of the meniscus, relative to the manometer level at the bottom of the column, as functions of time. The slope of the former curve at a given instant is the rate of flow, and the height of the mensicus at the same instant is the potential difference.

A relationship between the total amount of water transmitted and the time of duration of the flow may be derived as follows. Let $Q$ be the instantaneous rate of flow at a given moment. If $a$ is the area of cross-section of the tube containing the upper meniscus and $l$ is the height of the meniscus above the level of water in the manometer at the base of the column which may be taken as constant, then

$$Q = -a(\mathrm{d}l/\mathrm{d}t) \qquad (19.1)$$

But $l$ is also the potential difference between the surface of the column and the point, distance $D$ below the surface, where the lower manometer is inserted. That is to say $D$ is the effective length of the column. Hence by Darcy's law,

$$Q = KAl/D \qquad (19.2)$$

where $A$ is the area of cross-section of the column. Hence from Equations (19.1) and (19.2),

$$-a(\mathrm{d}l/\mathrm{d}t) = KAl/D \qquad (19.3)$$

At moments of time, $t_0$ and $t$, the observed levels of the meniscus are $l_0$ and $l$ respectively, and by integrating Equation (19.3) between these limits one has

$$1\mathrm{n}(l_0/l) = (A/a)(K/D)(t-t_0) \qquad (19.4)$$

### 19.3 The oscillating permeameter

Both the constant head and variable head permeameters require the passage of water from an external source continuously through the pore space. When it is important to ensure that the nature of the porous material is not changed by this procedure, as for example when the stability of a structured soil must remain unchanged and therefore the transmitted solution must be in ionic equilibrium with the soil, it may be necessary to determine what the constitution of such a solution must be and make up stocks of it. This necessity may be avoided by using a constant volume of solution, as little as possible in excess of what is required to saturate the soil and provide a measurable head, and by alternating the flow to and fro. Such a solution will come into equilibrium with the soil without altering its ionic constitution excessively, and will remain in equilibrium for as long as the procedure of measurement of conductivity requires. An alternating permeameter of this kind has been designed by Childs and Poulovassilis (1960). One form of it is shown diagrammatically in Figure (19.4). The

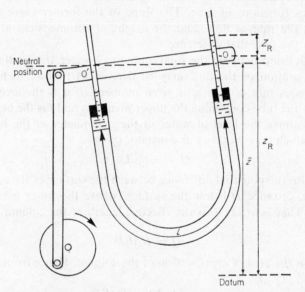

Figure 19.4 The oscillating permeameter.

sample is shown packed in a U-tube so that the surfaces are free and require no retaining membranes or filters. The saturating solution extends beyond both ends of the sample, so providing hydrostatic pressure contributions to the potential. The free liquid may extend into tubes, the bore of which is different from that of the tube which contains the sample, so providing a control of sensitivity of measurement as in the case of the falling head permeameter. The whole assembly is mounted on a frame which may be rocked about a horizontal axis by a geared motor turning a crankshaft and connecting rod, so that the time period $T$ of the simple harmonic motion may be freely chosen.

Let the apparatus be brought to rest in its neutral position and then let a reference mark $O$ be drawn on one limb, say the right-hand one, at the equilibrium position of the meniscus. This mark will be at some height $\bar{z}$ above an arbitrary level which may be chosen as the datum for the measurement of hydraulic potential. Since the apparatus is in static equilibrium at this stage, the meniscus in the left-hand limb will also be at this level, and a mark may be imagined to correspond to $O$ but need not in fact be made. On initiation of the oscillatory motion, the height $z_R$ of the reference mark $O$ will depend on the lapse of time according to the equation,

$$z_R = \bar{z} + z_0 \sin(2\pi t/T) \tag{19.5}$$

After the lapse of a sufficient length of time, the movement of the liquid through the porous column will have settled down to a simple harmonic motion in sympathy with the frequency of rocking, that is to say, with a time period $T$, and the meniscus on each side will rise and fall about the reference marks $O$ with this frequency. The sympathetic oscillation of the meniscuses will not, however, in general be in phase with the forcing frequency, and the amplitude of oscillation remains to be found.

If the height of the meniscus in the right-hand limb, relative to the reference mark $O$, is $Z_R$, then as is shown in Note 44, the oscillation is described by the equation,

$$Z_R = Z_0 \sin(2\pi t/T + \pi - \theta) \tag{19.6}$$

where $Z_0$ is the amplitude of the oscillation and the phase difference with respect to the forcing oscillation is $\pi - \theta$. The formula for $\theta$ is

$$\tan \theta = \frac{a\pi l}{KAT} \tag{19.7}$$

where $l$ and $A$ are respectively the length and the area of cross-section of the sample of the porous material in the column, and $a$ is the area of cross-section of each of the tubes in which the meniscuses move. The amplitude $Z_0$ is given by the equation,

$$Z_0 = \frac{z_0}{(1+\tan^2\theta)^{\frac{1}{2}}} \tag{19.8}$$

Figure (19.5) shows the result of an experiment in the form of a simultaneous plot of $z_R$, the oscillation of the limb, and of $Z_R$, the oscillation of the meniscus within the limb. The phase difference $\pi - \theta$ may be measured directly and the angle $\theta$ may be thus revealed, or alternatively, the amplitude $Z_0$ and the forcing amplitude $z_0$ may be measured and the value of $\theta$

derived indirectly by use of Equation (19.8). Either way, the value of $K$ then follows from Equation (19.7), since all the other parameters in the expression for tan $\theta$ are directly measurable. Thus one and the same experiment provides estimates of $K$ from two different measurements.

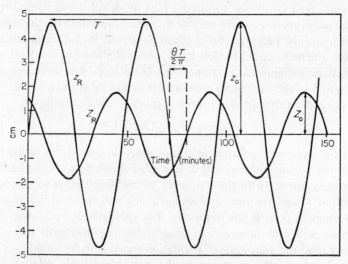

Figure 19.5 A typical set of observations from the oscillating permeameter.

The best accuracy of measurement is achieved when the value of $\theta$ turns out to be about $\pi/4$ or $45°$, and this may be achieved by a judicious choice of parameters under the control of the experimenter. At this value of $\theta$,

$$\tan \theta = -(a\pi l)/(KAT) = -1$$

The length of the time period $T$ and the ratio of the areas of cross-section, $a/A$, may be chosen freely, and a preliminary rough measurement of $K$ will indicate suitable values for the final accurate measurement.

### 19.4 The measurement of unsaturated hydraulic conductivity

Unsaturated porous materials are necessarily at a negative hydrostatic pressure, and consequently methods which require the maintenance of water standing on the exit and inflow faces, at positive pressure, are unsuitable. In principle, it should be possible to supply water to the inflow face and extract it from the outflow face via tensiometers maintained at a measured suction and to measure the potential difference between two separated points in the unsaturated column by other tensiometers, thus reproducing the features of the constant head permeameter with unsaturated soil under suction. This was in fact done in the first measurements of this nature by Richards (1931). The method is not, however, a good one, because in general the maintenance of a potential gradient causes the imposition of a suction gradient and, with it, a consequent gradient of moisture content and of hydraulic conductivity. Thus the potential gradient is not uniform for a constant rate of water movement, for the potential difference must be more concentrated across the layers of lower conductivity. The measured mean potential gradient therefore provides no more than a measure of the mean hydraulic conductivity of a column of varying moisture content which may be assigned a mean moisture content, but because of the sharply varying nature of the moisture content with suction and of the conductivity with moisture content, these mean values cannot be said to provide a measure of the hydraulic conductivity at a stated moisture content. For example, a thin layer at low moisture content would provide a very small contribution to the mean moisture content, but would dominate the overall hydraulic conductivity. The mean moisture content would approximate to that of the wetter parts of the column, but the overall conductivity would be dominated by that of the driest part.

The difficulty may be overcome by securing a uniform moisture content throughout the length of the sample column, or by endeavouring to measure the moisture content and potential gradient at a point. The latter course was pursued by Moore (1939), who plotted the suction distribution along the column by means of an array of tensiometers and the moisture distribution by sampling at the end of the experiment. In another experiment of this kind, Wyckoff and Botset (1936) used water in which gas was dissolved under pressure. As the pressure was reduced, the gas came out of solution in the pore space, thereby displacing liquid and causing a reduction of the degree of saturation. The moisture content was determined by measuring the electric conductivity of the unsaturated material while the potential gradient was measured by manometers. It cannot be said with certainty that the distribution of water and air in the pore space in such an experiment is of the kind that has been specified elsewhere in this book where the air and water are separately continuous and move independently. It may well be that the air is in the form of bubbles carried along by the water, and in fact the authors in their discussion envisage that this is the case. If so, the results of the experiment are not relevant to the movement of soil water in which the air in the pore space is in continuous equilibrium with the external atmosphere, although the curves of conductivity expressed as a function of the degree of saturation proved to be of the same kind as those which are the result of more valid kinds of experiments.

Columns of material at uniform moisture content may be obtained by infiltration procedures. It has been shown in Section 12.2 that if infiltration at a controlled rate takes place into a sufficiently long soil profile which drains to a water table, the upper part of the moisture profile is at a uniform moisture content and the hydraulic conductivity at this section is equal to the rate of infiltration. Thus one merely has to control and measure the rate of infiltration and to measure the moisture content when a steady state has been achieved. Childs and Collis-George (1950) used this method with columns of porous material which could be suspended at an angle to the vertical, so that the potential gradient might be varied at will. In this way they verified the applicability of Darcy's law in unsaturated materials as well as demonstrating the dependence of the conductivity on the moisture content.

It was pointed out in Section 12.7(b) that if a steady rate of infiltration is imposed at the surface of a deep soil profile in the absence of a water table, a condition is reached where a moisture profile of constant shape descends at a constant speed, and the upper part of this profile is then at uniform moisture content. Again the rate of infiltration is equal to the hydraulic conductivity at the moisture content which prevails at the surface, so that this also provides a method of demonstrating the form of the dependence of the hydraulic conductivity on the moisture content.

It was also shown in Section 12.7(b) that if the upper surface of a soil profile is maintained at a moisture content $C$, or at a constant suction appropriate to this moisture content, then the rate of infiltration settles down to a constant value equal to the hydraulic conductivity at the moisture content $C$ after sufficient time has elapsed for a moisture profile to develop with a finite depth near the surface at uniform moisture content. Again a pair of simple measurements gives directly the hydraulic conductivity and the moisture content at which it prevails.

It is, of course, impracticable to maintain a soil surface at any constant moisture content less than saturation, and if the alternative is adopted, namely to control the surface suction, this can only be done via a surface tensiometer plate which will itself absorb a part of the potential gradient in transmitting the liquid, so that the measured suction in the tensiometer chamber is not necessarily the same as that at the surface of the soil column. Nevertheless, it remains true that the surface layers of the soil column itself will ultimately achieve a state of uniform moisture content at the appropriate suction, although that suction will not be the same as that which is measured in the tensiometer chamber; and at that moisture content the hydraulic conductivity will be equal to the ultimate rate of infiltration. This technique is therefore available as a method of measuring the conductivity as a function of the moisture content. A variation of the method, in which the ultimate steady infiltration stage is not waited for but estimated by an extrapolation, has been reported by Youngs (1964).

Another type of measurement utilizes the theory of the non-steady state of flow from a sample on a suction table or in a pressure plate apparatus when the equilibrium is disturbed by a sudden change of the soil water suction. As presented by Gardner (1956), the analysis is simplified by the introduction of approximations. Gravity is ignored, which may well be warranted at higher suctions and lower moisture contents, so that the basic equation even for vertical flow is taken to be

$$\partial c / \partial t = \partial \{ D(\partial c / \partial z) \} / \partial z \qquad (19.9)$$

This is the form which Equation (11.22) takes when it is limited to the vertical direction in the absence of gravity. $D$ is here the diffusivity.

The sample is allowed to come to equilibrium at pore pressure $H_i$. Because gravity is ignored, it is assumed that the moisture content $c_i$ is uniformly distributed and is appropriate to this pore pressure, which is of course negative. At a recorded moment, the pore pressure at the face in contact with the membrane of the apparatus is changed to a new value $H_u$. In the case of the tension table, the tension is changed to this new value, while in the pressure plate apparatus, the air pressure in the chamber is increased by an equivalent amount. This difference is immaterial to the course of the anlaysis, which will be carried through on the former assumption.

The soil sample at once begins to lose water, and the loss, $Q$, is measured at recorded time intervals. Ultimately a new state of equilibrium is reached at moisture content $c_u$, the total water loss having been $Q_u$. If the area of cross-section of the sample is $A$ and the thickness $l$, then

$$Q_u = Al(c_i - c_u) \qquad (19.10)$$

From the curve of $Q$ as a function of time, the diffusivity $D$ and hence the conductivity $K$ are deduced as follows.

It must be assumed that the increment of suction in the stage of unsaturation from $c_i$ to $c_u$ is so small that $D$ may be regarded as constant throughout the thickness and the time interval. The basic equation then becomes, instead of Equation (19.9),

$$\partial c / \partial t = D \; \partial^2 c / \partial z^2 \qquad (19.11)$$

with the boundary conditions that at the extracting membrane, where $z$ is taken to be zero, the moisture content instantaneously takes the value $c_u$ at which it is maintained, while for vanishing $t$, namely the initial moment, the moisture content throughout the sample at all other values of $z$, up to the surface at $l$, is $c_i$. At the surface the rate of flow is zero, since there is neither infiltration nor evaporation, so that the gradient of pressure and therefore of moisture content vanishes. The solution of Equation (19.11) in these

P*

circumstances is given by Carslaw and Jaeger (1959) and is, with present nomenclature,

$$c - c_u = \frac{4(c_i - c_u)}{\pi} \sum_{n=1}^{n=\infty} \frac{1}{n} \left\{ e^{-\frac{(n\pi)^2 Dt}{(2l)^2}} \right\} \sin\frac{n\pi z}{2l} \tag{19.12}$$

The volume of water, $Q$, released up to time $t$, is

$$Q = A \int_0^l (c_i - c) \mathrm{d}z$$

On substitution of $c$ from Equation (19.12) and carrying out the elementary integration of the sin $(n\pi z/2l)$ terms, one has

$$Q = A(c_i - c_u)l \left( 1 - \frac{8}{\pi^2} \sum_1^\infty \frac{1}{n^2} e^{-\frac{(n\pi)^2 Dt}{(2l)^2}} \right)$$

The summation terms converge very rapidly for all but very short time intervals, so that if one neglects all but the first term and in addition makes use of Equation (19.10), the result is, in logarithmic form,

$$\ln\left(1 - \frac{Q}{Q_u}\right) = \ln\left(\frac{8}{\pi^2}\right) - \left(\frac{\pi}{2l}\right)^2 Dt \tag{19.13}$$

Thus a plot of $\ln(1 - Q/Q_u)$ as a function of elapsed time provides a straight line, the slope of which is measurable, and provides the value of $(\pi/2l)^2 D$ and thus of $D$ itself.

With $D$ known at each stage of the moisture content, the value of $K$ is determined from the relationship Equation (11.6),

$$D = K(\mathrm{d}H/\mathrm{d}c) = \frac{KAl(H_i - H_u)}{Q_u} \tag{19.14}$$

Two assumptions render the results of such measurements open to criticism. The first criticism is that the sample must rest on a membrane which itself absorbs a part of the measured pressure difference so that the pressure at the base is not the value $H_u$ as assumed, but a higher value (lower suction) which approaches $H_u$ as the rate of flow decreases during the stage. To revise the theory to take this into account leads to difficulties (Miller and Elrick, 1958). Secondly, the assumption of the constancy and uniformity of $D$ is hardly warranted in a stage during which a sufficiently large volume of water is withdrawn to be measured with accuracy, and an independent check that such uniformity and constancy in fact prevailed is not provided. It would seem that such doubts reduce the value of these methods in comparison with the steady-state methods to which such objections do not apply, but for the fact that it is claimed that soil samples which retain their natural structure can be studied by the outflow method. However, even this advantage may be exaggerated, since it is doubtful whether a sample can be extracted, even with an implement which allegedly preserves structure, without a disturbance which, small though it may be, may nevertheless modify the hydraulic conductivity profoundly.

## 19.5 Field measurements of saturated hydraulic conductivity; large scale well pumping tests

Situations in the field may be divided into those where there is ground water so that the soil is naturally already saturated, and those where the condition of saturation must be imposed artificially as part of the technique of the measurement. The former situations are dealt with by some form or other of artificial perturbation of the ground water and an observation of the flow of water caused by an imposed and measured difference of potential. The latter condition necessitates an observation of the infiltration rate from an artificially flooded surface.

Probably the simplest of all methods is the observation of the drawdown of the water table or of the piezometric surface in the neighbourhood of a pumped well of known diameter and depth of penetration. A true steady state with a stationary water table is not in principle attainable, but in practice the water table is drawn down to a sensibly stationary level in the near neighbourhood of the well fairly early in the pumping test, the distance at which the drawdown becomes appreciable then steadily increasing in the later stages. Thus, provided that one confines observation of the drawdown to distances within which the rate of fall of the water table contributes only a negligible fraction of the measured rate of pumping from the well, one may safely apply the steady-state formula Equation (14.14), of Section 14.6, namely

$$\Delta\phi = (Q/2\pi lK) \ln(r/r_W)$$

where $Q$ is the rate of pumping from the well, which has radius $r_W$ and penetrates a distance $l$ into the aquifer, while $\Delta\phi$ is the distance by which the water table at the well itself falls below that at the distance $r$ from the well axis. This method is commonly associated with the name of Thiem.

The problem of the transient stage of the flow to the well may be solved if it is assumed that the flow is radial. As with Thiem's method, this is strictly true of the flow in a confined aquifer which is completely penetrated by the well. It cannot be strictly true of even a completely penetrating well in groundwater under a free water table, since there is a cone of depression in the vicinity of the well, but provided that the drawdown is small compared with the well penetration, the assumption is not entirely unrealistic. It is in fact the characteristic assumption of the Dupuit-Forchheimer approximation.

If the well is pumped at a constant rate $\dot{Q}$, the drawdown $\Delta\phi$ at time $t$ at a distance $r$ from the well is, as is shown in Note 45,

$$\Delta\phi = \frac{\dot{Q}}{4\pi Kl}W(n) \tag{19.15}$$

where $W(n)$ is a function given by

$$W(n) = \int_n^\infty \frac{e^{-u}}{u}\,\mathrm{d}u \tag{19.16}$$

The parameter $n$ is the quantity,

$$n = \frac{r^2 Y}{4Klt} \tag{19.17}$$

In the case of pumped groundwater under a free surface, $\Delta\phi$ is the water table drawdown $\Delta Z$. $W$ is in fact, in all but sign convention, the exponential integral which is tabulated in Jahnke and Emde's Tables (1938), and reproduced by Peterson (1957). In a confined aquifer the specific yield, $Y$, is the change in the volume of water stored in the full thickness of the confined aquifer due to expansion when the aquifer potential is increased by a unit change of the pressure component.

In a pumping test $\Delta\phi$ is measured at specified values of $r$ and $t$, and it is evident from an inspection of Equations (19.15) to (19.17) that values of $K$ and $Y$ are not obtainable directly. Theis's (1935) method of handling the data is based on the fact that from Equation (19.15),

$$\log \Delta\phi = \log \frac{\dot{Q}}{4\pi Kl} + \log W(n) \tag{19.18}$$

and from Equation (19.17),

$$\log n = \log(Y/4Kl) + \log(r^2/t) \tag{19.19}$$

If then one plots a curve of $\log \Delta\phi$ against $\log (r^2/t)$ from the pumping test and another of $\log W(n)$ against $\log n$ from the tables of $W(n)$, then each point of the former is displaced from the corresponding point on the latter by the constant vertical distance $\log(\dot{Q}/4\pi Kl)$ and by the constant horizontal distance $\log (Y/4Kl)$. Thus one may draw the former curve as a transparency, and by manoeuvring it over the latter curve until the best fit is found, the magnitudes of the displacements are revealed. Thus from the vertical displacement of the horizontal axes, the value of $K$ is determined, since the remaining factors of $\log (\dot{Q}/4\pi Kl)$ are known parameters; while from the horizontal displacement of the vertical axes, $Y$ is determined once $K$ is known.

An alternative manipulation of the data, due to Cooper and Jacob (1946), relies on the expansion of $W(n)$ as a convergent series, of which all but the first two terms may be neglected for sufficiently small values of $r^2/t^2$. The resulting equation is, with the help of Equation (19.17),

$$\Delta\phi = (\dot{Q}/4\pi Kl)\{-0\cdot5772 - \ln(r^2 Y/4Klt)\}$$

or

$$\Delta\phi = (\dot{Q}/4\pi Kl)\ln(2\cdot25Klt/r^2 Y)$$
$$= (\dot{Q}/4\pi Kl)\ln(2\cdot25Kl/Y) + (\dot{Q}/4\pi Kl)\ln(t/r^2) \qquad (19.20)$$

Hence a plot of $\Delta\phi$ against $\ln(t/r^2)$ provides a curve which is a straight line over much of its length. The slope provides a measure of $K$ and the intercept on the axis of $\Delta\phi$ provides a measure of $Y$ when $K$ is thus determined.

Chow (1952) handled the same data in a different manner. The drawdown $\Delta\phi$ is again plotted against $\ln(t/r^2)$. The slope of this curve at a selected point is, from Equation (19.15),

$$\frac{d(\Delta\phi)}{d\left(\ln\dfrac{t}{r^2}\right)} = \frac{\dot{Q}}{4\pi Kl}\frac{dW(n)}{d\left(\ln\dfrac{t}{r^2}\right)} \qquad (19.21)$$

The full expansion of $W(n)$ is

$$W(n) = -0\cdot5772 - \ln n + n - \frac{n^2}{2\lfloor2} + \frac{n^3}{3\lfloor3} - \cdots$$

whence

$$\frac{dW(n)}{d\left(\ln\dfrac{t}{r^2}\right)} = -\frac{d(\ln n)}{d\left(\ln\dfrac{t}{r^2}\right)} + \frac{dn}{d\left(\ln\dfrac{t}{r^2}\right)} - \frac{\dfrac{n}{\lfloor2}dn}{d\left(\ln\dfrac{t}{r^2}\right)} + \frac{\dfrac{n^2}{\lfloor3}dn}{d\left(\ln\dfrac{t}{r^2}\right)} - \cdots \qquad (19.22)$$

From Equation (19.19),

$$\frac{d(\ln n)}{d\left(\ln\dfrac{t}{r^2}\right)} = \frac{\dfrac{1}{n}dn}{d\left(\ln\dfrac{t}{r^2}\right)} = -1$$

With these substitutions in Equation (19.22), one has

$$\frac{dW(n)}{d\left(\ln\dfrac{t}{r^2}\right)} = 1 - n + \frac{n^2}{2} - \frac{n^3}{3} \cdots$$

$$= e^{-n} \tag{19.23}$$

Thus Equation (19.21) becomes

$$\frac{d(\Delta\phi)}{d\left(\ln\dfrac{t}{r^2}\right)} = \frac{\dot{Q}}{4\pi K l} e^{-n} \tag{19.24}$$

Dividing Equation (19.15) by Equation (19.24) one arrives at the equation,

$$\Delta\phi \frac{d\left(\ln\dfrac{t}{r^2}\right)}{d(\Delta\phi)} = e^n W(n) = F(n) \tag{19.25}$$

Since $W(n)$ and $e^n$ are tabulated functions of $n$, it follows that a table can be compiled to give $F(n)$ as a function of $n$. Chow in fact presents curves instead of a table.

From the data of the pumping test both the drawdown itself and the slope of the curve of drawdown versus $\ln(t/r^2)$ are known at a chosen value of $t/r^2$, so that $F(n)$ may be computed from Equation (19.25). Thus from the table of $F(n)$, the value of $n$ is revealed as also is $W(n)$ from the table of that function. Then $K$ follows from Equation (19.15) with $W(n)$ known, and $Y$ follows from Equation (19.17) with $K$, $n$ and $t/r^2$ known.

A solution of the equations of flow of groundwater to a pumped well has been presented by Boulton (1954) in which he has not resorted to the Dupuit-Forchheimer assumption of radial flow, although certain other simplifying approximations are necessary, such as that the water table is nowhere disturbed very markedly and that both radial and vertical potential gradients at the water table are so small that their squares may be neglected. In terms of the symbols used here, the solution is

$$\Delta Z = (\dot{Q}/2\pi K l) V(\rho, \tau) \tag{19.26}$$

where $\rho$ and $\tau$ are the dimensionless quantities given by

$$\rho = r/l \tag{19.27}$$

$$\tau = Kt/Yl \tag{19.28}$$

$V(\rho, \tau)$ is a function of $\rho$ and $\tau$ which may be tabulated from the formula given by the author and for which a skeletal table is presented in his paper. The formula is

$$V(\rho,\tau) = \int_0^\infty \{J_0(\rho\lambda)/\lambda\}\{1 - e^{-\tau\lambda \tanh \lambda}\}d\lambda \qquad (19.29)$$

where $J_0(\rho\lambda)$ is the Bessel function of the first kind of zero order, tabulated in the British Association Mathematical Tables (1937). For values of $\tau$ exceeding about 5, $V$ turns out to be $\frac{1}{2}W(n)$ so that Boulton's formula then agrees with that of Theis. Equation (19.26) may be handled by the matching of curves just as proposed by Theis for Equation (19.15). By plotting log $V$ against log $\tau$ from the tables for a selected $\rho$, and log $\Delta Z$ against log $t$ from the pumping test for the value of $r$ which corresponds to $\rho$ from Equation (19.27), curves are obtained which should be of the same shape but relatively displaced. By matching them, the displacements may be measured. The displacement of log $\Delta Z$ relative to log $V$ is, from Equation (19.26), log $(\dot{Q}/2\pi Kl)$ and this permits the calculation of $K$. The displacement of log $\tau$ relative to log $t$ is, from Equation (19.28), equal to log $(K/Yl)$, and this provides the value of $Y$ when $K$ is known.

In all these derivations of formulas for the transient states of pumped wells, it has been assumed that the specific yield $Y$ is a constant which is characteristic of the aquifer. It has been shown in Section 12.11 that $Y$ can vary between wide limits for one and the same aquifer, according to the circumstances in which it measured, and Youngs and Smiles (1963) demonstrated this variability by applying Theis's analysis to pumping tests carried out in an experimental sand tank. It has not yet been shown how seriously this variability invalidates the analyses themselves.

## 19.6 Pumped boreholes and cavities of small diameter

The methods described in Section 19.5 are commonly associated with the pumping of high-yielding wells of substantial diameter. The trial well itself is often ultimately a service well, so that the cost of boring it is not to be debited wholly to the conductivity survey. When the survey is for a purpose in which the test well will play no part, such as for the design of a land drainage system, great expense is often not warranted and the test boreholes or cavities tend to be small and temporary. The disturbance of the water table is small and localized near the borehole, so that measurements of drawdown can be made with only low accuracy. Boreholes must usually be confined to relatively shallow depths of the order of a few metres, and observation of the water table is usually confined to the level of the water in the borehole itself.

A method due to Kirkham (1946) employs a cavity of restricted length created in the groundwater zone by driving a tube temporarily stopped at the lower end by a closely fitting rivet. When the tube is at the required depth, the rivet is driven out by hammering an internal plunger to a measured further depth. In this way, a cavity of known radius and length results, and is accessible through the tube, which is left in position.

After a period sufficient to enable the groundwater to rise in the tube to an equilibrium level, which measures the hydraulic potential at the cavity and at the same time the level of the water table if there is no substantial groundwater flow, the level is suddenly depressed by the rapid removal of a volume of water from the tube. During the ensuing period of recovery, the level of the water in the tube is recorded as a function of elapsed time. From the results, the hydraulic conductivity may be determined on the assumption that it is isotropic. The analysis of the situation is as follows.

Provided that not too much water has been pumped from the tube, it may be assumed (and indeed the assumption may be tested in certain cases of specified geometry, such as a spherical cavity) that the water table is not seriously disturbed during the period of recovery so that the problem may be treated as one of the flow of water in an aquifer of known limits between a plane equipotential (the water table) of known position and a cavity surface of known shape and position whose potential is measured by the level of water in the observation tube. Since the effective shape and size of the conducting body is thus assumed to be constant, the rate of flow to the cavity is simply proportional to the potential difference between the water table and the cavity and to the hydraulic conductivity of the aquifer. Thus if the lower extremity of the cavity is taken as the datum level and the height of the water table above it is $Z_{wt}$ while the height of the' water surface in the observation tube is $Z$, then $Z_{wt}$ and $Z$ are repectively the hydraulic potentials at the water table and the cavity, from Equation (9.3). Thus if $A$ is a constant of proportionality which takes account of the shape but not of the conductivity of the flow-net, the rate of flow $\dot{Q}$, from the imperceptibly descending water table to the cavity is

$$\dot{Q} = KA(Z_{wt} - Z) \tag{19.30}$$

This rate of flow produces a rate of rise of water in the observation tube, of which the radius is $r$, as given by the equation,

$$\dot{Q} = \pi r^2 dZ/dt \tag{19.31}$$

From Equation (19.30) and (19.31), one has

$$\pi r^2 \, dZ/dt = KA(Z_{wt} - Z)$$

the solution of which is

$$\pi r^2 \ln \frac{Z_{wt} - Z_1}{Z_{wt} - Z_2} = KAt \tag{19.32}$$

where $Z_1$ and $Z_2$ are respectively the levels of water in the observation tube at the beginning and end of an observation period of duration $t$. Thus if $A$ is known, the conductivity $K$ may be determined.

The factor $A$ may be determined by mathematical analysis in certain impracticable cases, but in general may be found by experiments with three-dimensional electric analogues. The groundwater itself may be simulated in an inverted form by a tank of electrolytically conducting solution of known conductivity at the bottom of which is a sheet of highly conducting metal at zero voltage to simulate the water table. The surface of the cavity is represented by a cylindrical electrode supported in the correct position by an insulating rod of the same diameter to represent the barrier of the observation tube. The size and position of this electrode must represent the cavity to the same scale as that at which the depth of solution in the tank represents the thickness of the groundwater. The resistance of the analogue is measured by any suitable alternating current method, the resistance being taken between the cavity analogue and the base sheet which represents the water table. Let the measured conductance between these points be $\Sigma$ when the specific conductivity of the solution is $\sigma$.

Had the electric analogue been at full scale, each linear dimension would have been multiplied by the factor $N$, where

$$N = r/a \qquad (19.33)$$

In this equation $a$ is the radius of the electrode which represent the cavity to the scale $1/N$. Had the same voltages been applied to the electrodes in the process of measuring the conductance, then each element of the three-dimensional flow-net would have had an unchanged voltage difference imposed across its ends, since the magnified flow-net retains the same number of elements. But each element would have had the cross-section increased in the ratio $N^2$ and the length increased in the ratio $N$, so that the current passing would have been increased by the overall ratio $N$. The number of current tubes between the water table surface and the cavity remaining unchanged, the total current passing would therefore have been increased in the ratio $N$, and thus this is the ratio of the increase of overall conductance between the water table surface and the cavity electrode. Thus if $\Sigma_N$ is the conductance which would have been measured had the analogue been at full scale, then, with the help of Equation (19.33),

$$\Sigma_N/\Sigma = N = r/a \qquad (19.34)$$

A comparison may now be made between Equation (19.30) and a corresponding equation which expresses the electric current $I$ which flows in a full scale analogue when a voltage $V$ is imposed.

In Equation (19.30) the constant $A$ is the shape factor which, together with the hydraulic conductivity $K$, relates the total water flowing between the water table and the cavity to the hydraulic potential difference. It is precisely the same shape factor which, together with the electrical conductivity, relates the current $I$, flowing in the full scale analogue, to the voltage $V$. The difference between the units in the two cases is accounted for by the difference between the definitions of the conductivities. Hence, one may write for the analogue,

$$I = \sigma \, AV$$
$$= \Sigma_N V$$

whence

$$\Sigma_N = \sigma \, A \qquad (19.35)$$

From Equation (19.34), the measured conductance of the small scale analogue may be related to $\Sigma_N$ and thence to $A$ from Equation (19.35), the result being,

$$A = (r/a)(\Sigma/\sigma) \qquad (19.36)$$

Curves have been presented by Luthin and Kirkham (1949) to indicate the value of $A$ for a range of radii and lengths of cavity. Smiles and Youngs (1965) concluded that these were subject to a small systematic error and published an amended table.

The conductivity which such a method measures is the conductivity of that part of the flow-net in which the greater part of the potential difference is concentrated, namely the part in the immediate vicinity of the cavity where the potential gradient is steepest. Thus the sample measured is located at a known and fairly well-defined depth, so that the method is particularly suited to measuring the variation of conductivity with depth where the variation is not too pronounced. If there should be layers of comparatively low conductivity and considerable extent, these would absorb a preponderant proportion of the total potential difference and would have a dominant effect on the measured conductivity.

Small wells and boreholes have been much used in a similar kind of way, namely by the observation of the rate of recovery of the well water level after the sudden extraction of a quantity of water. The various analyses of the situation with greater or less degrees of approximation have been reviewed by Luthin (1957). A formula proposed by Hooghoudt (1936) is

$$Kt\left(1+\frac{2l}{r}\right) = A \, \ln\frac{l-Z_1}{l-Z_2} \qquad (19.37)$$

when any impermeable layer that there may be is deep, and another is

$$\frac{2Ktl}{r} = A \, \ln\frac{l-Z_1}{l-Z_2} \qquad (19.38)$$

when the borehole penetrates to the impermeable layer. In each case, $l$ is the depth of the water in the borehole at equilibrium, $r$ is the radius of the bore, and $t$ is the time taken for the disturbed level to rise from height $Z_1$ to $Z_2$ in the bore. The factor $A$ is empirically determined and as reported seems to be somewhat variable. For homogeneous land it is ultimately quoted as being given by $rl/0.19$.

The approximations that are called for in the derivation of these formulas are so ruthless that there is very little of physical interest in the analysis, and it will not be described here. For example, the constant $A$ has the nature of a shape factor which relates the flow into the borehole to the difference of potential between the water table and the water level in the bore, as in the case of the cavity or piezometer method, notwithstanding that the surface of entry is, in part, an equipotential below the water level and, in part, a surface of seepage above the water level in the bore; and that the proportions of these two parts change as the level of water rises. Furthermore, the same shape factor $A$ is applied to that part of the water which enters through the bottom of the bore as to that part which enters through the vertical wall. The formulas are perhaps best regarded as prompted by a degree of physical analysis, but as being essentially empirically established.

Van Bavel and Kirkham (1948) developed a rigorous solution of the equations of flow of water into a borehole which penetrates the whole thickness of the groundwater to the impervious layer below. It is

$$dZ/dt = 1.6(lK/r)\Sigma \qquad (19.39)$$

where

$$\Sigma = \sum_{n=1}^{\infty} (-1)^{(n-1)/2} \cos (n\pi Z/2l) \frac{K_1(n\pi r/2l)}{K_0(n\pi r/2l)}$$

$$n = 1, 3, 5, \ldots .$$

Here $K_0$ and $K_1$ are Bessel functions of the second kind, of zero and first order respectively, which are tabulated in the British Association Mathematical Tables (1937). Thus a measurement of the level of water in the borehole as the time interval increases permits one to determine the simultaneous level and rate of rise of level at any chosen moment, and thus to calculate $K$.

When the borehole does not penetrate to the impervious bed, the factor $\Sigma$ in Equation (19.39) has been determined theoretically by Kirkham (1959) in a rather complicated form, worked out numerically for some selected geometries. Van Bavel and Kirkham (1948) have in addition presented some curves, based on determinations by means of an electric analogue, which relate $\Sigma$ to various proportions of borehole and depths of impermeable bed.

In all the above analyses it has been tacitly assumed that the soil is isotropic. If it is not, then the emergent value of $K$ will be compounded from the horizontal and vertical conductivities, $K_H$ and $K_V$ respectively. Childs (1952) has proposed the use of a two-well method to determine the value of the true horizontal conductivity, after which the piezometer experiments may be interpreted in such a way as to reveal the true vertical conductivity. Laboratory tests of the procedure have been described by Childs, Cole and Edwards (1953), and applications in the field have been further described by Childs, Collis-George and Holmes (1957).

In the two-well method a pair of boreholes, each of radius $r$, penetration $l$ into the groundwater and separation $2d$ between the axes, are bored by a suitable auger, and time is allowed for the water in them to reach equilibrium. Water is then pumped at a steady rate $\dot{Q}$ from one well and fed back into the other, so that after a further period of time a steady state is attained with the water level in one well raised by an amount equal to that of the depression of the level in the other, the difference between the two levels being $\Delta Z$. If the water table is not excessively disturbed by this pumping, the flow between the wells will be approximately in horizontal planes, except near the base of the system, but this end effect may be allowed for by information gained from hydraulic or electric analogue experiments, or may be eliminated by conducting experiments at different penetrations. It will be supposed that the rate $\dot{Q}$ is known to be horizontal over the length $l$.

In a horizontal plane the problem is the well-known one of the two-dimensional flow between a point source and a point sink, the well surfaces corresponding to a pair of circular equipotentials. The solution is presented in standard texts (see, for example, Smythe, 1950), and is

$$\dot{Q} = \frac{\pi K_H l \Delta Z}{\cosh^{-1}(d/r)} \tag{19.40}$$

All other factors in this equation being measurable, $K_H$ may be calculated.

The vertical conductivity is most accurately determined by a piezometer cavity which most accentuates the vertical flow direction, namely a cavity of zero length formed by the open end of the piezometer tube itself. As shown in Section 11.3, if the piezometer is driven into soil with vertical conductivity $K_V$ and horizontal conductivity $K_H$, the observations of water level as a function of time are the same as those which would have been noted had the soil been of uniform conductivity $K$, equal to $(K_H K_V)^{\frac{1}{2}}$, and had all vertical dimensions of the system been changed in the proportion $(K_H/K_V)^{\frac{1}{2}}$. Since the cavity is of zero length, this dimension remains unchanged in the transformed system, and a change of depth does not in general markedly affect the factor $A$ in the appropriately used Equation (19.32).

Hence this equation may be used with the untransformed $A$ factor as a first trial, since of course the transformed value cannot be known until the value of $K_V$ is known. Thus Equation (19.32) is used to determine $K$ where

$$K = (K_H K_V)^{\frac{1}{2}} \tag{19.41}$$

Since $K_H$ is known from the two-well experiment, $K_V$ follows at once. If it should prove that the ratio $(K_H/K_V)^{\frac{1}{2}}$ thus revealed affects the depth of the cavity in the transformed space sufficiently to modify the $A$ factor appreciably, the new $A$ factor thus indicated is used in Equation (19.32) to obtain a new value of $K_V$ from the same experimental observations, and in this way, by successive approximation, a value of $K_V$ is ultimately obtained which satisfies Equation (19.32) with an $A$ factor appropriate to the value of $(K_H/K_V)^{\frac{1}{2}}$.

A summary of experience with various field methods of measurement of the hydraulic conductivity at the site has been reported by Donnan (1959). All methods are described as difficult and subject to considerable errors. Some of the objections reported are associated with genuine difficulties due to inherent soil properties, such as instability, which makes it impossible to maintain a borehole or cavity of known dimensions. Some, however, are the result of an inadequate appreciation of the nature of hydraulic conductivity as a statistically averaged soil property. For example, where the conductivity is inherently a property of the soil structure and the distance between the structural fissures is on an average several inches, a fair sample must necessarily be large. In such circumstances the use of boreholes or cavities of the order of one inch in diameter, as is commonly reported, invites failure. The effect of the variability of conductivity on the variability and asymmetry of levels of water observed in the two-well method has been discussed by Childs, Collis-George and Holmes (1957).

## NOTES

**Note 44. The oscillation of liquid in the oscillating permeameter**
In the apparatus shown in Figure (19.4) let the meniscus in the right-hand limb be at a height $Z_R$ relative to the fixed reference mark $O$ on that limb. Then if $a$ is the area of cross-section of the tube in which the meniscus moves, the rate of entry of water out of the surface of the porous material into the tube is

$$dQ/dt = a \ dZ_R/dt \tag{N44.1}$$

The height of the reference mark relative to a fixed datum is, as given in Equation (19.5),

$$z_R = \bar{z} + z_0 \sin{(2\pi t/T)} \qquad \left\{ \begin{array}{l} \text{(N44.2)} \\ \text{(19.5)} \end{array} \right.$$

while the height of the corresponding mark on the left-hand side is, since the motion is symmetrical about the axis of rocking,

$$z_L = \bar{z} - z_0 \sin{(2\pi t/Z)} \qquad \text{(N44.3)}$$

When the meniscus in the right-hand limb is at a height $Z_R$ above the mark $O$, the meniscus in the left-hand limb must be at a depth $Z_R$ below the reference mark in that limb, since the total volume of water in the apparatus is constant, and it is assumed that the two limbs are of the same bore of tubing. Thus when the height of the right-hand meniscus relative to the fixed arbitrary datum is

$$\phi_R = z_R + Z_R \qquad \text{(N44.4)}$$

the height of the left-hand meniscus, again relative to the arbitrary datum, is

$$\phi_L = z_L - Z_R \qquad \text{(N44.5)}$$

By the definition of potential, Equation (9.3), $\phi_R$ and $\phi_L$ are the hydraulic potentials at the surfaces of the material in the right- and left-hand limbs respectively. Hence the potential on the right-hand side exceeds that on the left by the amount, $\Delta\phi$, which may be expressed by combining Equations (N44.4) and (N44.5) with Equations (N44.2) and (N44.3). The result is

$$\Delta\phi = \phi_R - \phi_L = 2\{z_0 \sin{(2\pi t/T)} + Z_R\} \qquad \text{(N44.6)}$$

If the length of the column of porous material, measured along its bent path, is $l$ and the cross-sectional area is $A$, then the equation which expresses Darcy's law is

$$\frac{dQ}{dt} = -\frac{\Delta\phi K A}{l} \qquad \text{(N44.7)}$$

Substituting from Equation (N44.1) for $dQ/dt$ and from Equation (N44.6) for $\Delta\phi$, one has, after a little rearrangement,

$$\frac{dZ_R}{dt} = -\frac{2KA}{al}[z_0 \sin{(2\pi t/T)} + Z_R] \qquad \text{(N44.8)}$$

This is the differential equation of the motion of the meniscus. One knows intuitively that the motion must be of the same time period $T$ as that of the forcing oscillation, but with a phase difference, say $\alpha$, and an amplitude, $B$, both of which have to be determined. The solution of Equation (N44.7) will therefore be of the form,

$$Z_R = B \sin{(2\pi t/T + \alpha)} \qquad \text{(N44.9)}$$

Accordingly,

$$\frac{dZ_R}{dt} = \frac{2\pi B}{T} \cos{(2\pi t/T + \alpha)} \qquad \text{(N44.10)}$$

With these substitutions for $Z_R$ and $dZ_R/dt$ in Equation (N44.8), one has

$$\frac{2\pi B}{T} \cos{(2\pi t/T + \alpha)} = -\frac{2KA}{al}[z_0 \sin{(2\pi t/T)} + B \sin{(2\pi t/T + \alpha)}]$$

Expansion of the circular functions of $2\pi t/T + \alpha$ results in the equation,

$$\frac{\pi B}{T}[\cos(2\pi t/T)\cos\alpha - \sin(2\pi t/T)\sin\alpha] =$$

$$= -\frac{KA}{al}[(z_0 + B\cos\alpha)\sin(2\pi t/T) + B\sin\alpha\cos(2\pi t/T)]$$

This equation must be satisfied at all times, and this can be so only if the coefficients of $\cos(2\pi t/T)$ on the left-hand side equal those on the right, and similarly the coefficients of $\sin(2\pi t/T)$ are equal. This condition gives rise to the pair of equations,

$$\frac{\pi\cos\alpha}{T} = -\frac{KA\sin\alpha}{al} \tag{N44.11}$$

and

$$B\left(\sin\alpha - \frac{KAT}{\pi al}\cos\alpha\right) = \frac{KATz_0}{\pi al} \tag{N44.12}$$

Equation (N44.11) yields directly,

$$\tan\alpha = -\frac{\pi al}{KAT} \tag{N44.13}$$

With this substitution in Equation (N44.12), that equation becomes

$$B = -z_0\cos\alpha$$

$$= -\frac{z_0}{(1+\tan^2\alpha)^{\frac{1}{2}}} \tag{N44.14}$$

It is convenient to avoid negative signs in the calculations by writing

$$\theta = -\alpha$$

so that, from Equation (N44.13),

$$\tan\theta = -\tan\alpha = \frac{\pi al}{KAT} \qquad \begin{cases} \text{(N44.15)} \\ \text{(19.7)} \end{cases}$$

The solution then becomes, from Equations (N44.9) and (N44.14),

$$Z_R = -\frac{z_0}{(1+\tan^2\theta)^{\frac{1}{2}}}\sin(2\pi t/T - \theta)$$

or, avoiding the negative sign,

$$Z_R = \frac{z_0}{(1+\tan^2\theta)^{\frac{1}{2}}}\sin(2\pi t/T + \pi - \theta) \tag{N44.16}$$

Thus the amplitude, $Z_0$, of the oscillation of the meniscus is given by

$$Z_0 = \frac{z_0}{(1+\tan^2\theta)^{\frac{1}{2}}} \qquad \begin{cases} \text{(N44.17)} \\ \quad \text{(19.8)} \end{cases}$$

and the equation of motion, Equation (N44.16), becomes

$$Z_R = Z_0 \sin\left(2\pi t/T + \pi - \theta\right) \qquad \begin{cases} \text{(N44.18)} \\ \quad \text{(19.6)} \end{cases}$$

### Note 45. The transient stage of a pumped well

First suppose that a confined aquifer of thickness $l$ is completely penetrated by a well which is steadily pumped at a rate $\dot{Q}$. By symmetry, the flow is everywhere radial and, at a distance $r$ from the well axis, one may write Darcy's law in the form,

$$v = -K\,d\phi/dr$$

Hence over the cylindrical equipotential surface of radius $r$ and length $l$, the total rate of flow $\dot{Q}_r$ towards the well is

$$\dot{Q}_r = -2\pi r l v = 2\pi l K r\, d\phi/dr \qquad \text{(N45.1)}$$

Similarly at the distance $r + \delta r$, the flow toward the well is

$$\dot{Q}_{r+\delta r} = 2\pi l K \left[ r\frac{d\phi}{dr} + \frac{d\left(r\dfrac{d\phi}{dr}\right)}{dr}\delta r \right] \qquad \text{(N45.2)}$$

As regards the annular cylinder with inner and outer radii of $r$ and $r+\delta r$ respectively, Equation (N45.2) represents the rate of entry of water and Equation (N45.1) the rate of outflow, so that the rate of storage $\dfrac{dS}{dt}$ is the difference. The storage in a confined saturated aquifer can only be due to an increase of pore space due to expansion of the aquifer when the hydrostatic pressure component of the potential is increased.

$$\frac{dS}{dt} = 2\pi l r\,\delta r\frac{dc}{d\phi}\frac{d\phi}{dt} \qquad \text{(N45.3)}$$

By appeal to Equations (N45.1) and (N45.2), the expression for the rate of storage in terms of the difference between the inflow and outflow rates is

$$\frac{dS}{dt} = 2\pi l K \left[ \frac{d\left(r\dfrac{d\phi}{dr}\right)}{dr} \right]\delta r \qquad \text{(N45.4)}$$

The increased storage in a column of unit cross-section and length $l$ per unit increase of potential is $l\,dc/d\phi$, and this is analogous to the specific yield $Y$ of groundwater per unit change of height of the water table, since the water table height is the potential at the specified distance according to the Dupuit-Forchheimer assumption. Hence from Equations (N45.3) and (N45.4), one has

$$\frac{Y}{lK}\frac{\partial\phi}{\partial t} = \frac{1}{r}\left[\frac{\partial\left(r\frac{\partial\phi}{\partial r}\right)}{\partial r}\right] \qquad (N45.5)$$

The solution of this equation, with the boundary condition that the rate of flow at the axis of the well is maintained constant at the value $\dot{Q}$, is identical in form with an equation discussed by Carslaw and Jaeger (1959) in connection with a problem of heat flow, and is

$$\Delta\phi = (\dot{Q}/4\pi lK)W(n) \qquad \begin{cases}(N45.6)\\(19.15)\end{cases}$$

where $\Delta\phi$ is the drawdown, relative to the undisturbed potential at large distances, at the distance $r$, and the function $W(n)$ is defined as

$$W(n) = \int_n^\infty \frac{e^{-u}}{u}\,du \qquad \begin{cases}(N45.7)\\(19.16)\end{cases}$$

In Equation (N45.7), $n$ is the quantity,

$$n = r^2 Y/4Klt \qquad \begin{cases}(N45.8)\\(19.17)\end{cases}$$

The pumping rate $\dot{Q}$ is in fact maintained at the finite well radius and not at the axis, but the well radius is small compared with reasonably large distances at which observations are taken and the value of $\dot{Q}$ would not vary very much with distance near the well except in the very early stages of pumping.

When the method is applied to groundwater under a free water table, the potential $\phi$ is identical with the water table height $Z$ and so is the aquifer thickness $l$, so that Equation (N45.5) takes the form,

$$\frac{Y}{K}\frac{\partial Z}{\partial t} = \frac{Z}{r}\left[\frac{\partial\left(r\frac{\partial\phi}{\partial r}\right)}{\partial r}\right] \qquad (N45.9)$$

In the right-hand side, the factor $Z/r$ may be written $(l+Z')/r$ where $Z'$ is the variable departure from the undisturbed groundwater thickness $l$. If now one postulates that only those cases are to be considered where $Z'$ is negligibly small in comparison with $l$, then Equation (N45.9) becomes identical with Equation (N45.5) with $Z'$ taking the place of the phreatic level $\phi$. Consequently Equation (N45.6) may be applied.

# REFERENCES

Alexander, L. T., Shaw, T. M. and Muckenhirn, R. J. (1936). Detection of freezing point by dielectric measurements. *Soil Sci. Soc. Am. Proc.*, **1**, 113-119.

Alway, F. J. and Clark, V. L. (1911). A study of the movement of water in a uniform soil under artificial conditions. *25th Ann. Rept. Nebraska Exp. Sta.*, 246-287.

Anderson, A. B. C. and Edlefsen, N. E. (1942a). Volume-freezing point relations observed with a new dilatometer technique. *Soil Sci.*, **54**, 221-232.

Anderson, A. B. C. and Edlefsen, N. E. (1942b). Laboratory study of the response of 2- and 4-electrode plaster of paris blocks as soil-moisture content indicators. *Soil Sci.*, **53**, 413-428.

Aronovici, V. S. and Donnan, W. W. (1946). Soil permeability as a criterion for drainage design. *Trans. Am. Geophys. Un.*, **27**, 95-101.

van Bavel, C. H. M. and Kirkham, D. (1948). Field measurement of soil permeability using auger holes. *Soil Sci. Soc. Am. Proc.*, **13**, 90-96.

Baver, L. D. (1938). Soil permeability in relation to non-capillary porosity. *Soil Sci. Soc. Am. Proc.*, **3**, 52-56.

Bendixen, T. W. and Slater C. S. (1947). Effect of the time of drainage on the measurement of soil pore space and its relation to permeability. *Soil Sci. Soc. Am. Proc.*, **11**, 35-42.

Bernal, J. D. and Fowler, R. H. (1933). A theory of water and ionic solution, with particular reference to hydrogen and hydroxyl ions. *J. Chem. Phys.*, **1**, 515-548.

Biswas, T. D., Nielsen, D. R. and Biggar, J. W. (1966). Redistribution of soil water after infiltration. *Water Resources Res.*, **2**, 513-524.

Boulton, N. S. (1954). The drawdown of the water table under non-steady conditions near a pumped well in an unconfined formation. *Proc. Instn. Civ. Eng.*, **III**, 3, 564-579.

Boumans, J. H. (1954). Referred to in Visser (1954).

Bouwer, H. (1959). Theoretical aspects of flow above the water table in tile drainage of shallow homogeneous soil. *Soil Sci. Soc. Am. Proc.*, **23**, 200-263.

Bouyoucos, G. J. (1927). The hydrometer as a new method for the mechanical analysis of soils. *Soil Sci.*, **23**, 343-353.

Bouyoucos, G. J. and Mick, H. H. (1940). An electrical resistance method for the continuous measurement of soil moisture under field conditions. *Mich. Agr. Exp. Sta. Tech. Bull.*, 172.

Bouyoucos, G. J. and Mick, H. H. (1941). Comparison of absorbent materials employed in the electrical resistance method of making a continuous measurement of soil moisture under field conditions. *Soil Sci. Soc. Am. Proc.*, **5**, 77-79.

Bouyoucos, G. J. and Mick, H. H. (1948). A fabric absorption unit for continuous measurement of soil moisture in the field. *Soil Sci.*, **66**, 217-232.

British Association Mathematical Tables, Vol. VI (1937), Cambridge University Press.

Bybordi, M. (1968). Moisture profiles in layered porous materials during steady state infiltration. *Soil Sci.* (in press).

Carman, P. C. (1937). Fluid flow through granular beds. *Trans. Instn. Chem. Engs.*, **15**, 150-166.

Carslaw, H. S. and Jaeger, J. C. (1959). *Conduction of Heat in Solids*, 2nd Edn., Oxford University Press.

Childs, E. C. (1940). The use of soil moisture characteristics in soil studies, *Soil Sci.*, **50**, 239-252.

Childs, E. C. (1943a). A note on electrical methods of determining soil moisture. *Soil Sci.*, **55**, 219-223.

Childs, E. C. (1943b). The water table, equipotentials and streamlines in drained land. *Soil Sci.*, **56**, 317-330.

Childs, E. C. (1945a). The water table, equipotentials and streamlines in drained land: II. *Soil Sci.*, **59**, 313-327.

Childs, E. C. (1945b). The water table, equipotentials and streamlines in drained land: III. *Soil Sci.*, **59**, 405-415.

Childs, E. C. (1946). The water table, equipotentials and streamlines in drained land: IV. Drainage of foreign water. *Soil Sci.*, **62**, 183-192.

Childs, E. C. (1947). The water table, equipotentials and streamlines in drained land: V. The moving water table. *Soil Sci.*, **63**, 361-376.

Childs, E. C. (1950). The equilibrium of rain-fed groundwater resting on deeper saline water: the Ghyben-Herzberg lens. *J. Soil Sci.*, **1**, 173-181.

Childs, E. C. (1952). Measurement of the hydraulic permeability of saturated soil *in situ*. 1. Principles of a proposed method. *Proc. Roy. Soc.*, **215A**, 525-535.

Childs, E. C. (1954). The space charge in the Gouy layer between two plane, parallel non-conducting particles. *Trans. Faraday Soc.*, **50**, 1356-1362.

Childs, E. C. (1957). The anistropic conductivity of soil. *J. Soil Sci.*, **8**, 42-47.

Childs, E. C. (1959). A treatment of the capillary fringe in the theory of drainage. *J. Soil Sci.*, **10**, 83-100.

Childs, E. C. (1960a). The non-steady state of the water table in drained land. *J. Geophys. Res.*, **65**, 780-782.

Childs, E. C. (1960b). A treatment of the capillary fringe in the theory of drainage. II. Modifications due to an impermeable sub-stratum. *J. Soil Sci.*, **11**, 923-304.

Childs, E. C. (1967). Soil moisture theory. *Advan. Hydrosci.*, **4**, 73-117.

Childs, E. C., Cole, A. H. and Edwards, D. H. (1953). Measurement of the hydraulic permeability of saturated soil *in situ*. II. *Proc. Roy. Soc.*, **216A**, 72-89.

Childs, E. C. and Collis-George, N. (1950). The permeability of porous materials. *Proc. Roy. Soc.*, **201A**, 392-405.

Childs, E. C., Collis-George, N. and Holmes, J. W. (1957). Permeability measurements in the field as an assessment of anisotropy and structure development. *J. Soil Sci.*, **8**, 27-41.

Childs, E. C. and Poulovassilis, A. (1960). An oscillating permeameter. *Soil Sci.*, **90**, 326-328.

Childs, E. C. and Poulovassilis, A. (1962). The moisture profile above a moving water table. *J. Soil Sci.*, **13**, 272-285.

Childs, E. C. and Youngs, E. G. (1961). A study of some three-dimensional field drainage problems. *Soil Sci.*, **92**, 15-24.

Chow, V. T. (1952). On the determination of transmissibility and storage co-efficients from pumping test data. *Trans. Am. Geophys. Un.*, **33**, 397-404.

Colding, A. (1873). Om lovene for vandets bevaegelse i jorden. K. Danske Vidensk. *Selks. Skr. 5 Raekke, Naturvidenskabelig og mathematisk afdeling*, **9**, 563-621.

Collis-George, N. and Youngs, E. G. (1958). Some factors determining water-table heights in drained homogeneous soils. *J. Soil Sci.*, **9**, 332-338.

Colman, E. A. and Hendrix, T. M. (1949). The fibreglass electrical soil-moisture instrument. *Soil Sci.*, **67**, 425-438.

Conway, V. M. and Millar, A. (1960). The hydrology of some small peat covered catchments in the northern Pennines. *J. Inst. Water Eng.*, **14**, 415-424.

Cooper, H. H., Jr. and Jacob, C. E. (1946). A generalized graphical method for evaluating formation constants and summarizing well field history. *Trans. Am. Geophys. Un.*, **27**, 526-534.

Crank, J. (1956). *The Mathematics of Diffusion*. Oxford University Press.

Crank, J. and Henry, M. E. (1949). Diffusion in media with variable properties. I. The effect of a variable diffusion coefficient on the rate of absorption and desorption. *Trans. Faraday Soc.*, **45**, 636-650.

Crank, J. and Henry, M. E. (1949). Diffusion in media with variable properties. II. The effect of a variable diffusion coefficient on the concentration-distance relationship in the non-steady state. *Trans. Faraday Soc.*, **45**, 119-130.

Croney, D., Coleman, J. D. and Black, W. P. M. (1958). Movement and distri-bution of water in soil in relation to highway design and performance. *Highway Res. Bd. Special Rept.*, **40**, 226-252.

Darcy, H. (1856). *Les fontaines publiques de la ville de Dijon*. Dalmont, Paris.

Davis, W. E. and Slater, C. S. (1942). A direct weighing method for sequent measurements of soil moisture under field conditions. *J. Am. Soc. Agron.*, **34**, 285-287.

Day, P. R. and Luthin, J. N. (1956). A numerical solution of the differential equation of flow for a vertical drainage problem. *Soil Sci. Soc. Am. Proc.*, **20**, 443-447.

van Deemter, J. J. (1950). Bijdragen tot de kennis van enige natuurkundige grootheden van de grond, II. Theoretische en numerieke behandeling van ontwaterings—en infiltratie-stromingsproblemen. *Versl. Landb. Ond.*, **56**, No. 7, Staatsdrukkerij, The Hague.

Donnan, W. W. (1959). Field experiences in measuring hydraulic conductivity for drainage design. *Agr. Engin.*, **40**, 270-273.

Dupuit, J. (1863). *Études théoriques et pratiques sur le mouvement des eaux*, Edn. 2. Dunod, Paris.

Enderby, A. J. (1955). The domain model of hysteresis. I. *Trans. Faraday Soc.*, **51**, 835-848.

Enderby, A. J. (1956). The domain model of hysteresis. II. *Trans. Faraday Soc.*, **52**, 106-120.

Engelund, F. (1951). Mathematical discussion of drainage problems. *Trans. Dan. Acad. Tech. Sci.*, **3**, 1-64.

Ernst, L. F. (1954). Private communication. (See also Visser (1954).)

Evans, R. C. (1964). *An Introduction to Crystal Chemistry*, 2nd Edn. Cambridge University Press.

Everett, D. H. and Whitton, W. I. (1952). A general approach to hysteresis. I. *Trans. Faraday Soc.*, **48**, 749-757.

Everett, D. H. and Smith, F. W. (1954). A general approach to hysteresis. II. *Trans. Faraday Soc.*, **50**, 187-197.

Everett, D. H. (1954). A general approach to hysteresis. III. *Trans. Faraday Soc.*, **50**, 1077-1096.

Everett, D. H. (1955). A general approach to hysteresis. IV. *Trans. Faraday Soc.*, **51**, 1511-1557.

Fair, G. M. and Hatch, L. P. (1933). Fundamental factors governing the stream-line flow of water through sand. *J. Amer. Water Works Assoc.*, **25**, 1551-1665.

Fancher, G. H., Lewis, J. A. and Barnes, K. B. (1933). Some physical charac-teristics of oil sands. *Min. Ind. Exp. Sta.*, *Pennsylvania State Coll. Bul.*, **12**. (See also Muskat (1937), p. 60.)

Fletcher, J. E. (1939). A dielectric method for determining soil moisture. *Soil Sci. Soc. Amer. Proc.*, **4**, 84-88.

Forchheimer, P. (1914). *Hydraulik*. Teubner, Leipzig and Berlin.

Fourier, J. B. J. (1822). *Théorie analytique de la chaleur*. Firmin Didot Père et Fils, Paris.

Gardner, W. R. (1956). Calculation of capillary conductivity from pressure plate outflow data. *Soil Sci. Soc. Am. Proc.*, **20**, 317-320.

Gardner, W. R. (1958). Some steady state solutions of the unsaturated moisture flow equation with application to evaporation from a water table. *Soil Sci.*, **85**, 228-232.

Gardner, W. R. and Kirkham, D. (1952). Determination of soil moisture by neutron scattering. *Soil Sci.*, **73**, 391-401.

Gardner, W. and Widtsoe, J. A. (1921). The movement of soil moisture. *Soil Sci.*, **11**, 215-232.

Ghyben, W., Baden and Drabbe, J. (1888-89). Nota in verband met de voor-genomen proefboring te Amsterdam. *Tijdschr. Koninkl. Inst. Ingen.*, **1**, 8-22.

Gonçalvez dos Santos, Jr., A. (1967). Thesis, Cambridge University.

Gouy, M. (1910). Sur la constitution de la charge électrique a la surface d'un électrolyte. *Ann. Phys.*, **9**, 457-468.

Green, W. H. and Ampt, G. A. (1911). Studies in soil physics. I. The flow of air and water through soils. *J. Agr. Sci.*, **4**, 1-24.

Gustafsson, Y. (1946). Untersuchungen über die Strömungsverhältnisse in gedräntem Boden. *Acta Agr. Suecana*, **2** (1), 1-157.

Heikurainen, L. (1964). Improvement of forest growth on poorly drained peat soils. *Intern. Rev. Forestry Res.*, **1**, 39-113.

Hendricks, S. B., Nelson, R. A. and Alexander, L. T. (1940). Hydration mechan-ism of the clay mineral montmorillonite saturated with various cations. *J. Am. Chem. Soc.*, **62**, 1457-1464.

Herzberg (Baurat) (1901). Die Wasserversorgung einiger Nordseebader. *J. fur Gasbeleucht and Wasserversorg.*, **44**, 815-819 and 842-844.

Hooghoudt, S. B. (1936). Bijdragen tot de kennis van eenige natuurkundige grootheden van den grond, 4. Bepaling van den doorlaatfactor van den grond met behulp van pompproeven (z.g. Boorgatenmethode). *Versl. Landb. Ond.*, **42** (13) B, 449-541. The Hague.

Hooghoudt, S. B. (1940). Bijdragen tot de kennis van eenige natuurkundige grootheden van den grond, 7. Algemeene beschouwing van het probleem van de detail ontwatering en de infiltratie door middel van parallel loopende drains, greppels, slooten en kanalen. *Versl. Landb. Ond.*, **46**, 515-707. The Hague.

Horton, R. E. (1940). An approach toward a physical interpretation of infiltration-capacity. *Soil Sci. Soc. Am. Proc.*, **5**, 399-417.

Irmay, S. (1956). Extension of Darcy's law to unsteady, unsaturated flow through porous media. *Sympos. Darcy, Intern. Assoc. Sci. Hydrol., Dijon*, **2**, 57-66.

Isherwood, J. D. (1959). Water-table recession in tile-drained land. *J. Geophys. Res.*, **64**, 795-804.

Jahnke, E. and Emde, F. (1938). *Funktionentafeln*, Edn. 3, p. 52. Teubner, Leipzig and Berlin.

Karplus, W. J. (1958). *Analog Simulation Solution of Field Problems*. McGraw-Hill, New York.

Kirkham, D. (1946). Proposed method for field measurement of permeability of soil below the water table. *Soil Sci. Soc. Am. Soc.*, **10**, 58-68.

Kirkham, D. (1957). Theory of land drainage. The ponded water case. In J. N. Luthin (Ed.), *Drainage of Agricultural Lands*. Am. Soc. Agron., Madison, Wisconsin.

Kirkham, D, (1959). Exact theory of flow into a partially penetrating well. *J. Geophys. Res.*, **64**, 1317-1327.

Kirkham, D. (1966). Steady-state theories for drainage. *J. Irr. and Drain. Div., Proc. Am. Soc. Civ. Engr.*, **92**, 19-39.

Kirkham, D. and van Bavel, C. H. M. (1948). Theory of seepage into auger holes. *Soil Sci. Soc. Am. Proc.*, **13**, 75-82.

Klute, A. (1952). A numerical method for solving the flow equation for water in unsaturated materials. *Soil Sci.*, **73**, 105-116.

Kornev, V. G. (1921-23). The absorbing power of soils and the principle of automatic self-irrigation of soils. *Zhur. Opytnoi. Argon.*, **22**, 105-111. See also abstract (1924) in *Soil Sci.*, **17**, 428-429.

Kostiakov, A. N. (1932). On the dynamics of the coefficient of water-percolation in soils and on the necessity for studying it from a dynamic point of view for purposes of amelioration. *Trans. 6th Comm. Intern. Soc. Soil Sci.*, Russian Part A, 17-21.

Kozeny, J. (1927). Über kapillare Leitung des Wassers im Boden. *Ber. Wien Akad.*, **136A**, 271-306.

List, E. J. (1964). The steady flow of precipitation to an infinite series of tile drains above an impervious layer. *J. Geophys. Res.*, **69**, 3371-3381.

Luthin, J. N. (Ed.). (1957). *Drainage of Agricultural Lands*. Am. Soc. Agron., Madison, Wisconsin.

Luthin, J. N. and Kirkham, D. (1949). A piezometer method for measuring permeability of soil *in situ* below a water table. *Soil Sci.*, **68**, 349-358.

Marshall, C. E. (1930). A new method of determining the distribution curve of a polydisperse colloidal systems. *Proc. Roy. Soc.*, **126A**, 427-439.

Marshall, T. J. (1958). A relation between permeability and size distribution of pores. *J. Soil Sci.*, **9**, 1-8.

Matano, C. (1932-33). On the relation between the diffusion-coefficient and concentrations of solid metals (the nickel-copper system). *Jap. J. Phys.*, **8**, 109-133. (See also Crank, J. (1956).)

Mecke, R. and Baumann, W. (1932). Das Rotationsschwingungsspektrum des Wasserdampes. *Phys. Zeit.*, **33**, 833-835.

Miller, E. E. and Elrick, D. E. (1958). Dynamic determination of the capillary conductivity extended for non-negligible membrane impedance. *Soil Sci. Soc. Am. Proc.*, **22**, 483-486.

Moore, R. E. (1939). Water conduction from shallow water tables. *Hilgardia*, **12**, 383-426.

Muskat, M. (1937). *The Flow of Homogeneous Fluids Through Porous Media.* McGraw-Hill, New York.

Navier, C. L. M. H. (1822). Memoir sur les lois du mouvement des fluids. *Mem. de l'Acad. des Sci.*, **6**, 389.

Nelson, W. R. and Baver, L. D. (1940). Movement of water through soils in relation to the nature of the pores. *Soil Sci. Soc. Am. Proc.*, **5**, 69-76.

Odén, S. (1915). Eine neue Methode zur mechanischen Bodenanalyse. *Int. Mitt. Bodenk.*, **5**, 257-311.

Ohm, G. S. (1827). *Die galvanische Kette mathematisch bearbeitet.* T. H. Riemann, Berlin.

van Olphen, H. (1950). Stabilisation of montmorillonite soils by chemical treatment. *Rec. Trav. Chim.*, **69**, 1308-1312.

Pauling, L. (1929). Principles determining structure of ionic crystals. *J. Am. Chem. Soc.*, **51**, 1010-1026.

Perrin, J. (1908). L'agitation moléculaire et le mouvement brownien. *Comptes rendus*, **146**, 967-970.

Peterson, D. F. (1957). Theory of land drainage: the theory of drainage by pumping from wells. In J. N. Luthin (Ed.), *Drainage of Agricultural Lands.* Am. Soc. Agron., Madison, Wisconsin.

Philip, J. R. (1955a). The concept of diffusion applied to soil water. *Proc. Nat. Acad. Sci. (India)*, **24A**, 93-104.

Philip, J. R. (1955b). Numerical solution of equations of the diffusion type with diffusivity concentration dependent. *Trans. Farad. Soc.*, **51**, 885-892.

Philip, J. R. (1957a). The theory of infiltration: II. The profile of infinity. *Soil Sci.*, **83**, 435-448.

Philip, J. R. (1957b). Numerical solution of equations of the diffusion type with diffusivity concentration-dependent: 2. *Australian J. Phys.*, **10**, 29-42.

Philip, J. R. (1957c). The theory of infiltration: 4. Sorptivity and algebraic infiltration equations. *Soil Sci.*, **84**, 257-264.

Philip, J. R. (1960). General method of exact solutions of the concentration-dependent diffusion equation. *Australian J. Phys.*, **13**, 1-12.

Philip, J. R. (1967). Sorption and infiltration in heterogeneous media. *Australian J. Soil. Res.*, **5**, 1-10.

Poiseuille, J. L. M. (1842). Recherches expérimentales sur le mouvement des liquides dans les tubes de très petits diamètres. *Comptes rendus*, **11**, 961-967; 1041-1048.

Polubarinova-Kochina, P. Ya. (1962). *Theory of groundwater movement.* J. M. R. de Wiest (Transl.). Princeton University Press, Princeton, New Jersey.

Poulovassilis, A. (1962). Hysteresis of pore water, an application of the concept of independent domains. *Soil Sci.*, **93**, 405-412.

Poulovassilis, A. (1969). The effect of hysteresis of pore water on the hydraulic conductivity. *J. Soil Sci.*, **20** (in press).

Richards, L. A. (1931). Capillary conduction of liquids through porous mediums. *Physics*, **1**, 318-333.

Richards, L. A. (1949). Methods of measuring soil moisture tension. *Soil Sci.*, **68**, 95-112.

Richards, L. A. and Weaver, L. R. (1943). Fifteen atmosphere percentage as related to the permanent wilting percentage. *Soil Sci.*, **56**, 331-339.

Rubin, J. (1967). Numerical method for analyzing hysteresis-affected post-infiltration redistribution of soil moisture. *Soil Sci. Soc. Am. Proc.*, **31**, 13-20.

Russell, E. W. (1943). The subdivision of the clay fraction in mechanical analysis. *J. Agr. Sci.*, **33**, 147-154.

Russell, M. B. and Richards, L. A. (1938). The determination of soil moisture energy relations by centrifugation. *Soil Sci. Soc. Am. Proc.*, **3**, 65-69.

van Schilfgaarde, J. (1957). Theory of land drainage. In, J. N. Luthin (Ed.), *Drainage of Agricultural Lands*. Am. Soc. Agron., Madison, Wisconsin, U.S.A.

Schofield, R. K. (1935). The pF of the water in soil. *Trans. 3rd Intern. Congr. Soil Sci.*, **2**, 37-48.

Schofield, R. K. (1938). Pore-size distribution as revealed by the dependence of suction (pF) on moisture content. *Trans. 1st Comm. Intern. Soc. Soil Sci.*, A, 38-45.

Schofield, R. K. and Botelho da Costa, J. V. (1935). The determination of the pF at permanent wilting and at the moisture equivalent by the freezing point method. *Trans. 3rd Intern. Congr. Soil Sci.*, **1**, 6-10.

Schofield, R. K. and Samson, H. R. (1952). The deflocculation of kaolinite suspensions and the accompanying change-over from positive to negative chloride adsorption. *Clay Min. Bull.*, **2**, 45-51.

Shaw, B. and Baver, L. D. (1939). An electrothermal method for following moisture changes of the soil *in situ*. *Soil Sci. Soc. Am. Proc.*, **4**, 78-83.

Shaw, C. F. (1927). The normal moisture capacity of soils. *Soil Sci.*, **23**, 303-317.

Slater, C. S. and Bryant, J. C. (1946). Comparison of four methods of soil moisture measurement. *Soil Sci.*, **61**, 131-155.

Smiles, D. E. and Youngs, E. G. (1965). Hydraulic conductivity determinations by several field methods in a sand tank. *Soil Sci.*, **99**, 83-87.

Smith, R. M., Browning, D. R. and Pohlman, G. G. (1944). Laboratory percolation through undisturbed soil samples in relation to pore-size distribution. *Soil Sci.*, **57**, 197-213.

Smythe, W. R. (1950). *Static and Dynamic Electricity*, 2nd Edn. McGraw Hill, New York.

Southwell, R. V. (1946). *Relaxation Methods in Engineering Science*. Oxford University Press.

Staple, W. J. (1966). Infiltration and redistribution of water in vertical columns of loam soil. *Soil Sci. Soc. Am. Proc.*, **30**, 553-558.

Stokes, G. G. (1845). On the theories of the internal friction of fluids in motion and the equilibrium and motion of elastic solids. *Proc. Camb. Phil. Soc.*, **1**, 16-18. *Camb. Trans.*, **8**, 287.

Svedberg, T. and Rinde, H. (1923a). Determination of the distribution of size of particles in disperse systems. *J. Am. Chem. Soc.*, **45**, 943-954.

Svedberg, T. and Nichols, J. B. (1923b). Determination of the size and distribution of size of particles by centrifugal methods. *J. Am. Chem. Soc.*, **45**, 2910-2917.

Svedberg, T. and Rinde H. (1924). The ultracentrifuge, a new instrument for the determination of size and distribution of size of particles in amicroscopic colloids. *J. Am. Chem. Soc.*, **46**, 2677-2693.

Swanson, C. L. W. and Peterson, J. B. (1942). The use of the micrometric and other methods for the valuation of soil structure. *Soil Sci.*, **53**, 173-185.

Q

Swartzendruber, D. and Huberty, M. R. (1958). Use of infiltration equation parameters to evaluate infiltration differences in the field. *Trans. Amer. Geophys. Union*, **39**, 84-93.

Terzaghi, K. (1925). *Eng. News Record*, **95**, 832-836.

Theis, C. V. (1935). The relation between the lowering of the piezometric surface and the rate of duration of discharge of a well using groundwater storage. *Trans. Amer. Geophys. Un.*, **16**, 519-524.

Thiem, G. (1906). *Hydrologische Methoden*. Gebhardt, Leipzig.

Thiessen, P. A. (1942). Wechselseitige Adsorption von Kolloiden. *Z. Elektrochem.*, **48**, 675-681.

Todd, D. K. (1959). *Ground Water Hydrology*. John Wiley and Sons, Inc., New York.

Topp, G. C. and Miller, E. E. (1966). Hysteretic moisture characteristics and hydraulic conductivities for glass-bead media. *Soil Sci. Soc. Am. Proc.*, **30**, 156-162.

Vachaud, G. (1966). Essai d'analyse de la redistribution après l'arrêt d'une infiltration dans une colonne horizontale de sol non saturé. *Comptes rendus*, **A262**, 839-842.

Veihmeyer, F. J. (1927). Some factors affecting the irrigation requirements of deciduous orchards. *Hilgardia*, **2**, 125-291.

Verwey, E. J. W. and Niessen, K. F. (1939). The electrical double layer at the interface of two liquids. *Phil. Mag.*, (7) **28**, 435-446.

Verwey, E. J. W. and Overbeek, J. Th. G. (1948). *Theory of the Stability of Lyophobic Colloids*. Elsevier, Amsterdam.

Visser, W. C. (1954). Tile drainage in the Netherlands. *Neth. J. Agr. Sci.*, **2**, 69-87.

de Vries, D. A. (1952). Het warmtegeleidingsvermogen van grond. *Mededelingen van de Landbouwhogeschool te Wageningen*, **52(1)**, 1-73.

Warkentin, B. P., Bolt, G. H. and Miller, R. D. (1957). Swelling pressure of montmorillonite. *Soil Sci. Soc. Am. Proc.*, **21**, 495-497.

Weatherburn, C. E. (1924). *Advanced Vector Analysis*. Bell, London.

Wedernikov, V. V. (1936). Sur la solution du problème à deux dimensions du courant stationnaire des eaux souterraines à surface libre. *Comptes rendus*, **202**, 1155-1157.

Wedernikov, V. V. (1937). Über die Sickerung und Grundwasserbewegung mit freier Oberfläche. *Zeit. f. angew. Math. und Mech.*, **17**, 155-168.

Wedernikov, V. V. (1939). Sur la théorie du drainage. *Comptes rendus U.S.S.R.*, **23**, 335-337.

Wesseling, J. (1957). *Enige aspecten van de waterbeheersing in landbouwgronden*. Thesis, Landbouwhogeschool te Wageningen.

Wilson, R. E. (1921). Humidity control by means of sulfuric acid solutions, with critical compilation of vapor pressure data. *J. Ind. Eng. Chem.*, **13**, 326-331.

Wyckoff, R. D. and Botset, H. G. (1936). The flow of gas-liquid mixtures through unconsolidated sands. *Physics*, **7**, 325-345.

Wyllie, M. R. J. and Rose, W. D. (1950). Some theoretical considerations related to the quantitative evaluation of the physical characteristics of reservoir rock from electric log data. *J. Petr. Tech.*, *Petr. Trans. Amer. Inst. Min. Eng.*, **189**, 105-118.

Wyllie, M. R. J. and Spangler, M. B. (1952). Application of electrical resistivity measurements to problems of fluid flow in porous media. *Bull. Am. Assoc. Petrol. Geol.*, **36**, 359-403.

Youngs, E. G. (1957). Moisture profiles during vertical infiltration. *Soil Sci.*, **84**, 283-290.

Youngs, E. G. (1958). Redistribution of moisture in porous materials after infiltration. *Soil Sci.*, **86**, 117-125; 202-207.

Youngs, E. G. (1960). The drainage of liquids from porous materials. *J. Geophys. Res.*, **65**, 4025-4030.

Youngs, E. G. (1964). An infiltration method of measuring the hydraulic conductivity of unsaturated porous materials. *Soil Sci.*, **97**, 307-311.

Youngs, E. G. (1965). Horizontal seepage through unconfined aquifers with hydraulic conductivity varying with depth. *J. Hydrol.*, **3**, 283-296.

Youngs, E. G. (1966). Horizontal seepage through unconfined aquifers with non-uniform hydraulic conductivity. *J. Hydrol.*, **4**, 91-97.

Youngs, E. G. and Smiles, D. E. (1963). The pumping of water from wells in unconfined aquifers: a note on the applicability of Theis's formula. *J. Geophys. Res.*, **68**, 5905-5907.

Zunker, F. (1933). Die Durchlässigkeit des Bodens. *Trans. VIth Comm. Intern. Soc. Soil Sci. Groningen,* **B**, 18-43.

# Author Index

471

# Subject Index